AF356138

Biomass Derived Heterogeneous and Homogeneous Catalysts, 2nd Edition

Editors

José María Encinar Martín
Sergio Nogales Delgado

Basel • Beijing • Wuhan • Barcelona • Belgrade • Novi Sad • Cluj • Manchester

Editors

José María Encinar Martín
Department of Chemical Engineering and Physical-Chemistry
University of Extremadura
Badajoz
Spain

Sergio Nogales Delgado
Department of Applied Physics
University of Extremadura
Badajoz
Spain

Editorial Office
MDPI
St. Alban-Anlage 66
4052 Basel, Switzerland

This is a reprint of articles from the Special Issue published online in the open access journal *Catalysts* (ISSN 2073-4344) (available at: www.mdpi.com/journal/catalysts/special_issues/8UZ64Y6D95).

For citation purposes, cite each article independently as indicated on the article page online and as indicated below:

Lastname, A.A.; Lastname, B.B. Article Title. *Journal Name* **Year**, *Volume Number*, Page Range.

ISBN 978-3-7258-1414-5 (Hbk)
ISBN 978-3-7258-1413-8 (PDF)
doi.org/10.3390/books978-3-7258-1413-8

Cover image courtesy of Juan Carlos Aldana Sánchez

Contents

About the Editors

José María Encinar Martín

Prof. Dr. José María Encinar is an experienced researcher devoted to various research areas, with a focus on biomass conversion. He has developed his career at University of Extremadura, where he has taught numerous professionals and published multiple research works, while leading the Department of Chemical Engineering and Physical-Chemistry.

Sergio Nogales Delgado

Sergio Nogales Delgado is a multidisciplinary scientist, with more than 15 years of experience in different research fields such as minimally processed fruits and vegetables, biomass conversion and its environmental impacts, biodiesel and biolubricant production, and biogas steam reforming, among others. He has published more than 50 articles in peer-reviewed journals, participating in multiple international conferences. This is the second reprint edited in MDPI to which he has contributed.

Preface

In the near future, catalysts will have a vital role in the sustainable implementation of new technologies, contributing to enhanced efficiency and a circular economy, among other factors.

This reprint focuses on the use of heterogeneous and homogeneous catalysts in biomass conversion for multiple purposes such as energy and chemical production. Thus, a wide range of research is included in this collection, providing interesting and innovative findings about new and sustainable catalysts for these purposes. We hope this reprint provides a valuable complement to your research.

José María Encinar Martín and Sergio Nogales Delgado
Editors

 *catalysts*

Editorial

Editorial: Biomass Derived Heterogeneous and Homogeneous Catalysts, 2nd Edition

José María Encinar Martín [1] and Sergio Nogales-Delgado [2,*]

[1] Department of Chemical Engineering, University of Extremadura, Avda. De Elvas s/n, 06006 Badajoz, Spain
[2] Department of Applied Physics, University of Extremadura, Avda. De Elvas s/n, 06006 Badajoz, Spain
* Correspondence: senogalesd@unex.es

1. Introduction

There are plenty of challenges related to the current energy situation. For instance, conventional energy sources, mainly derived from petroleum or oil, are continuously decreasing due to different factors such as population growth, continuous industrialization (including developing areas), and unsustainable transportation models. On the other hand, energy dependency due to traditional energy sources is a real menace for those countries or regions without any alternatives, as geopolitical factors are becoming increasingly present in international settings, making prices unstable in many sectors, such as energy, raw materials, etc. [1].

As a consequence, this situation is a global concern, and many of the above-mentioned problems have sought to be avoided through the use of sustainable alternatives, such as the implementation of green chemistry or circular economy policies, where biorefineries can play an important role [2]. In that sense, the production of biofuel and bioproducts that can present properties compared to equivalent compounds derived from oil can contribute to the implementation of these facilities. However, there are also some challenges that new research should address, such as the possible competition between food and land use or the impacts of biomass processing [3]. In these cases, the valorization of waste could be interesting, and the use of catalysts could alleviate some of the related problems by making these processes more efficient and allowing new processes to reduce the environmental impact of these practices.

Even though these processes are very different (covering a large number of natural raw materials and wastes, including endless opportunities for their valorization), they share some common points. For instance, new research focuses on innovative sources, such as the use of oilseed crop residues [4], lignocellulose [5,6], or microalgae [7], to produce biofuels and biochemicals in different ways, such as using thermochemical (including, for instance, hydrothermal gasification [8]), physicochemical, or chemical processes [9]. On the other hand, the conversion of biomass could offer products with very competitive properties compared to traditional ones (as in the case of alcohols obtained from furfural), demonstrating their added value [10]. In any case, these processes should present high atom economy and energy efficiency, which can be achieved via the use of catalysts, mainly in order to reduce the activation energies of chemical reactions. Thus, homogeneous and heterogeneous catalysts are used, including enzymes (as in the case of immobilized lipases for biodiesel production [11,12]). In that sense, new sustainable catalysts are obtained, for instance, from wastes to produce biochar (a bioproduct by itself with many uses [13]) or nanocatalysts [14] that could act as catalysts if they are suitably prepared [15] for biofuel and bioproduct synthesis.

2. An Overview of Published Articles

Considering the above, this Special Issue aims to cover new trends in the use of catalysts for biofuel and bioproduct generation, addressing the above-mentioned challenges.

Citation: Encinar Martín, J.M.; Nogales-Delgado, S. Editorial: Biomass Derived Heterogeneous and Homogeneous Catalysts, 2nd Edition. *Catalysts* **2024**, *13*, 339. https://doi.org/10.3390/catal14060339

Received: 17 May 2024
Accepted: 21 May 2024
Published: 23 May 2024

As inferred in the Introduction, there is a wide variety of alternatives that can be used to make biofuel and bioproduct generation competitive and efficient through the use of catalysts, and this Special Issue only mentions a sample of the possibilities proposed by the scientific community.

Some studies have focused on the valorization of waste and the introduction of innovative catalysts, as in the case of the work carried out by Hanif et al. (Contribution 1), who used different waste plant oils to produce biodiesel through transesterification using a novel nano-catalyst (Li-TiO$_2$/feldspar). The use of heterogeneous catalysts is a challenging and interesting approach in this process, which could considerably reduce the number of post-treatments to purify the final biodiesel. However, lower yields compared to homogeneous catalysts are usually obtained. In this study, high conversions (above 90%) were obtained at relatively low temperatures (50–60 °C) and typical catalyst concentrations (2%), requiring high methanol/oil ratios (usually 10:1). The properties of the biodiesel that was obtained complied, in general, with standard requirements. The same authors of the previous study (Contribution 2) produced biodiesel from waste oils using novel nano-magnetic CaO/Fe$_2$O$_3$/feldspar catalyst, the use of which allowed high-yield ranges (93.6–99.9%) at low temperatures (40 °C) for 2 h, with methanol/oil ratios ranging from 5:1 to 10:1. Equally, the use of environmentally-friendly and reusable catalysts is the main subject of the work carried out by Khosa et al. (Contribution 3), where the synthesis of combined CaO nanocomposite and cellulose nanocrystals was carried out to produce biodiesel from waste cooking oil, with high conversion levels and the possibility of regeneration of the catalyst (up to three cycles without any changes in catalytic performance).

In that sense, biodiesel production can present a starting point for a biorefinery concept, for instance, to produce biolubricants. Thus, Nogales et al. (Contribution 4) carried out a review study centered around the factors that affect catalytic performance during biolubricant production, and they found that it contributed to high conversions of raw materials (including residues like waste cooking oil), obtaining a final product with interesting and versatile properties depending on the operating conditions.

One recurring byproduct obtained in many syntheses covered in this Special Issue (especially concerning biodiesel production from fatty acids) is glycerol, which can be used in different reactions to produce energy or a wide range of products. For instance, in the study carried out by Cornejo et al. (Contribution 5), glycerol was used to produce higher tert-butyl glycerol esters through acid-catalyzed etherification with tert-butanol, employing *p*-toluensulfonic acid as a catalyst.

In the same way, there are other wastes that can be valorized. For instance, Olivares-Marín et al. (Contribution 6) optimized the hydrothermal carbonization of a waste derived from the removal of an invasive floating plant (water hyacinth) by adjusting Al and Fe content during this process. Thus, an interesting porous carbon was obtained, presenting a feasible alternative for the environmental management of this invasive species. In the same way, the valorization potential of biomass containing glucose (for instance, in agricultural or forest residues) via isomerization to fructose and subsequent dehydration to obtain 5-hydroxymethylfurfural (HMF) was studied by David et al. (Contribution 7). In this study, an effective catalytic consortium was used with the following operating conditions: CX$_4$SO$_3$H/NbCl$_5$ (5 wt%/7.5 wt%) using water/NaCl and MIBK (1:3) at 150 °C for 17.5 min. The resulting catalyst was successfully reused up to seven times, keeping the HMF yield constant. Considering HMF as the cornerstone for the synthesis of biofuels and fine chemicals, Mitra et al. (Contribution 8) used hybrid phosphonates for 5-HTC production from carbohydrates derived from biomass, obtaining yields exceeding 90% through microwave-assisted reactions. Parralejo Alcobendas et al. (Contribution 9) also offered a study where the role of nanoparticles was important in biogas production from pepper waste and pig manure, increasing the methane production rate. For energy exploitation, biogas steam reforming is an important resource that can be used to produce hydrogen, the catalytic process of which is essential to improve its efficiency. As explained by Nogales et al. (Contribution 10), there are different key factors that clearly affect catalytic steam reforming, like the presence of H_2S, coke deposition, or sintering, and the synthesis of innovative and resistant catalysts is required for this purpose.

To sum up, the studies that are included in this Special Issue are diverse, and they focus on different aspects like catalyst optimization or the search for new raw materials, including a wide range of wastes that can be used to valorize them.

3. Conclusions

In conclusion, the studies covered in this Special Issue pointed out some interesting ideas, like the following:

- The latest research has focused on waste valorization as a means to produce both energy and high-value products, with interesting properties that can be used to replace traditional energy sources, such as petrol-based products;
- The use of catalysts has made these different processes more competitive, presenting an interesting starting point for the implementation of such technologies at an industry level in a biorefinery context.
- Operating conditions have an important impact on catalytic performance and the quality parameters of final products.
- New challenges should be addressed, like an increase in the life cycle of catalysts or the requirement of exergy analyses to obtain a real grasp of the above-mentioned processes.

Conflicts of Interest: The authors declare no conflicts of interest.

List of Contributions

- **Contribution 1**: Hanif, M.; Bhatti, I.A.; Shahzad, K.; Hanif, M.A. Biodiesel Production from Waste Plant Oil over a Novel Nano-Catalyst of Li-TiO_2/Feldspar. *Catalysts* **2023**, *13*, 310. https://doi.org/10.3390/catal13020310.
- **Contribution 2**: Hanif, M.; Bhatti, I.A.; Hanif, M.A.; Rashid, U.; Moser, B.R.; Hanif, A.; Alharthi, F.A. Nano-Magnetic CaO/Fe_2O_3/Feldspar Catalysts for the Production of Biodiesel from Waste Oils. *Catalysts* **2023**, *13*, 998. https://doi.org/10.3390/catal13060998.
- **Contribution 3**: Khosa, S.; Rani, M.; Saeed, M.; Ali, S.D.; Alhodaib, A.; Waseem, A. A Green Nanocatalyst for Fatty Acid Methyl Ester Conversion from Waste Cooking Oil. *Catalysts* **2024**, *14*, 244. https://doi.org/10.3390/catal14040244.
- **Contribution 4**: Nogales-Delgado, S.; Encinar, J.M.; González, J.F. A Review on Biolubricants Based on Vegetable Oils through Transesterification and the Role of Catalysts: Current Status and Future Trends. *Catalysts* **2023**, *13*, 1299. https://doi.org/10.3390/catal13091299.
- **Contribution 5**: Cornejo, A.; Reyero, I.; Campo, I.; Arzamendi, G.; Gandía, L.M. Acid-Catalyzed Etherification of Glycerol with Tert-Butanol: Reaction Monitoring through a Complete Identification of the Produced Alkyl Ethers. *Catalysts* **2023**, *13*, 1386. https://doi.org/10.3390/catal13101386.
- **Contribution 6**: Olivares-Marin, M.; Román, S.; Ledesma, B.; Álvarez, A. Optimizing Al and Fe Load during HTC of Water Hyacinth: Improvement of Induced HC Physico-chemical Properties. *Catalysts* **2023**, *13*, 506. https://doi.org/10.3390/catal13030506.
- **Contribution 7**: David, G.F.; Delgadillo, D.M.E.; Castro, G.A.D.; Cubides-Roman, D.C.; Fernandes, S.A.; Lacerda Júnior, V. Conversion of Glucose to 5-Hydroxymethylfurfural Using Consortium Catalyst in a Biphasic System and Mechanistic Insights. *Catalysts* **2023**, *13*, 574. https://doi.org/10.3390/catal13030574.
- **Contribution 8**: Mitra, R.; Malakar, B.; Bhaumik, A. Organically Functionalized Porous Aluminum Phosphonate for Efficient Synthesis of 5-Hydroxymethylfurfural from Carbohydrates. *Catalysts* **2023**, *13*, 1449. https://doi.org/10.3390/catal13111449.
- **Contribution 9**: Parralejo Alcobendas, A.I.; Royano Barroso, L.; Cabanillas Patilla, J.; González Cortés, J. Pretreatment and Nanoparticles as Catalysts for Biogas Production Reactions in Pepper Waste and Pig Manure. *Catalysts* **2023**, *13*, 1029. https://doi.org/10.3390/catal13071029.

- **Contribution 10**: Nogales-Delgado, S.; Álvez-Medina, C.M.; Montes, V.; González, J.F. A Review on the Use of Catalysis for Biogas Steam Reforming. *Catalysts* **2023**, *13*, 1482.

References

1. Vakulchuk, R.; Overland, I.; Scholten, D. Renewable Energy and Geopolitics: A Review. *Renew. Sustain. Energy Rev.* **2020**, *122*, 109547. [CrossRef]
2. Palmeros Parada, M.; Osseweijer, P.; Posada Duque, J.A. Sustainable Biorefineries, an Analysis of Practices for Incorporating Sustainability in Biorefinery Design. *Ind. Crops Prod.* **2017**, *106*, 105–123. [CrossRef]
3. Gallezot, P. Catalytic Conversion of Biomass: Challenges and Issues. *ChemSusChem* **2008**, *1*, 734–737. [CrossRef] [PubMed]
4. Yong, K.J.; Wu, T.Y. Second-Generation Bioenergy from Oilseed Crop Residues: Recent Technologies, Techno-Economic Assessments and Policies. *Energy Convers. Manag.* **2022**, *267*, 115869. [CrossRef]
5. Ma, R.; Xu, Y.; Zhang, X. Catalytic Oxidation of Biorefinery Lignin to Value-Added Chemicals to Support Sustainable Biofuel Production. *ChemSusChem* **2015**, *8*, 24–51. [CrossRef] [PubMed]
6. Pattnaik, F.; Tripathi, S.; Patra, B.R.; Nanda, S.; Kumar, V.; Dalai, A.K.; Naik, S. Catalytic Conversion of Lignocellulosic Polysaccharides to Commodity Biochemicals: A Review. *Environ. Chem. Lett.* **2021**, *19*, 4119–4136. [CrossRef]
7. Kowthaman, C.N.; Senthil Kumar, P.; Arul Mozhi Selvan, V.; Ganesh, D. A Comprehensive Insight from Microalgae Production Process to Characterization of Biofuel for the Sustainable Energy. *Fuel* **2022**, *310*, 122320. [CrossRef]
8. Elliott, D.C. Catalytic Hydrothermal Gasification of Biomass. *Biofuels Bioprod. Biorefining* **2008**, *2*, 254–265. [CrossRef]
9. Hansen, S.; Mirkouei, A.; Diaz, L.A. A Comprehensive State-of-Technology Review for Upgrading Bio-Oil to Renewable or Blended Hydrocarbon Fuels. *Renew. Sustain. Energy Rev.* **2020**, *118*, 109548. [CrossRef]
10. Iriondo, A.; Agirre, I.; Viar, N.; Requies, J. Value-Added Bio-Chemicals Commodities from Catalytic Conversion of Biomass Derived Furan-Compounds. *Catalysts* **2020**, *10*, 895. [CrossRef]
11. Pasha, M.K.; Dai, L.; Liu, D.; Du, W.; Guo, M. Biodiesel Production with Enzymatic Technology: Progress and Perspectives. *Biofuels Bioprod. Biorefining* **2021**, *15*, 1526–1548. [CrossRef]
12. Encinar, J.M.; González, J.F.; Sánchez, N.; Nogales-Delgado, S. Sunflower Oil Transesterification with Methanol Using Immobilized Lipase Enzymes. *Bioprocess Biosyst. Eng.* **2019**, *42*, 157–166. [CrossRef] [PubMed]
13. Kumar, M.; Xiong, X.; Sun, Y.; Yu, I.K.M.; Tsang, D.C.W.; Hou, D.; Gupta, J.; Bhaskar, T.; Pandey, A. Critical Review on Biochar-Supported Catalysts for Pollutant Degradation and Sustainable Biorefinery. *Adv. Sustain. Syst.* **2020**, *4*, 1900149. [CrossRef]
14. Nasrollahzadeh, M.; Soheili Bidgoli, N.S.; Shafiei, N.; Soleimani, F.; Nezafat, Z.; Luque, R. Low-Cost and Sustainable (Nano)Catalysts Derived from Bone Waste: Catalytic Applications and Biofuels Production. *Biofuels Bioprod. Biorefining* **2020**, *14*, 1197–1227. [CrossRef]
15. Kang, K.; Nanda, S.; Hu, Y. Current Trends in Biochar Application for Catalytic Conversion of Biomass to Biofuels. *Catal. Today* **2022**, *404*, 3–18. [CrossRef]

catalysts

Article

Biodiesel Production from Waste Plant Oil over a Novel Nano-Catalyst of Li-TiO$_2$/Feldspar

Maryam Hanif [1], Ijaz Ahmad Bhatti [1], Khurram Shahzad [2],* and Muhammad Asif Hanif [3],*

[1]   Department of Chemistry, University of Agriculture, Faisalabad 38040, Pakistan
[2]   Centre of Excellence in Environmental Studies, King Abdulaziz University, Jeddah 21589, Saudi Arabia
[3]   Nano and Biomaterials Lab, Department of Chemistry, University of Agriculture, Faisalabad 38040, Pakistan
*   Correspondence: ksramzan@kau.edu.sa (K.S.); muhammadasifhanif@uaf.edu.pk (M.A.H.)

Abstract: A novel Li-impregnated TiO$_2$ catalyst loaded on feldspar mineral (Li-TiO$_2$/feldspar) was synthesized via a wet impregnation method and was characterized using X-ray diffraction (XRD), scanning electron microscopy (SEM), and Fourier transform infrared (FTIR) analysis. Using these techniques, it was possible to confirm the catalyst's structural organization with a high crystallinity. This catalyst was used in the transesterification of five waste plant oils of *Citrullus colocynthis* (bitter apple), *Pongamia pinnata* (karanja), Sinapis arvensis (wild mustard), *Ricinus communis* (castor) and *Carthamus oxyacantha* (wild safflower). The catalytic tests were performed at temperatures ranging from 40 to 80 °C, employing a variable methanol/ester molar ratio (5:1, 10:1, 15:1, 20:1 and 25:1) and different catalyst concentrations (0.5%, 1%, 1.5%, 2% and 2.5%) relative to the total reactants mass. Conversion of 98.4% of fatty acid methyl esters (FAMEs) was achieved for *Pongamia pinnata* (karanja). The main fatty acids present in bitter apple, karanja, wild mustard, castor and wild safflower oils were linoleic acid (70.71%), oleic acid (51.92%), erucic acid (41.43%), ricinoleic acid (80.54%) and linoleic acid (75.17%), respectively. Li-TiO$_2$/feldspar produced more than 96% for all the feedstocks. Fuel properties such as iodine value (AV), cetane number (CN), cloud point (CP), iodine value (IV), pour point (PP) and density were within the ranges specified in ASTM D6751.

Keywords: Li-TiO$_2$; catalyst; feldspar; biodiesel; wild mustard

Citation: Hanif, M.; Bhatti, I.A.; Shahzad, K.; Hanif, M.A. Biodiesel Production from Waste Plant Oil over a Novel Nano-Catalyst of Li-TiO$_2$/Feldspar. *Catalysts* **2023**, *13*, 310. https://doi.org/10.3390/catal13020310

Academic Editor: Sergio Nogales Delgado

Received: 11 December 2022
Revised: 7 January 2023
Accepted: 18 January 2023
Published: 31 January 2023

1. Introduction

The world's primary sources of energy are fossil fuels, but due to the limited reservoirs remaining and the production of huge amounts of greenhouse gases during combustion, they are non-viable energy sources [1]. The emissions resulting from fossil fuel combustion adversely affect both the environment and human health [2]. Fossil fuels are continuous sources of emissions. Biodiesel (BD) is a useful alternative to fossil fuels [3]. BD comprises mono-alkyl esters of long-chain fatty acids (FA) [4]. Biodiesel is a renewable and biodegradable clean-burning fuel with low exhaust emissions.

The increasing interest in BD production is due to its ability to use unlimited feedstocks. High yield and low production cost are most significant aspects of ideal feedstocks. Usually, the raw material cost covers ~60–80% of the total production cost of BD [5]. The best feedstocks for biodiesel production have a low cost, high oil contents and are regionally available [6]. Different edible oils and microorganisms such as bacteria, microalgae, yeast, and fungi can be used for biodiesel production [7]. Edible plants as feedstocks for biodiesel are not good candidates for biodiesel production, as this will result in increased food prices [8]. The manufacturing cost has increased 70–92% because of the rise in edible oils' cost, thus stimulating biodiesel manufacturing from non-edible oils [9]. The use of non-edible oils removes the food verses fuel debate related to biodiesel's production from edible oils [10]. Moreover, non-edible oils are more efficient, economical, and environmentally friendly, and the reduce the deforestation rate, as they are easily available in wastelands that are not appropriate for food crops. Recently, non-edible oils have been considered as potential feedstocks for BD production. *Pongamia pinnata*, usually known as karanja, is a medium-sized glabrous perennial tree that grows in littoral regions of Australia and Southeastern Asia. The yield is 8–24 kg of oilseed per tree [11]. *Carthamus oxyacantha* is a 1.5 m-tall, spiny-leaved annual herb, belonging to the family Asteraceae, generally known as wild safflower. This species is not eaten by livestock. It also reduces the yield of cereal crops [12]. It grows on any land and has no production costs prior to harvesting operation. It is resistant to harsh environmental conditions and dry climates [13]. *Citrullus colocynthis* plant oil was discovered as a new option for biodiesel production. Its seeds contain 47% oil. *Sinapis arvensis* L. (wild mustard) is a wild plant with high oil yields. It grows in calcareous soils. *Ricinus communis* (castor plant) can also be explored as a potential resource for BD production. Its seeds contain 40–60% oil.

All vegetable oils are very viscous, and their viscosities are 10–20 times higher than that of diesel fuel [14]. To overcome the challenges associated with vegetable oils, an effective method is the conversion of these oils into FAME [11]. The most common and widely accepted method for biodiesel production is transesterification. Transesterification is a reaction between triglyceride from vegetable oils/animal fats and alcohol using a catalyst [15]. Transesterification requires normal conditions and provides the best quality and efficiency of converted fuel [16]. Biodiesel and glycerol are produced during transesterification [17]. Glycerol, a valuable byproduct of this reaction, is used in various industries [18].

The overall cost of the process will increase and requires a long time for transesterification in the absence of a catalyst [1]. Homogenous and heterogeneous catalysis are conventionally carried out to produce biodiesel. Heterogeneous catalysts are better than homogeneous catalysts because they are easier to use, experience less contamination of the product and co-products, their separation from the medium is easier, and they option of regeneration and reuse [19]. Some drawbacks of these catalysts include the requirement for a greater quantity and availability of surface area, which results in reduced overall catalytical activity. Hence, heterogenous nano-catalysts could play a prominent role not only in producing higher yields of biodiesel, but also a better quality in a shorter reaction time. TiO_2 has gained popularity as a heterogenous catalyst because of its mesoporous structure, providing it with a larger surface area and several additional properties such as environmental friendliness, durability and low cost as compared to conventional nanomaterials [20]. Lithium doping of TiO_2 increases its transesterification ability significantly [19].

The catalytic supports could play an effective role in the production and separation of catalysts from the reaction mixtures. The primary role of the catalytic support is to provide stability to small metal catalyst particles [21]. The catalyst supports not only provide support to catalysts, but are also helpful in increasing their surface area [22]. The most used catalytic supports are zirconium dioxide, aluminum oxides, alumino silicates, magnesium oxide, silica gel, and titanium oxide. The use of clay materials has some advantages as compared to other supports, such as operational simplicity, high selectivity, low cost, and reusability [23]. Feldspar comprises more than half of the Earth's crust and is the most abundant mineral. Feldspar minerals consist of tectosilicates. Tectosilicates are silicate minerals containing silicon ions linked by shared oxygen ions to form a 3D network. Hence, feldspar provides a 3D surface area for catalyst anchoring [24]. The current study is focused on biodiesel synthesis using a novel Li-impregnated TiO_2 catalyst loaded on feldspar minerals.

The current study is focused on biodiesel production from waste plant oils of *Citrullus colocynthis, Pongamia pinnata, Sinapis arvensis, Ricinus communis,* and *Carthamus oxyacantha,* utilizing a Li-incorporated titanium oxide catalyst supported on feldspar minerals (Li-TiO_2/feldspar). The catalytical support was also prepared from waste or low-cost materials including clay minerals. Lithium-doped TiO_2 supported on feldspar has been used as a low-cost novel catalyst for the conversion of waste seed oils into biodiesel.

2. Results and Discussion

2.1. Seed Oil Yield (%)

Oil was extracted from ground seeds of five different nonedible feedstocks such as *Pongamia pinnta* (karanja), *Sinapis arvensis* (wild mustard), *Carthamus oxyacantha* (wild safflower), *Ricinus communis* (castor oil) and *Citrullus colocynthis* (bitter apple) by means of the screw press method. The percentage oil yield of the five different plants using 20 kg of seeds of each feedstock is shown in Table 1. The oil content of the extracted castor oil, karanja, wild mustard, wild safflower, and bitter apple seeds before the conversion to biodiesel (crude oil) was 39.2%, 37.05%, 32.5%, 29.55%, and 17.95%, respectively. The percentage oil yield was calculated using Equation (1).

$$Yield\ \% = \frac{Weight\ of\ oil}{Weight\ of\ seeds} \times 100 \tag{1}$$

Table 1. Percentage oil yield of five different seeds.

Seeds Type	Seed Weight (kg)	Oil Weight (kg)	Yield (%)
Karanja	20	7.41	37.05
Wild mustard	20	6.50	32.5
Wild safflower	20	5.91	29.55
Castor oil	20	7.84	39.2
Bitter apple	20	3.59	17.95

2.2. X-ray Diffraction (XRD) Analysis

Feldspar has a main sample peak of $KAlSi_3O_8$ in XRD spectra, as shown in Figure 1a. Intense sharp diffraction peaks were observed in XRD at a low angle position, $2\theta = 26.63°$, which shows mesopores with a uniform diameter. The diffraction peaks at 2θ of 20.86°, 36.54°, 39.49°, 40.28°, 42.44°, 45.78°, 50.14°, 54.87°, 55.318°, 59.95°, 60.12°, 64.06°, 67.73°, 68.12°, 68.3° were identified, showing the single phase of a highly ordered structure. The crystal size was calculated from the Debye–Scherrer equation. The average size of feldspar crystals was 41.83 nm.

Since the catalyst with 20% Li content provided the highest conversion yield, the XRD pattern of 20% Li-TiO_2 supported on feldspar is shown in Figure 1b. The diffraction peaks were assigned to the Li_2TiO_3 phase at 2θ of 20.65°, 36.38°, 39.2°, 54.68° and 59.80°. Li_2O diffraction peaks were not detected for Li-loaded samples, as also noted previously [25]. Feldspar showed intense sharp diffraction peaks at $2\theta = 26.65°$, 50°, 67.96°. The average crystallite size of the Li-TiO_2/feldspar catalyst was determined as 19 nm by means of the Scherrer equation, and thus the catalyst exhibited a nanoparticle character. The higher particle size has a lower surface area.

2.3. Fourier Transform Infrared Specroscop (FTIR) Analysis

The FTIR spectrum of feldspar is shown in Figure 2a. Feldspars are igneous aluminosilicate minerals. The absorption peaks shown by pure feldspar at 775.3, ~776, 693, 1080 cm^{-1} were due to the stretching vibration of Si–O–Si, the SiO_2 bearing bond, the bending vibrations of silicon-oxygen, and the stretching vibrations of Si–O–Al, respectively. H–O–H bending vibrations were observed at 2102 cm^{-1} and 2318 cm^{-1} [26,27].

The FTIR spectrum of 20% Li-TiO_2 supported on feldspar is shown in Figure 2b. The bands at 1507 cm^{-1} and 1457 cm^{-1} are assigned to the specific characteristic of the Li–O–Ti bond [28]. The small band at the wavelength of 1090 cm^{-1} can be ascribed to the presence of aluminosilicates. The peaks at 2322 cm^{-1}, 2342 cm^{-1} and 2372 cm^{-1} were due to H–O–H bending vibration.

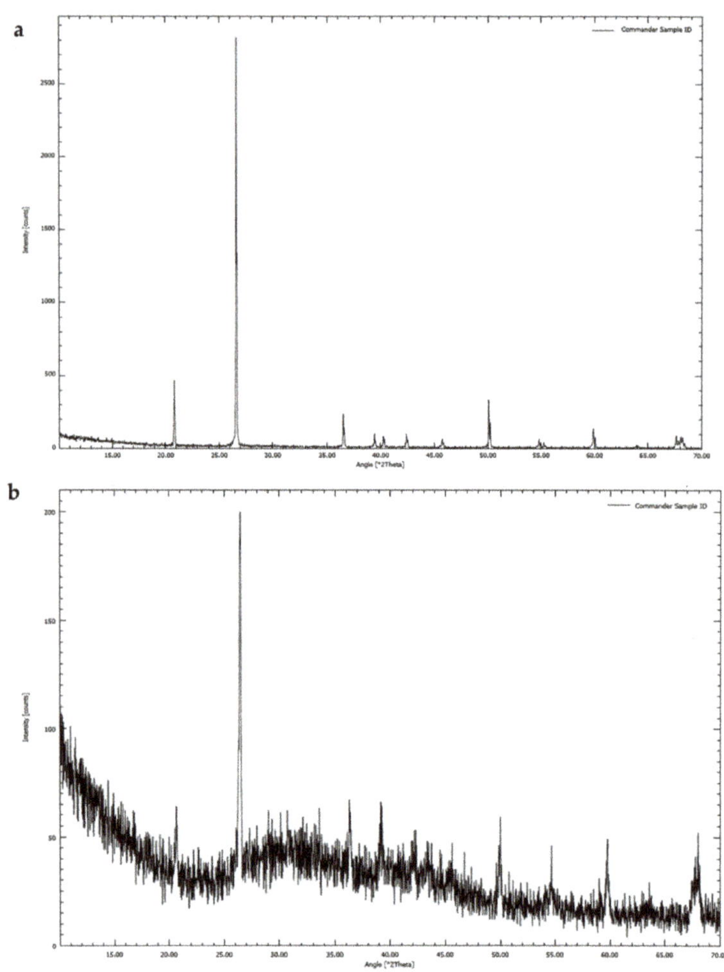

Figure 1. XRD of (**a**) feldspar (catalyst support) and (**b**) 20% Li-TiO$_2$/feldspar.

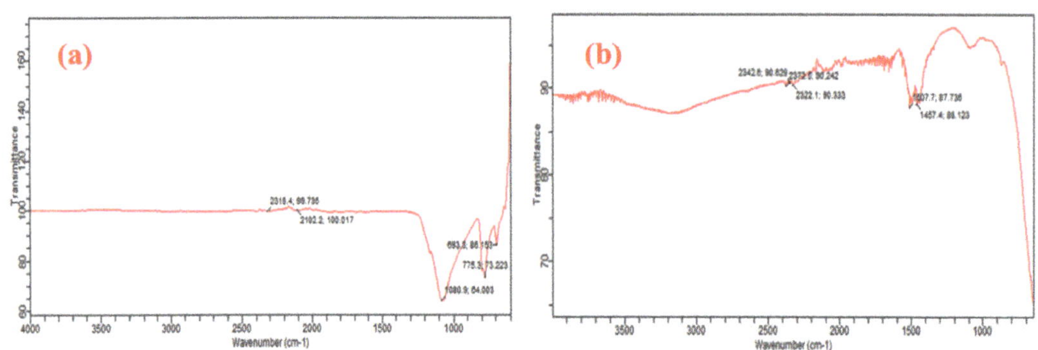

Figure 2. FTIR of (**a**) feldspar and (**b**) 20% Li-TiO$_2$/feldspar.

2.4. Scanning Electron Microscopy (SEM) with Energy Dispersive X-ray Analysis (EDX) Analysis

SEM/EDX for feldspar nanoparticles can be seen in Figure 3. The obtained micrographs of samples were of high crystallinity and non-uniform size. The K-feldspar showed a mesoporous lamellar structure. Due to its multiplicity and complex interfaces, a crossed-lamellar structure provides a high surface area [29,30].

(a)

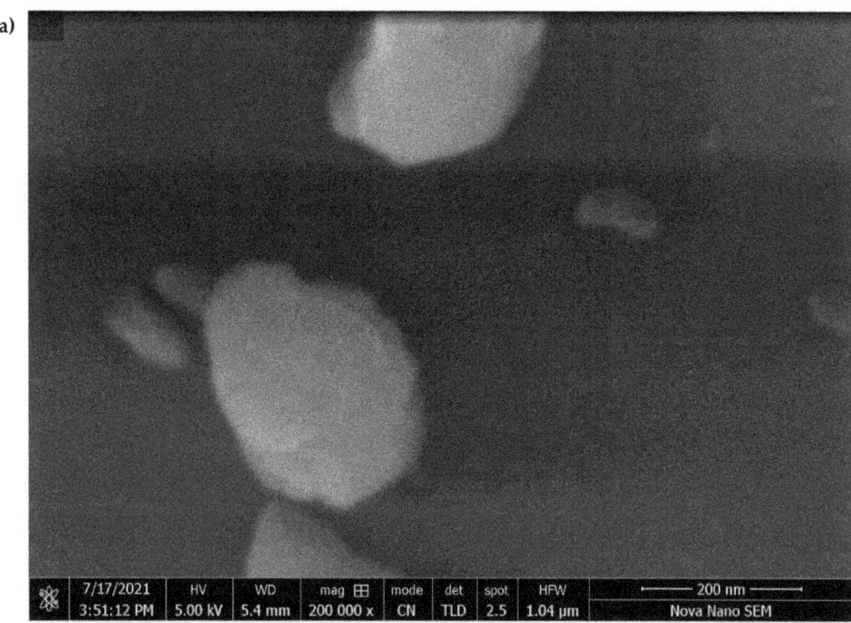

(b)

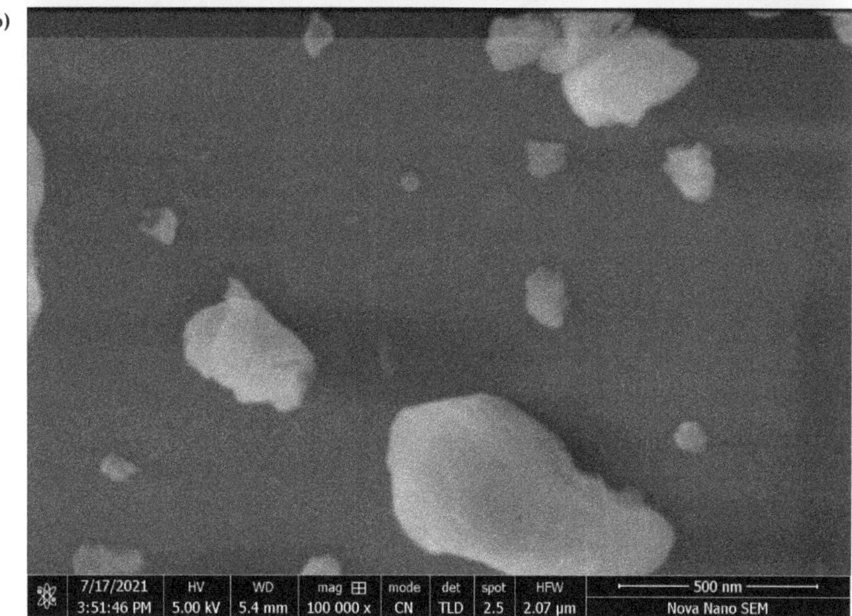

Figure 3. *Cont.*

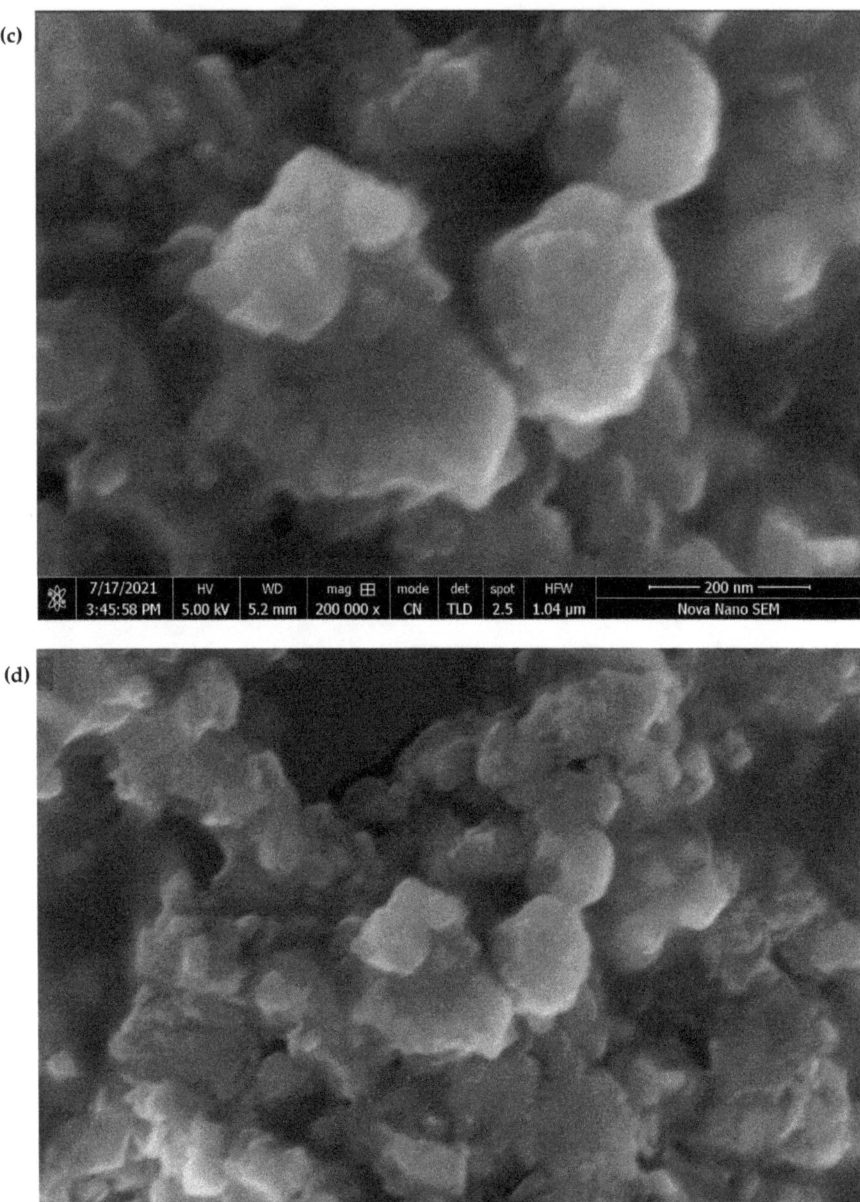

Figure 3. SEM images of (**a**) feldspar at 200 nm, (**b**) feldspar at 500 nm, (**c**) Li-TiO$_2$/feldspar at 200 nm, (**d**) Li-TiO$_2$/feldspar at 500 nm.

The SEM/EDX images of Li-TiO$_2$ supported on feldspar are displayed in Figure 3. SEM image shows the presence of embedded grains without sharp boundaries. The agglomeration of small grains to form large grains can be clearly observed in the images. This resulted in the appearance of small nano holes on the surface of the composite [31]. The dark spots represent the porosity in the sample. Porous materials increase the specific surface area and thus increase the catalytic activity. The EDX spectrum for Li-TiO$_2$ supported on feldspar was obtained in order to confirm the chemical composition of the sample. The spectrum depicts (Figure 4) the presence of all of the elements, i.e., O, Si, Ti, Fe and Al, with weight percentages of 57.75%, 18.77%, 5.68%, 1.22% and 0.75%, respectively. All of the elements in the sample were detected, except for lithium (Li), due to the inability of the EDX technique to detect small-Z elements [32].

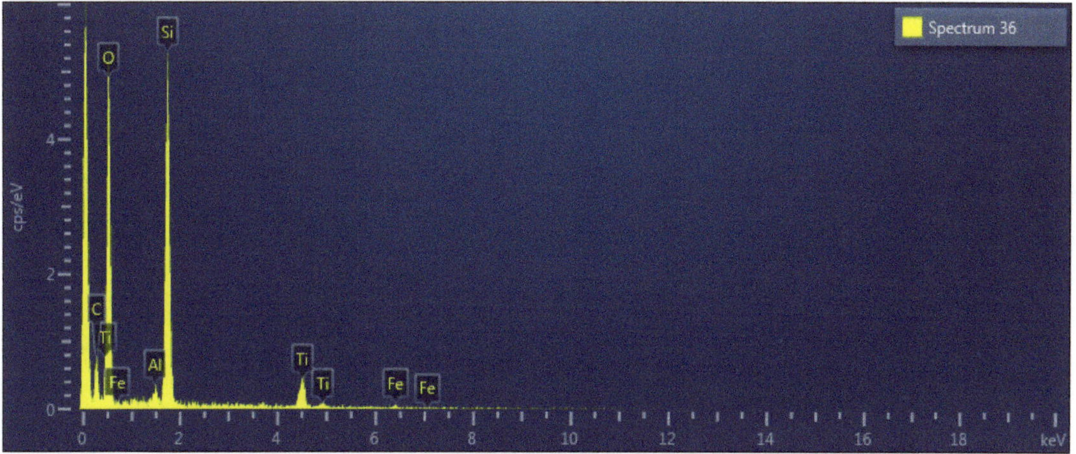

Figure 4. EDX of 20% Li-TiO$_2$/feldspar.

2.5. Effect of the Li to TiO$_2$ Percentage Weight Ratio

The weight ratio of the Li to TiO$_2$ percentage is shown in Table 2. The results show that impregnation with 20 wt% Li exhibited the highest conversion under the reaction conditions of a 5:1 methanol:oil ratio, for a reaction time of 120 min, at 40 °C and with a 1% catalyst concentration.

Table 2. Biodiesel yield (%) using Li-TiO$_2$/feldspar nanocatalysts.

Catalyst	Biodiesel Yield (%)				
	Karanja	**Wild Mustard**	**Castor**	**Wild Safflower**	**Bitter Apple**
10% Li/TiO$_2$/feldspar	85.0 ± 0.5	81.0 ± 0.8	76.6 ± 0.2	83.0 ± 0.9	78.2 ± 0.2
20% Li/TiO$_2$/feldspar	86.1 ± 0.8	89.1 ± 0.5	87.4 ± 0.3	89.7 ± 0.3	88.3 ± 0.9
30% Li/TiO$_2$/feldspar	84.4 ± 0.8	83.5 ± 0.8	80.0 ± 0.8	79.1 ± 0.3	80.3 ± 0.7
40% Li/TiO$_2$/feldspar	85.4 ± 0.4	80.2 ± 0.4	84.1 ± 0.6	83.0 ± 0.3	82.4 ± 0.5

2.6. Optimization of Process Parameters

Bitter apple, karanja, wild mustard, castor, and wild safflower oil transesterification was tested under different reaction conditions to obtain the maximum yield. Since the 20 wt% loading amount of lithium provided the highest production value, all investigations were further conducted with 20% Li-TiO$_2$/feldspar catalyst.

2.6.1. Effect of Catalyst Concentration

The biodiesel yield depends upon the catalyst concentration. The effect of selected supported nanocatalysts on methyl ester yield from bitter apple, karanja, wild mustard, castor, and wild safflower oils is depicted in Tables 3–7. The influence of catalyst amount was examined by changing the catalyst amount within a range of 0.5–2.5% (wt. of catalyst/wt. of oil). The other operational conditions were kept constant (temperature (40 °C), time (120 min) and methanol to oil ratio (5:1)) using five different feedstocks.

Table 3. Yield of biodiesel from karanja oil with changes in reaction parameters.

Feedstock	Conc. of Catalyst (%)	Methanol to Oil Ratio	Temperature (°C)	Reaction Time (min)	Biodiesel Yield (%)
	0.5				84.5 ± 0.8
	1.00				86.1 ± 0.7
	1.5	5:1	40	120	86.0 ± 0.6
	2.00				89.3 ± 0.5
	2.5				87.0 ± 0.4
		10:1			91.7 ± 0.6
	2.00	15:1	40	120	88.0 ± 0.4
		20:1			86.5 ± 0.9
Karanja		25:1			83.7 ± 0.7
			50		98.4 ± 0.9
	2.00	10:1	60	120	93.4 ± 0.5
			70		90.0 ± 0.6
			80		87.0 ± 0.8
				30	87.1 ± 0.1
	2.00	10:1	50	60	89.3 ± 0.6
				90	90.0 ± 0.7
				150	93.1 ± 0.7

Table 4. Yield of biodiesel from wild mustard oil with changes in reaction parameters.

Feedstock	Conc. of Catalyst (%)	Methanol to Oil Ratio	Temperature (°C)	Reaction Time (min)	Biodiesel Yield (%)
	0.5				86.0 ± 0.5
	1.00				89.1 ± 0.3
	1.5	5:1	40	120	87.1 ± 0.9
	2.00				85.0 ± 0.5
	2.5				84.2 ± 0.7
		10:1			87.1 ± 0.8
	1.00	15:1	40	120	85.3 ± 0.4
		20:1			83.6 ± 0.5
Wild mustard		25:1			81.5 ± 0.3
			50		94.7 ± 0.8
	1.00	5:1	60	120	89.0 ± 0.6
			70		84.0 ± 0.7
			80		80.0 ± 0.4
				30	88.2 ± 0.2
	1.00	5:1	50	60	91.2 ± 0.9
				90	96.7 ± 0.7
				150	90.5 ± 0.6

In the presence of 20% Li-TiO$_2$/feldspar, bitter apple, castor, wild mustard, and wild safflower oil provided the highest yield at 1%. The highest biodiesel yield from bitter apple, castor, wild mustard, and wild safflower was 88.3 ± 0.4%, 87.4 ± 0.4%, 89.1 ± 0.3% and 89.7 ± 0.3%, respectively.

Table 5. Yield of biodiesel from wild safflower oil with changes in reaction parameters.

Feedstock	Conc. of Catalyst (%)	Methanol to Oil Ratio	Temperature (°C)	Reaction Time (min)	Biodiesel Yield (%)
Wild safflower	0.5	5:1	40	120	86.1 ± 0.6
	1.00				89.7 ± 0.3
	1.5				87.2 ± 0.4
	2.00				85.3 ± 0.2
	2.5				85.0 ± 0.9
	1.00	10:1	40	120	88.0 ± 0.7
		15:1			86.3 ± 0.8
		20:1			84.5 ± 0.5
		25:1			82.7 ± 0.7
	1.00	5:1	50	120	96.5 ± 0.3
			60		92.5 ± 0.6
			70		90.0 ± 0.8
			80		89.5 ± 0.7
	1.00	5:1	50	30	91.4 ± 0.6
				60	94.2 ± 0.7
				90	96.5 ± 0.8
				150	95.0 ± 0.9

Table 6. Yield of biodiesel from bitter apple oil with changes in reaction parameters.

Feedstock	Conc. of Catalyst (%)	Methanol to Oil Ratio	Temperature (°C)	Reaction Time (min)	Biodiesel Yield (%)
Bitter apple	0.5	5:1	40	120	87.0 ± 0.5
	1.00				88.3 ± 0.4
	1.5				86.0 ± 0.7
	2.00				85.0 ± 0.6
	2.5				83.9 ± 0.4
	1.00	10:1	40	120	89.2 ± 0.7
		15:1			86.5 ± 0.6
		20:1			83.4 ± 0.9
		25:1			80.7 ± 0.8
	1.00	10:1	50	120	96.2 ± 0.8
			60		90.0 ± 0.3
			70		87.0 ± 0.5
			80		84.0 ± 0.6
	1.00	10:1	50	30	89.2 ± 0.6
				60	91.4 ± 0.7
				90	94.5 ± 0.8
				150	93.6 ± 0.2

However, karanja provided the highest biodiesel yield (89.3 ± 0.5%) at a catalyst concentration of 2%. The overall results of the effect of catalyst concentration on biodiesel yield are quite promising and indicate that a certain level of catalyst concentration is required to obtain the maximum biodiesel yield. The optimized concentration level was not only dependent upon the type of catalyst/support used, but was also dependent on the type of oil under investigation. Catalysts at low concentrations could not effectively drive reactions. At high catalyst concentrations, a viscous emulsion formed in the reaction mixture, which restricted the effective mass transfer of the reactants onto the active surface of catalyst, causing diffusion and saturation, thus hindering the interaction. The result is the overall reduced product yield [33].

Table 7. Yield of biodiesel from castor oil with changes in reaction parameters.

Feedstock	Conc. of Catalyst (%)	Methanol to Oil Ratio	Temperature (°C)	Reaction Time (min)	Biodiesel Yield (%)
	0.5				86.1 ± 0.8
	1.00				87.4 ± 0.4
	1.5	5:1	40	120	85.0 ± 0.6
	2.00				84.1 ± 0.9
	2.5				83.4 ± 0.6
		10:1			88.3 ± 0.8
	1.00	15:1	40	120	90.0 ± 0.4
		20:1			87.7 ± 0.8
Castor oil		25:1			85.0 ± 0.6
			50		91.0 ± 0.7
	1.00	15:1	60	120	96.1 ± 0.3
			70		89.0 ± 0.4
			80		87.0 ± 0.6
				30	90.5 ± 0.5
	1.00	15:1	60	60	93.1 ± 0.7
				90	96.1 ± 0.6
				150	94.5 ± 0.8

2.6.2. Effect of Methanol to Oil Ratio

The effect of methanol to oil ratio was studied at five different levels including 5:1, 10:1, 15:1, 20:1 and 25:1 (methanol:oil) for the 20% Li-TiO$_2$/feldspar catalyst to optimize biodiesel production from bitter apple, karanja, wild mustard, castor and wild safflower oils. The methanol/oil molar ratio is one of the most vital factors affecting biodiesel yield. Although the theoretically required methanol/oil molar ratio is 3:1, it is commonplace to carry out the transesterification reaction with an extra amount of alcohol to shift the equilibrium to the fatty acid methyl ester (FAME) side. The optimum molar ratio of alcohol is very important to reduce the production cost of biodiesel. At a lower alcohol molar ratio, the conversion of triglycerides into FAME will not be complete. On the other hand, a very high molar ratio may also decrease biodiesel yield, as methanol can cause emulsification of the polar hydroxyl groups present in glycerol, which is a byproduct of biodiesel production. This emulsification process hinders forward reactions and favors backward reactions. As a result, a decrease in the biodiesel yield is observed. An optimized amount of alcohol (a slight excess amount) is required to keep the transesterification reaction in the forward direction, as the transesterification reaction is reversible in nature [34]. The effect of the methanol to oil ratio on biodiesel production is summarized in Tables 3–7.

The optimum methanol to oil ratio was 10:1 for karanja and bitter apple using 20% Li/TiO$_2$/feldspar. The highest biodiesel yield obtained from karanja and bitter apple was 91.7 ± 0.6 and 89.2 ± 0.7, respectively, while wild mustard and wild safflower provided the maximum biodiesel yield (89.1 ± 0.3 and 89.7 ± 0.3%, respectively) at a 5:1 methanol to oil ratio.

It can be noted from the obtained results that castor oil requires a 15:1 methanol to oil ratio, which is relatively higher than the other oils used in the current study. The different methanol to oil ratios optimized for various oils depends upon the viscosity of the oil as well as the viscosity of the oil/methanol mixture formed after the addition of the catalyst. Castor oil is already known to have a higher viscosity than many other vegetable oils [35]. Thus, high amounts of methanol are required for the proper conversion of the reactant to biodiesel [36].

A longer separation time was required for the separation of the biodiesel from the water layer for a high molar ratio. This is due to the fact that the one hydroxyl group present in the methanol can work as an emulsifier to increase emulsion. Therefore, increasing the molar ratio of methanol to oil beyond 5:1 (in the case of wild mustard and wild safflower),

10:1 (in the case of karanja and bitter apple) and 15:1 (in the case of castor oil) was not suitable for increasing the biodiesel yield, but it made the recovery of ester difficult and increased the methanol recovery cost [36]. In addition to the already described viscosity factor, the variations in the FFA and water contents in the different types of oil also play a vital role in the optimization of the methanol to oil ratio. The variation in the FFA contents of some plant oils causes the used catalyst to react differently towards the different oils [37].

2.6.3. Effect of Temperature

Transesterification can be conducted at temperatures from room temperature to a temperature that is close to methanol's boiling point. However, transesterification reactions generally occur at high speed at elevated temperatures with shortened reaction periods. The importance of the reaction temperature in the transesterification of oils can be clearly observed from the much-boosted yields obtained in the present study after elevating reaction temperature. To study the effect of temperature on ester content and the yield of the transesterification, experiments were conducted at 40–80 °C by keeping the volume and other reaction variables constant. It is clear from the results shown in Tables 3–7 that the increase in the reaction temperature increases biodiesel yield. The increase in temperature accelerated the transesterification reaction, not only by decreasing viscosity but also by increasing mass transfer and initiating the more effective collision between molecules participating in the chemical reaction [36]. In the present study, karanja, bitter apple, wild mustard, and wild safflower oils showed the highest biodiesel yield at 50 °C, while castor oil provided the highest biodiesel yield at 60 °C. The higher optimized temperature to produce biodiesel from castor oil can be related to the higher viscosity of castor oil than other oils. Higher temperature could lead to a decrease in the viscosity of castor oil, making the reaction between oil and methanol molecules more favorable for the more effective conversion of triglycerides. The rise in temperature also favors the relative miscibility of the non-polar oil phase to polar alcoholic media to increase the speed of the chemical reaction [38]. However, increasing the temperature beyond the optimized temperature resulted in a decreased biodiesel yield due to the greater possibility of side reactions that can take place at higher temperatures along with a higher loss of methanol from the reaction mixture due to greater evaporation [39].

2.6.4. Reaction Time

The transesterification reaction is an equilibrium reaction, and the reaction time is a crucial variable to obtain the maximum biodiesel yield in an optimized reaction period. If a too-short reaction time is selected, there will not be complete conversion of reactants into products and the purification cost will also be higher. On the other hand, if a too-long reaction time is selected, there will be higher production cost and the product could also undergo decomposition or side reactions. To determine the optimum reaction time to produce biodiesel from oils (bitter apple; karanja; wild mustard; castor; and wild safflower oils), the reactions were performed at different reaction times (30 to 150 min) for the 20% Li-TiO$_2$/feldspar catalyst. The impact of time on the transesterification reactions performed on the different oil samples is presented in Tables 3–7.

The bitter apple, karanja, castor, and wild safflower oils provided the maximum methyl ester yield after 120 min. The highest biodiesel yield for bitter apple, karanja, castor, and wild safflower oils was 96.2 ± 0.8%, 96.4 ± 0.9%, 96.1 ± 0.3%, and 96.5 ± 0.3%, respectively, while the wild mustard oil provided the best biodiesel yield of 96.7 ± 0.7% after 90 min. Transesterification reaction is an equilibrium reaction, and there is always the possibility of a reverse reaction if reaction times are longer. Longer time periods are required for oils with higher saturated fatty acid contents [37–48].

2.7. Evaluation of Fuel Quality Parameters

The structural features of each fatty ester present in the fatty acid methyl esters determine its physicochemical properties such as density, acid value, iodine value, saponification value, cold flow properties (cloud and pour points) and ignition quality. The fuel characteristics of the alkyl esters synthesized from bitter apple, karanja, wild mustard, castor and wild safflower were evaluated according to ASTM (D6751) standard methods and compared with diesel (Table 8). This standard identifies the parameters that the pure biodiesel must meet before being used. Density is a very important fuel property because the precise fuel amount is required for proper combustion. High-density biodiesel could experience incomplete combustion, while on the other hand, low-density biodiesel fuels are highly volatile. The densities of biofuels depend on the feedstock's nature, the biodiesel's synthesis method and the structural features of methyl ester [40]. The densities of biodiesel produced from bitter apple, karanja, wild mustard, castor and safflower oils were 0.85, 0.89, 0.84, 0.87 and 0.86 g/mL, respectively.

Table 8. Fuel properties of FAME produced from various sources.

Fuel Parameters	Bitter Apple	Karanja	Wild Mustard	Castor Oil	Wild Safflower	Diesel ASTM D975	ASTM D6751 Limits
Density (g/mL)	0.85	0.89	0.84	0.87	0.86	0.85	Not specified
Cloud point (°C)	1.3	2.0	−2	1.2	0.2	−15–5	−15 to 10
Pour point (°C)	−4.2	−1.6	−4.2	−4.1	−4.0	−35–15	Not specified
Acid value (mg KOH/g)	0.43	0.17	0.43	0.27	0.41	-	0.50 max
Iodine value (g I_2/100 g)	96.31	85.1	76.77	86.04	78.64	-	Not specified
Saponification value (mg KOH g^{-1} oil)	190.92	176.03	185.05	187.01	179.74	-	Not specified
Cetene number	53.21	58.15	58.52	56.12	58.97	40–55	47 minimum

An important criterion to determine a biofuel's quality is the fuel's low temperature behavior, as it could solidify in filters and pipelines of the engine and may cause different problems such as fuel undernourishment and delayed ignition. The cloud point (CP) is the temperature at which it appears hazy or cloudy. The pour point (PP) is the temperature at which the lubricating oil ceases to flow [41]. From the obtained results, it was observed that the cloud and pour points measured for the produced FAME from bitter apple, karanja, wild mustard, castor and wild safflower oils using the 20% Li-TiO$_2$/feldspar nanocatalyst lie in the range which was prescribed by the ASTM.

The acid value (AV) is the amount of potassium hydroxide (in milligrams) required for the neutralization of the organic acids per gram of fat [42]. The AVs of the synthesized biodiesels from bitter apple, karanja, wild mustard, castor, and wild safflower oils in the presence of the 20% Li-TiO$_2$/feldspar catalyst were 0.43, 0.17, 0.43, 0.27 and 0.41 mg KOH g^{-1}, respectively (Table 8). The iodine value (IV) is measured as the amount of iodine (in grams) that is adsorbed by 100 g of oil or biodiesel [43]. The iodine values of the biodiesels synthesized from bitter apple, karanja, wild mustard, castor and wild safflower oils were 96.31, 85.1, 76.77, 86.04, and 78.64 g I_2/100 g, respectively.

The conversion of fat/oil/lipid by means of a reaction with aqueous alkali into alcohol and soap is called the saponification reaction [42]. The saponification values of the biodiesels produced from bitter apple, karanja, wild mustard, castor and wild safflower oils were 190.92, 176.03, 185.05, 187.01 and 179.74, respectively. The cetane number (CN) reflects the ignition delay period [41]. The cetene numbers of the biodiesels produced from bitter apple, karanja, wild mustard, castor and wild safflower oils were 53.21, 58.15, 58.52, 56.12, and 58.97, respectively.

2.8. Fatty Acid Profile

A biodiesel is a mixture of long-chain FAs, with the number of C atoms present in the chain varying from 14 to 22 [44]. The main fatty acids appearing in bitter apple [48], karanja [45], wild mustard, castor [47], and wild safflower [46] oils were linoleic acid (70.71%), oleic acid (51.92%), erucic acid (41.43%), ricinoleic acid (80.54%), and linoleic acid (75.17%), respectively (Table 9).

Table 9. The chemical composition of used oils.

Sr.No.	Fatty Acid	Molecular Formula	Fatty Acid Amount (%)				
			Karanja Oil	Wild Mustard Oil	Wild Safflower Oil	Castor Oil	Bitter Apple Oil
1	Capric acid	$C_{10}H_{20}O_2$	0.11	0.15	0.13	0.12	0.07
2	Lauric acid	$C_{12}H_{24}O_2$	0.22	0.12	0.09	0.08	0.06
3	Myristic acid	$C_{14}H_{28}O_2$	0.93	0.18	0.16	0.11	0.13
4	Palmitic acid	$C_{16}H_{32}O_2$	10.33	3.63	7.73	1.30	8.35
5	Margaric acid	$C_{17}H_{34}O_2$	0.09	0.05	0.06	0.07	0.01
6	Linolenic acid	$C_{18}H_{30}O_2$	3.15	0.09	0.32	1.57	0.17
7	Linoleic acid	$C_{18}H_{32}O_2$	11.03	15.75	75.17	7.65	70.71
8	Oleic acid	$C_{18}H_{34}O_2$	51.92	23.11	12.98	5.83	9.96
9	Ricinoleic acid	$C_{18}H_{34}O_3$	-	-	-	80.54	0
10	Stearic acid	$C_{18}H_{36}O_2$	4.66	1.15	0.89	1.43	8.29
11	Eicosanoic acid	$C_{20}H_{40}O_2$	9.76	12.83	0.11	0.18	0.03
12	Arachidic acid	$C_{20}H_{40}O_2$	0.96	0.07	0.76	0.21	0.11
13	Erucic acid	$C_{22}H_{42}O_2$	-	41.43	-	-	0.17
14	Behenic acid	$C_{22}H_{44}O_2$	4.36	0.09	0.43	0.17	0.07
15	Lignoceric acid	$C_{24}H_{48}O_2$	2.12	1.12	0.32	0.15	1.13

3. Materials and Methods

3.1. Chemicals and Reagents

Analytical grade methanol (99%), Wijs reagent, ethanol (99.5%), titanium dioxide, lithium nitrate, sodium sulphate (anhydrous), hydrochloric acid (37%), potassium hydroxide, sodium hydroxide (98%), petroleum ether, starch, Wijs solution, sodium thiosulphate, phenolphthalein, and potassium iodide were purchased from Sigma-Aldrich (Lahore, Pakistan).

3.2. Materials and Oil Extraction

Citrullus colocynthis (bitter apple), *Pongamia pinnata* (karanja), *Sinapis arvensis* (wild mustard), *Ricinus communis* (castor oil) and *Carthamus oxyacantha* (wild safflower) seeds were collected after obtaining permission from Head of Department (HOD,) Department of Chemistry, UAF, Pakistan. The plant materials, including bitter apple, karanja, wild mustard, castor and wild safflower, were identified by Dr. Mansoor Hameed, UAF, and the sample voucher specimen numbers were 21-R-001, 21-R-002, 21-R-003, 21-R-004 and 21-R-005, respectively. All of the experimental research and field studies on plants were conducted in compliance with the standard rules. After the sample collection, seeds were cleaned to remove all of the foreign particles, such as dirt, chaff, stones, dust, and immature broken seeds. Seeds were extracted from kernels. Extracted seeds were crushed and ground with the help of a pestle and mortar. Oil was extracted using an automatic screw press machine. The screw press machine compressed the seeds between the main screw and travelling cones for the extraction of oil from the seeds. This machine separated the oil and non-oily solid (cake). The extracted oil was further purified using a high-speed centrifuge at 5000 rpm followed by filtration with a vacuum filtration assembly. By using the vacuum filtration assembly, the oil was filtered off to remove impurities and solid particles.

3.3. Catalyst Preparation

The Li-incorporated TiO_2 nano-catalysts were synthesized via a wet impregnation process. We took 10 g of TiO_2 nanoparticles and dissolved them in 20 mL of distilled water and heated this solution at room temperature with slow stirring. Subsequently, the desired concentration of lithium nitrate was added dropwise. Additionally, the resultant slurry was mixed for 4 h at room temperature using a thermostatic magnetic stirrer and heated overnight in an oven at 120 °C, and the obtained sample was calcined at 600 °C in a muffle furnace for 4 h. Similarly, a series of Li-impregnated TiO_2 nanoparticles with different amounts of Li (10, 20, 30 and 40 wt%) were synthesized.

In order to resolve the problem faced by using a catalyst without a support for the production of biodiesel, the clay mineral feldspar was used to prepare the supported catalysts in the present study. The nano-catalysts of Li-TiO_2 with different amounts of Li (10%, 20%, 30%, 40%) were supported on feldspar. For the synthesis of supported catalysts, 0.5 g of each prepared nanocatalyst and 0.75 g of feldspar support were dissolved into distilled water to create a uniform paste. The mixture was then dried in an oven at 150 °C for half an hour to eliminate the moisture. The prepared supported nano-catalysts were ground with the help of a pestle and mortar. A nano-sieve was used to separate the nanoparticles from bigger particles.

3.4. Characterization

X-ray diffraction was performed to determine the nanocatalysts' structure by using a Shimadzu model 6000 Power X-ray diffractometer (Shimadzu Corp., Kypoto, Japan). The obtained diffraction peaks were compared with standard compounds described in the Joint Committee on Powder Diffraction Standards (JCPDS) databank. The average crystal size of Li-TiO_2/feldspar was determined using the Debye–Scherrer equation.

FTIR (Agilent technologies, Santa Clara, CA, USA) analysis was carried out to identify the functional groups present in Li-TiO_2/feldspar nanocatalysts. The morphological structure and elemental composition of nanoparticles were determined using SEM (NOVA NANOSEM-450, Thermo Fisher Scientific, New York, NY, USA) and EDX (Nova 450), respectively. The fatty acid composition of oils was determined by means of gas chromatography equipped with a flame ionization detector.

3.5. Transesterification Process

In the presence of nanocatalysts, biodiesel was produced by means of transesterification of bitter apple, karanja, wild mustard, castor, and wild safflower oils using methanol. Several reversible and successive steps are involved in transesterification. Biodiesel is the major product of the reaction and floats on top, while the by-product glycerol is present at the bottom. Different concentrations of lithium (10–40% Li/TiO_2)-doped titanium oxide nanocatalysts supported on feldspar were used for transesterification.

The methanol to oil molar ratio (5:1, 10:1, 15:1, 20:1 and 25:1), catalyst amount (0.5%, 1%, 1.5%, 2% and 2.5%), reaction temperature (40; 50; 60, 70; and 80 °C), and reaction time (30, 60, 90, 120 and 150 min) were optimized during the present study to obtain the maximum biodiesel yield. After the completion of the reaction, the prepared biodiesel was centrifuged at 1500 rpm for 20 min for the removal of the remaining nanocatalysts from the biodiesel. Hot distilled water in an excess amount was used to wash the produced biodiesel and to remove surplus methanol from the biodiesel. The cleared transparent liquid was obtained after proper washing of the biodiesel. The biodiesel % yield was calculated by applying Equation (2).

$$Process\ yield\ (\%) = \frac{Pure\ biodiesel\ (g)}{Oil\ used\ (g)} \times 100 \tag{2}$$

3.6. Quality Parameters of Biodiesel

The biodiesel quality parameters such as density, acid value, pour point (PP), iodine value (IV), cetane number (CN), saponification value (SV), and cloud point (CP) were determined using methods and equations given in the literature [6].

4. Conclusions

The aim of the present research was to compare the yield of biodiesel from different non-edible oils such as *Citrullus colocynthis* (bitter apple), *Pongamia pinnata* (karanja); *Sinapis arvensis* (wild mustard), *Ricinus communis* (castor oil), and *Carthamus oxyacantha* (wild safflower) using a novel lithium-impregnated titanium oxide catalyst supported on feldspar minerals (Li-TiO$_2$/feldspar) and to assess the efficiency of the synthesized nanocatalysts under mild reaction conditions. A series of lithium-impregnated titanium oxide nanocatalysts with different amounts of Li (10%, 20%, 30%, 40%) supported on the clay mineral feldspar (Li-TiO$_2$/feldspar) were investigated, conducting the transesterification process for biodiesel production. However, the study showed that 20 wt% of Li showed the highest activity for FAME formation. The characterization of support-loaded catalysts was performed using XRD, SEM/EDX and FTIR analysis. The major fatty acids present in bitter apple, karanja, wild mustard, castor, and wild safflower oils were linoleic acid (70.71%), oleic acid (51.92%), erucic acid (41.43%), ricinoleic acid (80.54%), and linoleic acid (75.17%), respectively. One of the aims of the present study was to look for a universal transesterification catalyst that could effectively convert feedstock oils of quite variable compositions with good efficiency, and the produced Li-TiO$_2$/feldspar is found to be such a catalyst.

Author Contributions: M.H.: Conceived and designed the analysis, Investigation, Collected the data, Writing—original draft; I.A.B.: Supervision, Conceptualization, Data curation; K.S.: Formal analysis, Writing—review & editing; M.A.H.: Data curation, Validation, Writing—review & editing. All authors have read and agreed to the published version of the manuscript.

Funding: The Deanship of Scientific Research (DSR) at King Abdulaziz University, Jeddah, Saudi Arabia has funded this project, under grant no (KEP-1-155-42).

Data Availability Statement: There is no data associated with it.

Conflicts of Interest: The authors declare no conflict of interest.

References

1. Mahlia, T.; Syazmi, Z.; Mofijur, M.; Abas, A.P.; Bilad, M.; Ong, H.C.; Silitonga, A. Patent landscape review on biodiesel production: Technology updates. *Renew. Sustain. Energy Rev.* **2020**, *118*, 109526. [CrossRef]
2. Chidambaranathan, B.; Gopinath, S.; Aravindraj, R.; Devaraj, A.; Krishnan, S.G.; Jeevaananthan, J. The production of biodiesel from castor oil as a potential feedstock and its usage in compression ignition Engine: A comprehensive review. *Mater. Today Proc.* **2020**, *33*, 84–92. [CrossRef]
3. Silitonga, A.S.; Shamsuddin, A.H.; Mahlia, T.M.I.; Milano, J.; Kusumo, F.; Siswantoro, J.; Dharma, S.; Sebayang, A.H.; Masjuki, H.H.; Ong, H.C. Biodiesel synthesis from Ceiba pentandra oil by microwave irradiation-assisted transesterification: ELM modeling and optimization. *Renew. Energy* **2020**, *146*, 1278–1291. [CrossRef]
4. Ueki, Y.; Saiki, S.; Hoshina, H.; Seko, N. Biodiesel fuel production from waste cooking oil using radiation-grafted fibrous catalysts. *Radiat. Phys. Chem.* **2018**, *143*, 41–46. [CrossRef]
5. Srivastava, N.; Srivastava, M.; Gupta, V.K.; Manikanta, A.; Mishra, K.; Singh, S.; Singh, S.; Ramteke, P.; Mishra, P. Recent development on sustainable biodiesel production using sewage sludge. *3 Biotech* **2018**, *8*, 245. [CrossRef]
6. Hanif, M.; Bhatti, H.N.; Hanif, M.A.; Rashid, U.; Hanif, A.; Moser, B.R.; Alsalme, A. A Novel Heterogeneous Superoxide Support-Coated Catalyst for Production of Biodiesel from Roasted and Unroasted Sinapis arvensis Seed Oil. *Catalysts* **2021**, *11*, 1421. [CrossRef]
7. Lee, J.-C.; Lee, B.; Ok, Y.S.; Lim, H. Preliminary techno-economic analysis of biodiesel production over solid-biochar. *Bioresour. Technol.* **2020**, *306*, 123086. [CrossRef]
8. Abomohra, A.E.-F.; Elsayed, M.; Esakkimuthu, S.; El-Sheekh, M.; Hanelt, D. Potential of fat, oil and grease (FOG) for biodiesel production: A critical review on the recent progress and future perspectives. *Prog. Energy Combust. Sci.* **2020**, *81*, 100868. [CrossRef]

9.	Nair, A.S.; Al-Bahry, S.; Gathergood, N.; Tripathi, B.N.; Sivakumar, N. Production of microbial lipids from optimized waste office paper hydrolysate, lipid profiling and prediction of biodiesel properties. *Renew. Energy* **2020**, *148*, 124–134. [CrossRef]
10.	Sangwan, S.; Rao, D.; Sharma, R. A review on *Pongamia pinnata* (L.) Pierre: A great versatile leguminous plant. *Nat. Sci.* **2010**, *8*, 130–139.
11.	Demirbas, A.; Bafail, A.; Ahmad, W.; Sheikh, M. Biodiesel production from non-edible plant oils. *Energy Explor. Exploit.* **2016**, *34*, 290–318. [CrossRef]
12.	Sadia, H.; Ahmad, M.; Zafar, M.; Sultana, S.; Azam, A.; Khan, M.A. Variables effecting the optimization of non edible wild safflower oil biodiesel using alkali catalyzed transesterification. *Int. J. Green Energy* **2013**, *10*, 53–62. [CrossRef]
13.	Khanahmadzadeh, A.; Almasi, M.; Meighani, H.M.; Mohamad, A.; Borghei, J.A. Assessment of Wild Safflower oil Methyl Ester as Potential Alternative Fuel. *J. Appl. Environ. Biol. Sci.* **2011**, *1*, 325–328.
14.	Aburas, H.; Bafail, A.; Demirbas, A. The pyrolizing of waste lubricating oil (WLO) into diesel fuel over a supported calcium oxide additive. *Pet. Sci. Technol.* **2015**, *33*, 226–236. [CrossRef]
15.	Supamathanon, N.; Wittayakun, J.; Prayoonpokarach, S. Properties of Jatropha seed oil from Northeastern Thailand and its transesterification catalyzed by potassium supported on NaY zeolite. *J. Ind. Eng. Chem.* **2011**, *17*, 182–185. [CrossRef]
16.	Gashaw, A.; Getachew, T.; Teshita, A. A Review on biodiesel production as alternative fuel. *J. Prod. Ind.* **2015**, *4*, 80–85.
17.	Mishra, V.K.; Goswami, R. A review of production, properties and advantages of biodiesel. *Biofuels* **2018**, *9*, 273–289. [CrossRef]
18.	Hanif, M.A.; Nisar, S.; Akhtar, M.N.; Nisar, N.; Rashid, N. Optimized production and advanced assessment of biodiesel: A review. *Int. J. Energy Res.* **2018**, *42*, 2070–2083. [CrossRef]
19.	Hanif, M.; Bhatti, I.A.; Zahid, M.; Shahid, M. Production of biodiesel from non-edible feedstocks using environment friendly nano-magnetic Fe/SnO catalyst. *Sci. Rep.* **2022**, *12*, 16705. [CrossRef]
20.	Carlucci, C.; Degennaro, L.; Luisi, R. Titanium Dioxide as a Catalyst in Biodiesel Production. *Catalysts* **2019**, *9*, 75. [CrossRef]
21.	Sankar, M.; He, Q.; Engel, R.V.; Sainna, M.A.; Logsdail, A.J.; Roldan, A.; Willock, D.J.; Agarwal, N.; Kiely, C.J.; Hutchings, G.J. Role of the support in gold-containing nanoparticles as heterogeneous catalysts. *Chem. Rev.* **2020**, *120*, 3890–3938. [CrossRef]
22.	Shuit, S.H.; Yee, K.F.; Lee, K.T.; Subhash, B.; Tan, S.H. Evolution towards the utilisation of functionalised carbon nanotubes as a new generation catalyst support in biodiesel production: An overview. *RSC Adv.* **2013**, *3*, 9070–9094. [CrossRef]
23.	Xue, B.; Guo, H.; Liu, L.; Chen, M. Preparation, characterization and catalytic properties of yttrium-zirconium-pillared montmorillonite and their application in supported Ce catalysts. *Clay Miner.* **2015**, *50*, 211–219. [CrossRef]
24.	Rahman, M.; Zaidan, U.H.; Basri, M.; Othman, S.S.; Rahman, R.N.Z.R.A.; Salleh, A.B. Modification of natural feldspar as support for enzyme immobilization. *J. Nucl. Relat. Technol. (Spec. Ed.)* **2009**, *6*, 25–42.
25.	Alsharifi, M.; Znad, H.; Hena, S.; Ang, M. Biodiesel production from canola oil using novel Li/TiO$_2$ as a heterogeneous catalyst prepared via impregnation method. *Renew. Energy* **2017**, *114*, 1077–1089. [CrossRef]
26.	Salimkhani, H.; Joodi, T.; Bordbar-Khiabani, A.; Dizaji, A.M.; Abdolalipour, B.; Azizi, A. Surface and structure characteristics of commercial K-Feldspar powders: Effects of temperature and leaching media. *Chin. J. Chem. Eng.* **2020**, *28*, 307–317. [CrossRef]
27.	Todea, M.; Turcu, R.; Frentiu, B.; Simon, S. FTIR and NMR evidence of aluminosilicate microspheres bioactivity tested in simulated body fluid. *J. Non-Cryst. Solids* **2016**, *432*, 413–419. [CrossRef]
28.	Basiron, N.; Sreekantan, S.; Akil, H.M.; Saharudin, K.A.; Harun, N.H.; Mydin, R.B.S.; Seeni, A.; Rahman, N.R.A.; Adam, F.; Iqbal, A. Effect of Li-TiO$_2$ nanoparticles incorporation in ldpe polymer nanocomposites for biocidal activity. *Nano-Struct. Nano-Objects* **2019**, *19*, 100359. [CrossRef]
29.	Ji, H.; Li, X.; Chen, D. *Cymbiola nobilis* shell: Toughening mechanisms in a crossed-lamellar structure. *Sci. Rep.* **2017**, *7*, 40043. [CrossRef]
30.	Park, K.; Kim, W.; Kim, H.-Y. Optimal lamellar arrangement in fish gills. *Proc. Natl. Acad. Sci. USA* **2014**, *111*, 8067–8070. [CrossRef]
31.	Lin, Y.-H.; Weng, C.-H.; Srivastav, A.L.; Lin, Y.-T.; Tzeng, J.-H. Facile synthesis and characterization of N-doped TiO$_2$ photocatalyst and its visible-light activity for photo-oxidation of ethylene. *J. Nanomater.* **2015**, *2015*, 807394. [CrossRef]
32.	Abbasi, S.A.; Rafique, M.; Mir, A.A.; Kearfott, K.J.; Ud-Din Khan, S.; Ud-Din Khan, S.; Khan, T.M.; Iqbal, J. Quantification of elemental composition of Granite Gneiss collected from Neelum Valley using calibration free laser-induced breakdown and energy-dispersive X-ray spectroscopy. *J. Radiat. Res. Appl. Sci.* **2020**, *13*, 362–372. [CrossRef]
33.	Kusumaningtyas, R.D.; Pristiyani, R.; Dewajani, H. A new route of biodiesel production through chemical interesterification of jatropha oil using ethyl acetate. *Int. J. ChemTech Res.* **2016**, *9*, 627–634.
34.	Verma, P.; Sharma, M. Comparative analysis of effect of methanol and ethanol on Karanja biodiesel production and its optimisation. *Fuel* **2016**, *180*, 164–174. [CrossRef]
35.	Keera, S.; El Sabagh, S.; Taman, A. Castor oil biodiesel production and optimization. *Egypt. J. Pet.* **2018**, *27*, 979–984. [CrossRef]
36.	Leung, D.; Guo, Y. Transesterification of neat and used frying oil: Optimization for biodiesel production. *Fuel Process. Technol.* **2006**, *87*, 883–890. [CrossRef]
37.	Hoque, M.E.; Singh, A.; Chuan, Y.L. Biodiesel from low cost feedstocks: The effects of process parameters on the biodiesel yield. *Biomass Bioenergy* **2011**, *35*, 1582–1587. [CrossRef]
38.	Ogbu, I.; Ajiwe, V. Biodiesel production via esterification of free fatty acids from *Cucurbita pepo* L. seed oil: Kinetic studies. *Int. J. Sci. Technol.* **2013**, *2*, 616–621.

39. De, A.; Boxi, S.S. Application of Cu impregnated TiO$_2$ as a heterogeneous nanocatalyst for the production of biodiesel from palm oil. *Fuel* **2020**, *265*, 117019. [CrossRef]

40. Zango, Z.U.; Kadir, H.A.; Imam, S.S.; Muhammad, A.I.; Abu, I.G. Optimization Studies for Catalytic Conversion of Waste Vegetable Oil to Biodiesel. *Am. J. Chem.* **2019**, *9*, 27–32.

41. Sakthivel, R.; Ramesh, K.; Purnachandran, R.; Shameer, P.M. A review on the properties, performance and emission aspects of the third generation biodiesels. *Renew. Sustain. Energy Rev.* **2018**, *82*, 2970–2992. [CrossRef]

42. Ismail, S.A.-e.A.; Ali, R.F.M. Physico-chemical properties of biodiesel manufactured from waste frying oil using domestic adsorbents. *Sci. Technol. Adv. Mater.* **2015**, *16*, 034602. [CrossRef] [PubMed]

43. Folayan, A.J.; Anawe, P.A.L.; Aladejare, A.E.; Ayeni, A.O. Experimental investigation of the effect of fatty acids configuration, chain length, branching and degree of unsaturation on biodiesel fuel properties obtained from lauric oils, high-oleic and high-linoleic vegetable oil biomass. *Energy Rep.* **2019**, *5*, 793–806. [CrossRef]

44. Nurdin, M.; Fatma, F.; Natsir, M.; Wibowo, D. Characterization of methyl ester compound of biodiesel from industrial liquid waste of crude palm oil processing. *Anal. Chem. Res.* **2017**, *12*, 1–9.

45. Islam, A.K.M.A.; Chakrabarty, S.; Yaakob, Z.; Ahiduzzaman, M.; Islam, A.K.M.M. Koroch (*Pongamia pinnata*): A Promising Unexploited Resources for the Tropics and Subtropics. In *Forest Biomass-From Trees to Energy*; IntechOpen: London, UK, 2021.

46. Murthy, I.; Anjani, K. Fatty acid composition in Carthamus species. In Proceedings of the 7th International Safflower Conference, Wagga Wagga, NSW, Australia, 3–6 November 2008.

47. Salimon, J.; Noor, D.A.M.; Nazrizawati, A.; Noraishah, A. Fatty acid composition and physicochemical properties of Malaysian castor bean *Ricinus communis* L. seed oil. *Sains Malays.* **2010**, *39*, 761–764.

48. Sabiri, M. Fatty acid and sterol constituents of *Citrullus colocynthis* (L.) Schard Seed oil. *RHAZES Green Appl. Chem.* **2019**, *7*, 57–60.

 *catalysts*

Article

Nano-Magnetic CaO/Fe$_2$O$_3$/Feldspar Catalysts for the Production of Biodiesel from Waste Oils

Maryam Hanif [1], Ijaz Ahmad Bhatti [1], Muhammad Asif Hanif [1], Umer Rashid [2,3,*], Bryan R. Moser [4], Asma Hanif [5] and Fahad A. Alharthi [6]

[1] Department of Chemistry, University of Agriculture, Faisalabad 38040, Pakistan; drmaryamhanifchemist@gmail.com (M.H.); ijazchem@gmail.com (I.A.B.); drmuhammadasifhanif@gmail.com (M.A.H.)

[2] Institute of Nanoscience and Nanotechnology (ION2), Universiti Putra Malaysia, Serdang 43400, Selangor, Malaysia

[3] Center of Excellence in Catalysis for Bioenergy and Renewable Chemicals (CBRC), Faculty of Science, Chulalongkorn University, Bangkok 10330, Thailand

[4] Bio-Oils Research Unit, National Center for Agricultural Utilization Research, United States Department of Agriculture, Agricultural Research Service, Peoria, IL 61604, USA; bryan.moser@usda.gov

[5] Department of Chemistry, Government College Women University, Faisalabad 38000, Pakistan; asmarana_hanif@yahoo.com

[6] Chemistry Department, College of Science, King Saud University, Riyadh 1145, Saudi Arabia; fharthi@ksu.edu.sa

* Correspondence: umer.rashid@upm.edu.my or dr.umer.rashid@gmail.com; Tel.: +60-3-97697393

Citation: Hanif, M.; Bhatti, I.A.; Hanif, M.A.; Rashid, U.; Moser, B.R.; Hanif, A.; Alharthi, F.A. Nano-Magnetic CaO/Fe$_2$O$_3$/Feldspar Catalysts for the Production of Biodiesel from Waste Oils. *Catalysts* **2023**, *13*, 998. https://doi.org/10.3390/catal13060998

Academic Editors: José María Encinar Martín and Sergio Nogales Delgado

Received: 18 April 2023
Revised: 30 May 2023
Accepted: 4 June 2023
Published: 13 June 2023

Abstract: Production of biodiesel from edible vegetable oils using homogenous catalysts negatively impacts food availability and cost while generating significant amounts of caustic wastewater during purification. Thus, there is an urgent need to utilize low-cost, non-food feedstocks for the production of biodiesel using sustainable heterogeneous catalysis. The objective of this study was to synthesize a novel supported nano-magnetic catalyst (CaO/Fe$_2$O$_3$/feldspar) for the production of biodiesel (fatty acid methyl esters) from waste and low-cost plant seed oils, including *Sinapis arvensis* (wild mustard), *Carthamus oxyacantha* (wild safflower) and *Pongamia pinnata* (karanja). The structure, morphology, surface area, porosity, crystallinity, and magnetization of the nano-magnetic catalyst was confirmed using XRD, FESEM/EDX, BET, and VSM. The maximum biodiesel yield (93.6–99.9%) was achieved at 1.0 or 1.5 wt.% catalyst with methanol-to-oil molar ratios of 5:1 or 10:1 at 40 °C for 2 h. The CaO/Fe$_2$O$_3$/feldspar catalyst retained high activity for four consecutive cycles for conversion of karanja, wild mustard, and wild safflower oils. The effective separation of the catalyst from biodiesel was achieved using an external magnet. Various different physico-chemical parameters, such as pour point, density, cloud point, iodine value, acid value, and cetane number, were also determined for the optimized fuels and found to be within the ranges specified in ASTM D6751 and EN 14214, where applicable.

Keywords: biodiesel; nano-magnetic catalyst; feldspar; transesterification; waste oils

1. Introduction

Energy plays a critical role in economic growth and modern society. Moreover, economic development greatly depends on the long-term availability, security, and affordability of energy. Over the last 30 years there has been a dramatic increase in world-wide energy demand due to the emergence of developing economies and continued population growth [1]. The ever-increasing use of nonrenewable energy is not sustainable due to the irregular geographic distribution of resources as well as environmental, climatic, geopolitical, and economic concerns associated with their continued use [2]. In addition, the combustion of fossil fuels releases greenhouse gases that renders nonrenewable energy unsustainable and environmentally damaging [3].

Biodiesel represents a renewable, sustainable, and environmentally friendly alternative to petroleum diesel for combustion in compression-ignition (diesel) engines. Biodiesel is composed of fatty acid methyl (or ethyl) esters prepared from plant oils, waste lipids, used cooking oils, or other triglycerides and is typically prepared by alkaline-catalyzed transesterification. Biodiesel is a clean burning fuel with low exhaust emissions, high biodegradability, and minimal heteroatom residues. Biodiesel is an important alternative energy source because of its safe and renewable nature. Biodiesel has significantly reduced emissions of sulfates, particulate matter, unburnt hydrocarbons, and polycyclic aromatic hydrocarbons relative to petroleum diesel. In addition, net CO_2 emissions are not increased through the combustion of biodiesel because greenhouse gases are recycled as CO_2 by plants during photosynthesis [4]. Biodiesel is more environmentally friendly because it significantly reduces net CO_2 emissions versus petroleum diesel and is biodegradable and non-toxic [5].

In the biodiesel industry, the choice of raw material is extremely vital, as 80% of the cost to prepare biodiesel is associated with feedstock cost [6]. Edible feedstocks compete with food uses, deplete freshwater supplies, and crucially require large areas of fertile land [7]. In addition, production costs have increased 70–92% due to escalate in the cost of edible oils. These factors have stimulated interest in biodiesel production from non-edible oils [8]. Lately, low-cost waste oils have been under consideration as a potential feedstock alternative for BD production. In the current study, biodiesel was produced from waste/low-cost plant seed oils, including *Carthamus oxyacantha* (wild safflower), *Sinapis arvensis* (wild mustard) and *Pongamia pinnata* (karanja).

Predominantly, homogeneous catalysts are normally used for the industrial production of biodiesel due to their high catalytic activities and low cost. However, these catalysts cannot be reused, and the removal of dissolved homogeneous catalysts is tedious. Additionally, numerous processing steps are necessary, such as neutralization, washing, and dehydration. Thus, the major disadvantages of homogeneous catalysis are wastewater generation, extensive post-processing, and incompatibility with low-quality, low-cost feedstocks [9].

Heterogeneous catalysts have important advantages over homogeneous catalysts because they catalyze important chemical reactions associated with green chemistry that are environmentally friendly and economically favorable [10]. However, they often require high reaction temperatures and pressures, and catalytic performance is generally lower than homogeneous catalysts. Furthermore, coking, leaching, poisoning, and sintering may result in the deactivation of heterogeneous catalysts over time.

Nanomaterials have served as ideal supports for the preparation of solid acid catalysts because they possess large surface areas and can minimize mass transfer resistance [11]. Unfortunately, for small-size solid catalysts, their recovery and reutilization through the commonly used filtration or centrifugation approaches is difficult, especially with high-viscosity reaction systems (generally soybean oil has high viscosity). In such cases, a large amount of catalyst is lost during separation. Magnetic heterogeneous nano-catalysts are of interest because they have a higher surface area, which imparts greater catalytic activity relative to lower-surface-area solid catalysts. In addition, magnetic catalysts can be easily separated from reaction mixtures without the generation of large amounts of caustic wastewater [12,13].

Various common ferromagnetic materials, such as Fe_3O_4 and γ-Fe_2O_3, can be used as magnetic catalysts or catalyst supports. After modification and functionalization, they not only retain their magnetic properties but can also be quickly and easily separated when an external magnetic field is applied. Ferrites have considerable porosity, good chemical properties, high contact area, and high thermal stability, culminating in a more effective catalyst for the production of biodiesel [11]. Agglomeration in magnetite synthesis is very difficult to avoid, because of its magnetism. Thus, the selection of a suitable catalytic support has been of immense interest. Typically, a solid support with a high surface area is used to attach catalysts. Supports prevent the sintering and agglomeration of small catalyst particles, revealing additional surface area [14]. Most catalytic supports are zirconium dioxide, aluminum oxides, alumino silicates, magnesium oxide, silica gel, and titanium oxide [15]. Clay materials, compared to other supports, have the advantages of high quantity, low-priced, environmental compatibility, high selectivity, reusability, and operational simplicity [16].

The primary objective of this study was the synthesis of the novel supported nano-magnetic catalyst CaO/Fe_2O_3/Feldspar to produce biodiesel from waste/low-cost plant seed oils, including *Carthamus oxyacantha* (wild safflower), *Sinapis arvensis* (wild mustard) and *Pongamia pinnata* (karanja). The novel catalytic support which is low-cost feldspar (a clay mineral) was also prepared and characterized. Other objectives included the optimization of the transesterification reaction and characterization of the resulting biodiesel and catalytic materials.

2. Results and Discussion

2.1. Characterization of CaO/Fe₂O₃ and CaO/Fe₂O₃/feldspar Nano-Magnetic Catalyst

2.1.1. XRD Analysis

The XRD patterns of the nano-magnetic catalyst CaO/Fe_2O_3/feldspar and CaO/Fe_2O_3 are depicted in Figure 1. The results indicated that the CaO/Fe_2O_3/feldspar has several peaks at different angles, which indicated a crystalline structure. The prominent peaks were observed at $2\theta = 30°$, $32.34°$, $35.55°$, $37.36°$, $43.5°$, $53.61°$, $63.15°$, and $64.27°$, which was attributed to the crystalline phases of (220), (111), (311), (200), (422), (511), (440), and (311), respectively (ICSD File No. 00-006-0711 and 00-032-0168) [17] [Maryam et al. 2021]. These indicated that Fe_2O_3 and CaO were present in the structure of the catalyst [16]. The intense sharp diffraction peaks within $2\theta = 25–30°$ and the specifically intense peak at $27.4°$ suggested the presence of feldspar [18], which possessed a uniform structure and pore diameter. The narrow peaks showed the single phase of a highly ordered structure, which was indicated in the FESEM analysis as well (Figure 2). The crystal size of CaO/Fe_2O_3/feldspar nanoparticles calculated from the Scherrer equation was 45.73 nm.

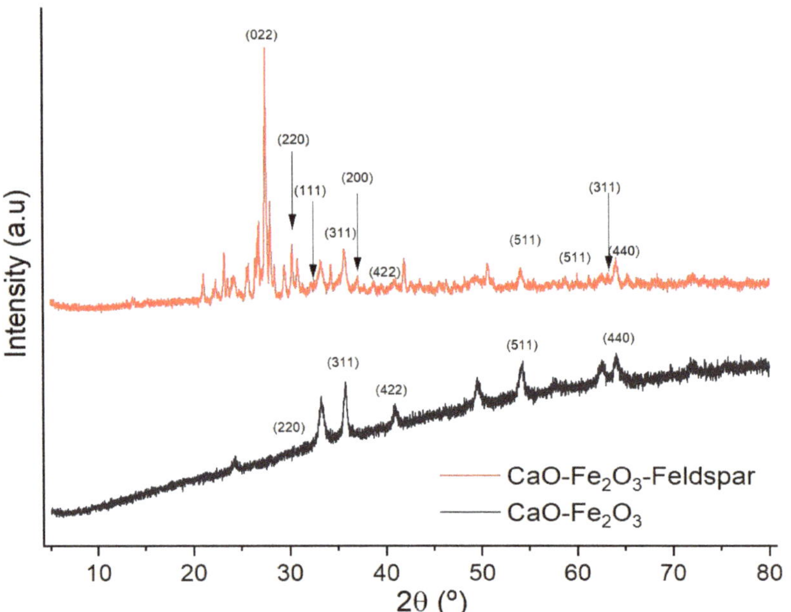

Figure 1. XRD analysis of CaO/Fe_2O_3 and CaO/Fe_2O_3/feldspar.

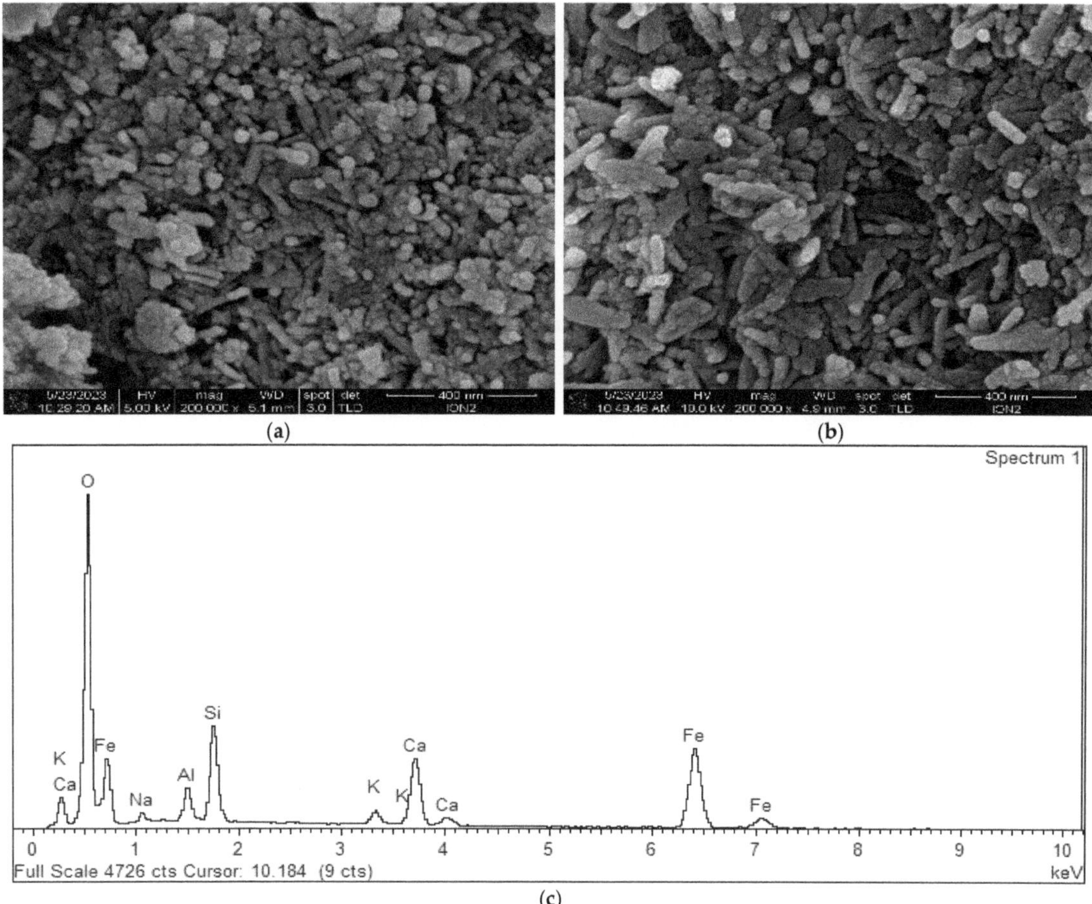

Figure 2. FESEM images of (**a**) CaO/Fe_2O_3, (**b**) CaO/Fe_2O_3/feldspar and (**c**) EDX analysis of CaO/Fe_2O_3/feldspar.

2.1.2. FESEM and EDX Analysis

Figure 2a,b depict different surface roughness, particle sizes, and arrangement of particles, which is very important for the conversion of oil into biodiesel. Figure 2a shows the agglomeration and coral-like shape, whereas the CaO-Fe_2O_3 composite attached to the surface of feldspar (Figure 2b) indicates a morphology of well-arranged small platelets laced with pores. The surface morphology by SEM thus confirmed the successful synthesis of the CaO-Fe_2O_3 nano-catalyst supported on the feldspar, which was also indicated by the EDX analysis. A close inspection of the spectra revealed an exact wt.% of O (43.46%), Na (0.99%), Al (2.18%), Si (6.69%), K (1.34%), Ca (8.66%), and Fe (36.68%) elements in the CaO/Fe_2O_3/feldspar catalyst, as presented in Table 1. Based on the EDX analysis, the CaO-Fe_2O_3 was rich in Fe with a 59.82 wt.% composition. The large amount of O (37.12 wt.%) and Ca (3.06 wt.%) indicated that the CaO/Fe_2O_3 was pure. The results confirmed that the CaO-Fe_2O_3 was efficiently supported on the feldspar matrix without distorting the elemental ratio. The quantitative data derived from the EDX analysis for the CaO/Fe_2O_3/feldspar is presented in Figure 2c.

Table 1. EDX analysis of CaO/Fe_2O_3 and $CaO/Fe_2O_3/feldspar$.

Catalyst	Element (wt.%)							
	O	Na	Al	Si	K	Ca	Fe	Total
CaO/Fe_2O_3	37.12	-	-	-	-	3.06	59.82	100
$CaO/Fe_2O_3/feldspar$	43.46	0.99	2.18	6.69	1.34	8.66	36.68	100

2.1.3. Porosity and Surface Area Analysis

Results indicating the surface area, pore volume, and average pore size of the CaO/Fe_2O_3 and $CaO/Fe_2O_3/feldspar$ are presented in Table 2. The CaO/Fe_2O_3 displayed a surface area of 68.680 m^2/g, pore volume of 0.189 cm^3/g, and average pore size of 7.670 nm. After the synthesis of the $CaO/Fe_2O_3/feldspar$, the surface area was 19.523 m^2/g with a pore volume of 0.060 cm^3/g and an average pore size of 7.911 nm. After the doping of feldspar, the surface area and pore volume of the $CaO/Fe_2O_3/feldspar$ catalyst were higher than those of K-feldspar [17]. When the CaO/Fe_2O_3 was doped with feldspar, the surface area was reduced due to a blockage of pores by feldspar. Figure 3 illustrates the nitrogen adsorption–desorption isotherms of the CaO/Fe_2O_3 and $CaO/Fe_2O_3/feldspar$. The N_2 adsorption–desorption isotherms exhibited a typical type IV adsorption–desorption isotherm of $CaO/Fe_2O_3/feldspar$ [19]. The mesoporous pore size of the CaO/Fe_2O_3 and $CaO/Fe_2O_3/feldspar$ were consistent, as shown in Figure 3.

Table 2. Surface property analysis of CaO/Fe_2O_3 and $CaO/Fe_2O_3/feldspar$.

Catalyst	Surface Area (m^2/g)	Pore Volume (cm^3/g)	Average Pore Size (nm)
CaO/Fe_2O_3	68.680	0.189	7.670
$CaO/Fe_2O_3/feldspar$	19.523	0.060	7.911

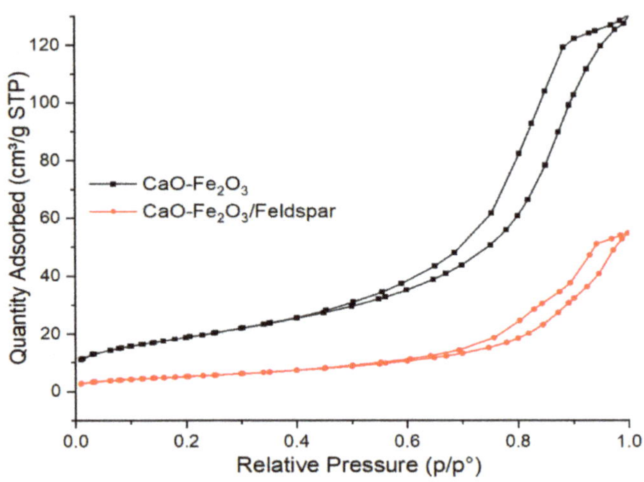

Figure 3. N_2 adsorption–desorption isotherms of CaO/Fe_2O_3 and CaO/Fe_2O_3 /feldspar.

2.1.4. VSM Analysis

The magnetic properties of the $CaO/Fe_2O_3/feldspar$ catalyst were determined using VSM, as presented in Figure 4. The VSM results revealed that the M-H loop of the produced nano-magnetic catalysts had a magnetization of 22.15 emu/g. The VSM curve for the $CaO/Fe_2O_3/feldspar$ in the absence of hysteresis curves is representative of normal super-paramagnetic behavior at room temperature. These results confirmed that the iron-containing feldspar catalysts can be easily separated from the reaction mixture by using an external magnet.

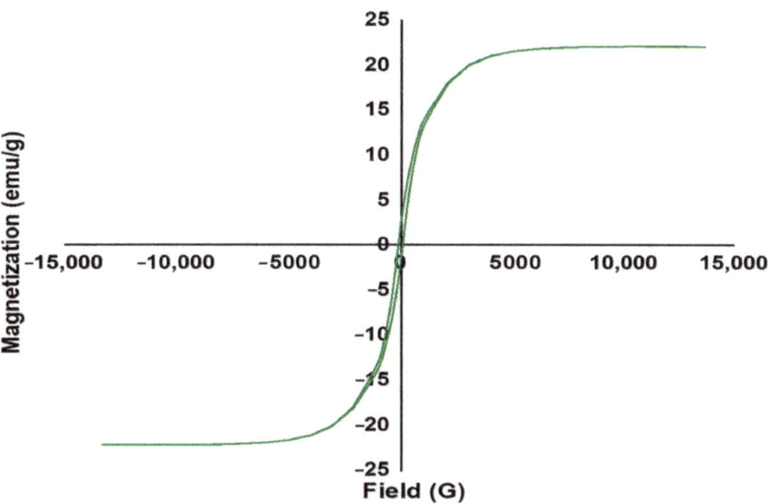

Figure 4. VSM analysis of CaO/Fe$_2$O$_3$/feldspar.

2.2. Optimization of Transesterification Reaction Process Parameters

The impact of two important operational reaction parameters i.e., catalyst concentration and methanol-to-oil molar ratio on the biodiesel yield of non-edible oils (wild safflower, wild mustard and karanja) in the presence of the CaO/Fe$_2$O$_3$/feldspar nano-magnetic catalyst is summarized in Table 3. The other reaction parameters were fixed at: a reaction temperature of 40 °C and a reaction time of 2 h. For karanja and wild mustard, the optimal catalyst concentration was 1.0%, which produced a maximum ester content of 89.1 ± 0.8% and 93.6 ± 0.3%, respectively. For wild safflower, a CaO/Fe$_2$O$_3$/feldspar concentration of 1.5% was adequate to complete the conversion of triglycerides into methyl esters. The utmost yield of biodiesel was obtained from wild safflower was 99.3 ± 0.7 at 1.5% catalyst concentration. Wild mustard and wild safflower oils gave highest biodiesel yields of 93.6 ± 0.3 and 99.3 ± 0.7, respectively, at a 5:1 methanol-to-oil molar ratio. Karanja oil conferred the highest biodiesel yield of 99.9 ± 0.9% at a 10:1 methanol-to-oil molar ratio. Certain levels of catalyst and methanol were required to obtain the maximum BD yield. The optimized catalyst concentration amount not only depended upon the category of catalyst used but was also dependent on the type of the feedstock oil under investigation. In addition, low catalyst levels were not sufficient to give a high conversion to methyl esters. However, increasing the catalyst concentration or the methanol-to-oil molar ratio ahead of the optimized levels resulted in decreased yields of biodiesel due to the formation of a viscous emulsion in the reaction mixture [20].

2.3. Proposed Mechanism of CaO/Fe$_2$O$_3$/Feldspar Catalyst for Transesterification

The initial step in the transformation of fatty oil into biodiesel involves the removal of a proton from methanol by the basic catalyst, resulting in the formation of a methoxide anion and a catalyst that has been protonated as shown in Figure 5. The methoxide anion then performs a nucleophilic attack on the carbonyl group of the triglyceride, leading to the formation of a tetrahedral intermediate. This intermediate subsequently produces a methyl ester and the corresponding anion of the diglyceride. The catalyst is restored in situ through proton abstraction from the catalyst by the diglyceride anion. Following the same mechanism, diglycerides and monoglycerides are consecutively converted into biodiesel and glycerol.

Table 3. Optimization reaction parameters for karanja, wild mustard, and wild safflower seed oils using CaO/Fe_2O_3/feldspar nano-magnetic catalyst.

Feedstock	Catalyst Conc. (%)	Methanol-to-Oil Ratio	Biodiesel Yield (%)
Karanja	0.5	5:1	85.0 ± 0.5
	1	5:1	89.1 ± 0.8
	1.5	5:1	87.5 ± 0.7
	2	5:1	86.0 ± 0.6
	2.5	5:1	85.6 ± 0.5
	1	10:1	99.9 ± 0.9
	1	15:1	88.1 ± 0.7
	1	20:1	85.6 ± 0.8
	1	25:1	83.7 ± 0.5
Wild mustard	0.5	5:1	85.0 ± 0.7
	1	5:1	93.6 ± 0.3
	1.5	5:1	84.7 ± 0.9
	2	5:1	82.9 ± 0.5
	2.5	5:1	81.8 ± 0.8
	1	10:1	85.4 ± 0.6
	1	15:1	83.2 ± 0.3
	1	20:1	80.5 ± 0.9
	1	25:1	79.8 ± 0.4
Wild safflower	0.5	5:1	85.0 ± 0.5
	1	5:1	85.9 ± 0.7
	1.5	5:1	99.3 ± 0.7
	2	5:1	86.5 ± 0.5
	2.5	5:1	85.0 ± 0.6
	1.5	10:1	86.0 ± 0.3
	1.5	15:1	84.5 ± 0.9
	1.5	20:1	82.3 ± 0.8
	1.5	25:1	80.4 ± 0.6

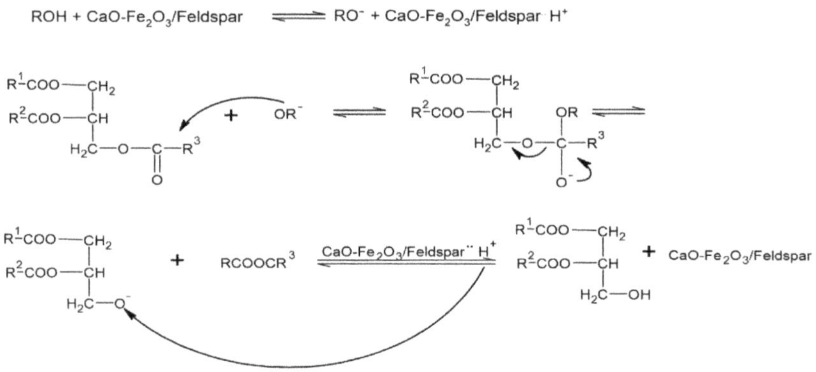

Figure 5. The proposed mechanism of CaO/Fe_2O_3/feldspar catalyst for transesterification of karanja, wild mustard, and safflower seed oil to biodiesel.

2.4. Comparison of Catalytic Activity with Published Literature Reported on Magnetic Catalysts

Table 4 summarizes results obtained from previous studies on the utilization of magnetic catalysts for the production of biodiesel. For example, Hazmi et al. [21] obtained a high FAME yield of 98.60% when applying a methanol-to-oil molar ratio of 12:1, although a higher reaction temperature of 75 °C and higher amount of catalyst loading (4 wt.%) were utilized in comparison to the current work. In another study, Ibrahim et al. [22] reported that an 18:1 methanol-to-oil molar ratio with 3 wt.% catalyst loading gave a FAME yield of 98.30%. Xie et al. [23] observed a 99.20% conversion rate; however, this was at a high methanol-to-oil molar ratio and reaction time of 8 h compared to the present study (2 h). Krishnan et al. [24] obtained a FAME yield of 87.32% with a catalyst loading of 10 wt.%, which was very high in comparison to the present study. Ali et al. [25] also achieved a higher esters yield but the reaction conditions were higher than our study (Table 4). Overall, the FAME yield is primarily impacted by the methanol-to-oil ratio and catalyst loading. Previous studies also noted a typical increase in biodiesel production when the reaction time and temperature were increased [21,23].

Table 4. Comparison study of different magnetic catalysts for biodiesel production.

Types of Catalyst	Experimental Reaction Conditions	Esters Yield (%)	References
RHC/K$_2$O-20%/Fe-5%	Methanol-to-oil ratio = 12:1, temp. = 75 °C, time = 4 h, catalyst loading = 4 wt.%	98.60	[21]
CaO-Fe$_2$O$_3$/AC	Methanol-to-oil ratio = 18:1, temp. = 65 °C, time = 3 h, catalyst loading = 3 wt.%	98.30	[22]
Fe$_3$O$_4$/MCM-41 composites	Methanol-to-oil ratio = 25:1, time = 8 h, catalyst loading = 3 wt.%	99.20	[23]
EFB supported magnetic solid acid catalyst	Catalyst loading = 10 wt.%	87.32	[24]
Magnetically recyclable CuFe$_2$O$_4$	Methanol-to-oil ratio = 18:1, temp. = 60 °C, time = 0.5 h, catalyst loading = 3 wt.%	90.24	[25]
CaO/Fe$_2$O$_3$/feldspar	Methanol-to-oil ratio = 10:1, catalyst loading = 1 wt.% (karanja oil)	99.9	Present study
	Methanol-to-oil ratio = 5:1, catalyst loading = 1 wt.% (wild mustard oil)	93.6	
	Methanol-to-oil ratio = 5:1, catalyst loading = 1.5 wt.% (safflower oil)	99.3	

2.5. CaO/Fe$_2$O$_3$/Feldspar Catalyst Reusability for Karanja, Wild Mustard, and Safflower Seed Oil

A catalytic test was carried out to assess the reusability of the CaO/Fe$_2$O$_3$/feldspar for repeated use without further treatment. The optimized set of conditions for each oil was chosen because it led to better performance in catalytic activity and comparison using the same catalyst. The reaction was performed five times and the respective ester yields were measured, as depicted in Figure 6.

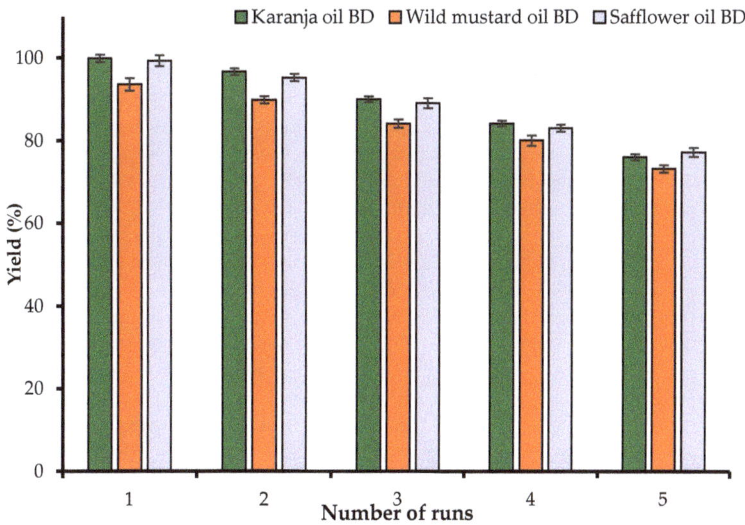

Figure 6. The reusability performance of CaO/Fe$_2$O$_3$/feldspar catalyst for karanja, wild mustard, and safflower seed oil.

The results indicated that after the fourth run of karanja, wild mustard, and safflower seed oils, ester yields of 76%, 73.2%, and 77.2% were obtained, respectively. It was found that the biodiesel yield was maintained at more than 80% when the catalyst was used consecutively four times. Thus, it can be concluded that the CaO/Fe_2O_3/feldspar catalyst does not merely show preferable catalytic performance throughout the transesterification reaction but is also highly suitable for repeated use for the studied oils. The results showed that the CaO/Fe_2O_3/feldspar catalyst has potential for industrial applications due to its effective reusability, absence of wastewater production, and easy removability from the reaction mixture using an external magnet.

2.6. Physicochemical Properties

The physicochemical properties of karanja, wild mustard, and wild safflower seed oil biodiesels using the optimized reaction conditions is summarized in Table 5. Biodiesel with a high density can contribute to the combustion issues while a low-density fuel could be substantially explosive. The American biodiesel standard, ASTM D6751, does not define a density limit [26]. The specified range of density of EN 14214:2003, the European biodiesel standard, lies between 0.86–0.90 g/mL. The densities of all the biodiesel samples were within the range specified in EN 14214 for karanja, wild mustard, and wild safflower. The acid value (AV) indicates the amount of free fatty acids present in biodiesel, with higher numbers indicating a higher acid content [27]. According to American (ASTM) and European (EU) standard requirements, the limit allowed level of AV in pure biodiesel is 0.50 mg KOH g^{-1} [28]. The AVs of the biodiesel fuels prepared from karanja, wild mustard, and wild safflower seed oils were 0.15, 0.46, and 0.15 mg KOH g^{-1}, respectively. The conversion of triglycerides by reaction with aqueous alkali into glycerol and soap is called the saponification reaction. The saponification value (SV) is the indication of the amount of saponifiable units per unit weight of oil. A low SV indicates a higher proportion of high-molecular-weight fatty acids in the fat/oil or vice versa [28]. The saponification values of karanja, wild mustard, and wild safflower oil biodiesels were 180.04, 175.04, and 196.58 mg KOH g^{-1} oil, respectively. The comparatively low SV for the wild mustard biodiesel was attributed to its high content of higher-molecular-weight erucic acid (Table 5). The iodine value (IV) is a measure of the degree of unsaturation in a fat or oil and is an important indicator of oxidative stability, as oxidative stability is dependent on the number and type of double bonds present in biodiesel. According to EN14214, a maximum IV of 120 g I_2/100 g is specified [29]. ASTM D6751 does not specify limits for IV. The IVs of the biodiesel produced from karanja, wild mustard, and wild safflower seed oils were 82.05, 86.65, and 69.44 g I_2/100 g, respectively. The low temperature operability of biodiesel is measured by CP and PP, which give an indication of when biodiesel will solidify and clog fuel filters and lines as ambient temperatures decrease. As seen in Table 5, the CPs and PPs of biodiesel produced from karanja, wild mustard, and wild safflower oils were in the range of 0.8–4.0 °C and −3.9, −1.1 °C, respectively. The cetane number (CN) is a dimensionless parameter that indicates ignition delay and fuel quality. Fuels with a higher CN auto ignite in shorter times after injection to the combustion chamber. Biodiesel CN increases with degree of saturation and chain length of fatty acids. The minimum specifications for CN in ASTM D6751 and EN 14214 are 47 and 51, respectively [30]. The CNs of karanja, wild mustard, and wild safflower seed oil biodiesels were 58.05, 57.98, and 58.44, respectively.

2.7. Fatty Acid Profile

The major fatty acids identified in karanja [31], wild mustard [32], and wild safflower [33] seed oils were oleic acid (51.92%), erucic acid (41.43%), and linoleic acid (75.17%), respectively, as shown in Table 6. Other fatty acids of significance included linoleic (11.03%) and palmitic (10.33%) acids in karanja oil, oleic (23.11%), linoleic (15.75%), and eicosanoic (12.83%) acids in wild mustard oil, and oleic acid (12.98%) in wild safflower oil. Lastly, all three oils contained low levels ($\leq$3.15%) of trienoic linolenic acid as well as shorter chain capric ($\leq$0.15%), lauric ($\leq$0.22%), and myristic ($\leq$0.93%) acids. IV can be calculated from the fatty acid profile following the American Oil Chemists' Society's official method Cd 1c-85. The calculated IV of karanja, wild mustard, and wild safflower oil biodiesels were 78.9, 81.0, and 141.5 g I_2/100 g, respectively. The higher value for safflower oil biodiesel was attributed to its high content of linoleic acid (75.17%) relative to the other methyl ester samples. Biodiesel fuels with higher levels of polyunsaturated fatty acid methyl esters will exhibit higher IV, as IV is a measure of total unsaturation within the sample.

Table 5. Comparison of fuel properties of biodiesel prepared from karanja, wild mustard, and wild safflower with ASTM D6751.

Fuel Parameters	Karanja	Wild Mustard	Wild Safflower	ASTM D6751 Limits
Density (g/mL)	0.85	0.86	0.88	Not specified
Cloud point (°C)	4.0	2.1	0.8	Report
Pour point (°C)	−1.1	−1.6	−3.9	Not specified
Acid value (mg KOH/g)	0.15	0.46	0.15	0.50 max
Iodine value (g I_2/100 g)	82.5	86.65	69.44	Not specified
Saponification value (mg KOH g^{-1} oil)	180.04	175.04	196.58	Not specified
Cetene number	58.05	57.98	58.44	47 minimum

Table 6. Fatty acid composition of karanja, wild mustard, and wild safflower seed oils.

Fatty Acid	Molecular Formula	Fatty Acid Amount (%)		
		Karanja Oil	Wild Mustard Oil	Wild Safflower Oil
Capric acid	$C_{10}H_{20}O_2$	0.11	0.15	0.13
Lauric acid	$C_{12}H_{24}O_2$	0.22	0.12	0.09
Myristic acid	$C_{14}H_{28}O_2$	0.93	0.18	0.16
Palmitic acid	$C_{16}H_{32}O_2$	10.33	3.63	7.73
Linolenic acid	$C_{18}H_{30}O_2$	3.15	0.09	0.32
Linoleic acid	$C_{18}H_{32}O_2$	11.03	15.75	75.17
Oleic acid	$C_{18}H_{34}O_2$	51.92	23.11	12.98
Stearic acid	$C_{18}H_{36}O_2$	4.66	1.15	0.89
Eicosanoic acid	$C_{20}H_{40}O_2$	9.76	12.83	0.11
Erucic acid	$C_{22}H_{42}O_2$	-	41.43	-

3. Materials and Methods

3.1. Materials and Chemical Reagent

P. pinnata (karanja), *S. arvensis* (wild mustard), and *C. oxyacantha* (wild safflower) seeds were collected from wild areas of Pakistan. Oil was extracted from the ground seeds (20 kg) using a VOSOCO automatic screw press machine. The extracted oil was filtered using a vacuum filtration assembly to remove residual solids. The acid values of karanja oil (3.75 mg KOH/g), wild mustard oil (2.30 mg KOH/g), and safflower seed oil (1.95 mg KOH/g) were determined before the transesterification reaction. Feldspar (clay mineral) was taken from the northern area of Pakistan. All chemicals used were of analytical grade and came from Sigma Aldrich.

3.2. Preparation of Supported Nano-Magnetic Catalyst

Calcium nitrate (0.05 g) was dissolved into 100 mL of deionized water, after which 3.5 g of ferric oxide was added. This solution was agitated for 10 min at 1500 rpm, followed by ultrasonication for 15 min to achieve a well-dispersed suspension. Then, a 2M solution of sodium hydroxide was added dropwise under constant stirring at 1500 rpm until a pH of 12 was reached. The suspension was then stirred overnight at 1500 rpm. The precipitates were washed with deionized water until a pH of 7 was obtained. The solid was then dried overnight in an oven at 80 °C. The product was then ground into a fine powder by a mortar and pestle. Finally, the powder underwent calcination at 550 °C for 1 h in a furnace. A total of 23.5 g of catalyst was produced from 25 g of starting material [34,35].

Initially, 0.5 g of CaO/Fe_2O_3 and 0.75 g of feldspar as a support were dissolved in distilled water via precipitation. The material was stirred for one hour at room temperature, filtered, washed three times with distilled water, and then dried for 0.5 h at 150 °C. The resulting powder underwent calcination at 550 °C for 1 h in a furnace. The supported nano-magnetic catalyst (CaO/Fe_2O_3/feldspar) was then ground using a mortar and pestle. Finally, nano-sieving was used to collect the nano-sized particles.

3.3. Characterization of Catalyst and Esters

X-ray diffraction (XRD) was used to obtain information regarding the structure, crystalline phase change, and degree of crystallinity of the magnetic feldspar catalyst. The structures of nano-catalysts were studied using a Rigaku SmartLab powder x-ray diffractometer (XRD, 40 kV, 30 mA, Rigaku Corporation, Tokyo, Japan) of Hypix-400 monochromatic $CuK\alpha$ ($\lambda = 1.5406$ Å), which radiated in the range of 5° to 60°. The crystalline phases of the CaO/Fe_2O_3 and CaO/Fe_2O_3/feldspar were determined.

The structural properties of nanoparticles were characterized using FESEM (JFC-1600, JEOL, Tokyo, Japan), while the elemental analysis for major elements in the samples were determined using an EDX spectrometer model EX-230BU from JEOL (Tokyo, Japan) with an emission current of 15.00 kV and 8.00 mm WD.

The total surface areas of the samples were measured using the Brunauer–Emmett–Teller (BET) method. The BET analyses were conducted using a Thermo–Finnigan Sorpmatic 1990 series (Thermo Finnigan LLC, San Jose, CA, USA) by using nitrogen adsorption-desorption analysis.

The magnetic properties of the nano-magnetic catalyst (CaO/Fe_2O_3/feldspar) were determined using a Lakeshore 7404 Series VSM instrument (Lake Shore Cryotronics, Westerville, OH, USA). During the test, the sample was placed on a rod in an external magnetic field and vibrated via connection with a vibrator. Furthermore, the magnetization as a function of the applied magnetic field (M–H loop) was measured. The electromotive induction method was used to quantify the induced electromotive force generated by flux changes.

The fatty acid compositions of the oils were determined by gas chromatography equipped with a flame ionization detector (GC-FID) utilizing an Agilent 7890A equipped with a capillary column (BPX-70, 60 m × 0.25 mm × 0.25 μm, Trajan Scientific, Victoria, Australia). The H_2 carrier gas was fixed at a flow rate of 40 mL/min and the temperature program was set to increase from 100 to 250 °C at 10 °C/min. Peaks were identified by comparison to reference standards.

3.4. Transesterification and Physiochemical Properties Evaluation

The extracted oil (50 g) was added to a 250 mL conical flask. Then, the waste oils, methanol and CaO/Fe_2O_3/feldspar nano-magnetic catalyst were each transesterified at 60 °C for 2 h on a temperature-controlled hotplate fitted with a magnetic stirrer. After the reaction, the magnetic catalyst was separated using an external magnet. The CaO/Fe_2O_3/feldspar catalyst and glycerol created during the transesterification reaction were separated from the biodiesel phase using separatory funnel. The biodiesel phase was washed with excess hot (75–80 °C) distilled water. The percentage biodiesel yield was calculated using Equation (1) [35].

$$Biodiesel\ yield\ (\%) = \frac{Weight\ of\ biodiesel\ (g)}{Weight\ of\ Oil\ (g)} \times 100 \tag{1}$$

The effect of different reaction parameters, such as the catalyst amount (0.5%, 1%, 1.5%, 2% and 2.5%) and methanol-to-oil molar ratio (5:1, 10:1, 15:1, 20:1 and 25:1), were optimized to give the highest yield of fatty acid methyl esters.

The physiochemical parameters of the resulting biodiesel, such as density, cloud point, (CP), pour point (PP), acid value (AV), cetane number (CN), iodine value (IV) and saponification value (SV), were measured according to the methods and instrumentation described previously [30].

3.5. CaO/Fe$_2$O$_3$/Feldspar Catalyst Reusability Test

The reusability experiments of the CaO/Fe$_2$O$_3$/feldspar catalyst in repeated transesterification reaction cycles were conducted for karanja, wild mustard, and wild safflower oils. The CaO/Fe$_2$O$_3$/feldspar catalyst was used under optimum reaction conditions without undergoing a regeneration process. After each run, the spent catalyst was washed using hexane then by methanol and dried at 100 °C for 6 h before further experiments. The produced biodiesel was analyzed by GC for fatty acid conversion of each oil.

3.6. Statistical Analysis

Each sample was analyzed in triplicate and data presented as mean $\pm$ SD.

4. Conclusions

The transesterification of waste and low-cost oils was investigated using a novel CaO/Fe$_2$O$_3$/feldspar catalyst. The morphology (FESEM), phase and crystallize size (XRD), surface area (BET), and magnetic properties (VSM) were measured. It was found that magnetic separation not only facilitated separation but also avoided material loss and increased fatty acid methyl ester yields. The highest biodiesel yield for karanja, wild mustard, and wild safflower oils were 99.9 ± 0.9, 93.6 ± 0.3, and $99.3 \pm 0.7\%$, respectively. These yields were obtained after optimization of the catalyst concentration and methanol-to-oil molar ratio. For the karanja and wild mustard oils, the optimal catalyst concentration was 1.0%, whereas 1.5% was the optimum amount for wild safflower oil. The wild mustard and wild safflower oils gave a maximum biodiesel yields at a 5:1 methanol-to-oil molar ratio. Karanja gave the highest yield at a 10:1 methanol-to-oil molar ratio. The reusability study depicted that the CaO/Fe$_2$O$_3$/feldspar catalyst is capable of good reusability (>70%) for karanja, wild mustard, and safflower seed oils. The resulting fuel properties of the optimized methyl esters indicated their acceptability for use as biodiesel fuels in compression-ignition engines. This information is valuable in biodiesel production from inexpensive feedstocks with high acid value. Overall, the present study clearly offered a cheap and environmentally beneficial technique for efficient catalysts synthesis and its efficient application for biodiesel production with higher yield. In the current investigation, it demonstrated that CaO/Fe$_2$O$_3$/feldspar nano-magnetic catalysts can be used as a low-cost and naturally safe material for producing biodiesel as compared to traditionally employed homogeneous catalysts.

Author Contributions: Conceptualization, M.H., M.A.H. and U.R.; methodology, M.H., U.R., M.A.H. and B.R.M.; software, U.R. and F.A.A.; validation, I.A.B., M.A.H., A.H., U.R. and B.R.M.; formal analysis, M.H. and A.H.; investigation, M.H., resources, I.A.B., U.R., F.A.A. and B.R.M.; data curation, M.H.; writing—original draft preparation, M.H., A.H., I.A.B. and M.A.H.; writing—review and editing, U.R., F.A.A. and B.R.M.; visualization, M.H., M.A.H. and U.R.; supervision, I.A.B.; project administration, I.A.B., F.A.A. and U.R.; funding acquisition, F.A.A. and B.R.M. All authors have read and agreed to the published version of the manuscript.

Funding: This research work obtained financial support under Geran Putra Berimpak from Universiti Putra Malaysia (Kod Project: UPM/800–3/ 3/1/GPB/2019/9679700). This research was funded (in part) by the U.S. Department of Agriculture, Agricultural Research Service.

Data Availability Statement: All the data is contained within the article.

Acknowledgments: Authors extend their thanks to the Researchers Supporting Project (RSP2023R160), King Saud University (Riyadh, Saudi Arabia).

Conflicts of Interest: The authors declare no conflict of interest.

Disclaimer: Mention of trade names or commercial products in this publication is solely for the purpose of providing specific information and does not imply recommendation or endorsement by the U.S. Department of Agriculture. USDA is an equal opportunity provider and employer.

References

1. Ali, A.; Mushtaq, A. Do biofuels really a need of the day: Approaches, mechanisms, applications and challenges. *Int. J. Chem. Biochem. Sci.* **2023**, *23*, 219–226.
2. Sanjid, A.; Masjuki, H.; Kalam, M.; Abedin, M.; Rahman, S.A. Experimental investigation of mustard biodiesel blend properties, performance, exhaust emission and noise in an unmodified diesel engine. *APCBEE Procedia* **2014**, *10*, 149–153. [CrossRef]
3. Siddique, H.S.; Nadeem, F.; Inam, S.; Kazerooni, E.A. Recent production methodologies and advanced spectroscopic characterization of biodiesel: A review. *Int. J. Chem. Biochem. Sci.* **2020**, *18*, 134–144.
4. Lokman, I.M.; Rashid, U.; Yunus, R.; Taufiq-Yap, Y.H. Carbohydrate-derived solid acid catalysts for biodiesel production from low-cost feedstocks: A review. *Catal. Rev. Sci. Eng.* **2014**, *56*, 187–219. [CrossRef]
5. Jamil, F.; Al-Haj, L.; Al-Muhtaseb, A.H.; Al-Hinai, M.A.; Baawain, M.; Rashid, U.; Ahmad, M.N.M. Current scenario of catalysts for biodiesel production: A critical review. *Rev. Chem. Eng.* **2018**, *34*, 267–297. [CrossRef]
6. Srivastava, N.; Srivastava, M.; Gupta, V.K.; Manikanta, A.; Mishra, K.; Singh, S.; Singh, S.; Ramteke, P.; Mishra, P. Recent development on sustainable biodiesel production using sewage sludge. *3 Biotech* **2018**, *8*, 245. [CrossRef]
7. Abomohra, A.E.-F.; Elsayed, M.; Esakkimuthu, S.; El-Sheekh, M.; Hanelt, D. Potential of fat, oil and grease (FOG) for biodiesel production: A critical review on the recent progress and future perspectives. *Prog. Energy Combust. Sci.* **2020**, *81*, 100868. [CrossRef]
8. Aslam, K.; Mushtaq, A.; Nadeem, F.; Ghnia, J.B.; Rafique, M.; Sillanpää, M. Economic feasibility of non-edible oils as biodiesel feedstock: A brief. *Int. J. Chem. Biochem. Sci.* **2020**, *18*, 145–150.
9. Irshad, F.; Mushtaq, A. Biomass-derived Materials and their Commercial Applications. *Int. J. Chem. Biochem. Sci.* **2020**, *23*, 202–211.
10. Ikram, M.M.; Hanif, M.A.; Khan, G.S.; Rashid, U.; Nadeem, F. Significant seed oil feedstocks for renewable production of biodiesel: A review. *Curr. Org. Chem.* **2019**, *23*, 1509–1516. [CrossRef]
11. Xie, W.; Li, J. Magnetic solid catalysts for sustainable and cleaner biodiesel production: A comprehensive review. *Renews* **2023**, *171*, 113017. [CrossRef]
12. Xie, W.; Wang, H. Immobilized polymeric sulfonated ionic liquid on core-shell structured Fe_3O_4/SiO_2 composites: A magnetically recyclable catalyst for simultaneous transesterification and esterifications of low-cost oils to biodiesel. *Renew. Energy* **2020**, *145*, 1709–1719. [CrossRef]
13. Xie, W.; Wan, F. Basic ionic liquid functionalized magnetically responsive Fe_3O_4@HKUST-1 composites used for biodiesel production. *Fuel* **2018**, *220*, 248–256. [CrossRef]
14. Xie, W.; Wang, H. Grafting copolymerization of dual acidic ionic liquid on core-shell structured magnetic silica: A magnetically recyclable Brönsted acid catalyst for biodiesel production by one-pot transformation of low-quality oils. *Fuel* **2021**, *283*, 118893. [CrossRef]
15. Hanif, M.A.; Nisar, S.; Rashid, U. Supported solid and heteropoly acid catalysts for production of biodiesel. *Catal. Rev.* **2017**, *59*, 165–188. [CrossRef]
16. Xue, B.; Guo, H.; Liu, L.; Chen, M. Preparation, characterization and catalytic properties of yttrium-zirconium-pillared montmorillonite and their application in supported Ce catalysts. *Clay Miner.* **2015**, *50*, 211–219. [CrossRef]
17. Keshavarz, M.; Papari, R.F.; Bulgariu, L.; Esmaeili, H. Synthesis of CaO/Fe_2O_3 nanocomposite as an efficient nanoadsorbent for the treatment of wastewater containing Cr (III). *Sep. Sci. Technol.* **2021**, *56*, 1328–1341. [CrossRef]
18. Ali, B.J.; Othman, S.S.; Harun, F.W.; Jumal, J.; Rahman, M.B.A. Immobilization of enzyme using natural feldspar for use in the synthesis of oleyl oleate. In *AIP Conference Proceedings*; AIP Publishing LLC: Melville, NY, USA, 2018; p. 030018.
19. Lokman, I.M.; Rashid, U.; Taufiq-Yap, Y.H. Production of biodiesel from palm fatty acid distillate using sulfonated-glucose solid acid catalyst: Characterization and optimization. *Chin. J. Chem. Eng.* **2015**, *23*, 1857–1864. [CrossRef]
20. Kusumaningtyas, R.D.; Pristiyani, R.; Dewajani, H. A new route of biodiesel production through chemical interesterification of jatropha oil using ethyl acetate. *Int. J. ChemTech Res.* **2016**, *9*, 627–634.
21. Hazmi, B.; Rashid, U.; Taufiq-Yap, Y.H.; Ibrahim, M.L.; Nehdi, I.A. Supermagnetic nano-bifunctional catalyst from rice husk: Synthesis, characterization and application for conversion of used cooking oil to biodiesel. *Catalysts* **2020**, *10*, 225. [CrossRef]
22. Ibrahim, N.A.; Rashid, U.; Hazmi, B.; Moser, B.R.; Alharthi, F.A.; Rokhum, S.L.; Ngamcharussrivichai, C. Biodiesel production from waste cooking oil using magnetic bifunctional calcium and iron oxide nanocatalysts derived from empty fruit bunch. *Fuel* **2022**, *317*, 123525. [CrossRef]
23. Xie, W.; Han, Y.; Wang, H. Magnetic $Fe_3O_4/MCM-41$ composite-supported sodium silicate as heterogeneous catalysts for biodiesel production. *Renew Energy* **2018**, *125*, 675–681. [CrossRef]
24. Krishnan, S.G.; Pua, F.; Syed Jaafar, S.N. Synthesis and characterization of local biomass supported magnetic catalyst for esterification reaction. *Mater. Today Proc.* **2020**, *31*, 161–165. [CrossRef]
25. Ali, R.M.; Elkatory, M.R.; Hamad, H.A. Highly active and stable magnetically recyclable $CuFe_2O_4$ as a heterogenous catalyst for efficient conversion of waste frying oil to biodiesel. *Fuel* **2020**, *268*, 117297. [CrossRef]
26. Saeed, A.; Hanif, M.A.; Hanif, A.; Rashid, U.; Iqbal, J.; Majeed, M.I.; Moser, B.R.; Alsalme, A. production of biodiesel from *Spirogyra elongata*, a common freshwater green algae with high oil content. *Sustainability* **2021**, *13*, 12737. [CrossRef]

27. Ijaz, B.; Hanif, M.A.; Rashid, U.; Zubair, M.; Mushtaq, Z.; Nawaz, H.N.; Choong, T.S.Y.; Nehdi, I.A. High vacuum fractional distillation (HVFD) approach for quality and performance improvement of *Azadirachta indica* biodiesel. *Energies* **2020**, *13*, 2858. [CrossRef]

28. Ismail, S.A.-E.A.; Ali, R.F.M. Physico-chemical properties of biodiesel manufactured from waste frying oil using domestic adsorbents. Science and technology of advanced materials. *Sci. Technol. Adv. Mater.* **2015**, *16*, 034602. [CrossRef]

29. Folayan, A.J.; Anawe, P.A.L.; Aladejare, A.E.; Ayeni, A.O. Experimental investigation of the effect of fatty acids configuration, chain length, branching and degree of unsaturation on biodiesel fuel properties obtained from lauric oils, high-oleic and high-linoleic vegetable oil biomass. *Energy Rep.* **2019**, *5*, 793–806. [CrossRef]

30. Sakthivel, R.; Ramesh, K.; Purnachandran, R.; Shameer, P.M. A review on the properties, performance and emission aspects of the third generation biodiesels. *Renew. Sustain. Energy Rev.* **2018**, *82*, 2970–2992. [CrossRef]

31. Islam, A.K.M.A.; Chakrabarty, S.; Yaakob, Z.; Ahiduzzaman, M.; Islam, A.K.M.M. Koroch (*Pongamia pinnata*): A promising unexploited resources for the tropics and subtropics. In *Forest Biomass-From Trees to Energy*; IntechOpen: London, UK, 2021.

32. Hanif, M.; Bhatti, H.N.; Hanif, M.A.; Rashid, U.; Hanif, A.; Moser, B.R.; Alsalme, A. A novel heterogeneous superoxide support-coated catalyst for production of biodiesel from roasted and unroasted *Sinapis arvensis* seed oil. *Catalysts* **2021**, *11*, 1421. [CrossRef]

33. Murthy, I.; Anjani, K. Fatty acid composition in *Carthamus* species. In Proceedings of the 7th International Safflower Conference, Wagga Wagga, Australia, 3–6 November 2008; Australian Oilseeds Federation: Sydney, Australia, 2007.

34. Ali, M.A.; Al-Hydary, I.A.; Al-Hattab, T.A. Nano-magnetic catalyst CaO-Fe$_3$O$_4$ for biodiesel production from date palm seed oil. *Bull. Chem. React. Eng. Catal.* **2017**, *12*, 460–468. [CrossRef]

35. Shi, M.; Zhang, P.; Fan, M.; Jiang, P.; Dong, Y. Influence of crystal of Fe$_2$O$_3$ in magnetism and activity of nanoparticle CaO@ Fe$_2$O$_3$ for biodiesel production. *Fuel* **2017**, *197*, 343–347. [CrossRef]

 *catalysts*

Article

A Green Nanocatalyst for Fatty Acid Methyl Ester Conversion from Waste Cooking Oil

Sadaf Khosa [1], Madeeha Rani [1], Muhammad Saeed [2], Syed Danish Ali [3], Aiyeshah Alhodaib [4],* and Amir Waseem [1],*

[1] Department of Chemistry, Quaid-i-Azam University, Islamabad 45320, Pakistan
[2] School of Chemistry, University of the Punjab, Lahore 54590, Pakistan
[3] Nanoscience and Technology Department, National Centre for Physics, Islamabad 44000, Pakistan
[4] Department of Physics, College of Science, Qassim University, Buraydah 51452, Saudi Arabia
* Correspondence: ahdieb@qu.edu.sa (A.A.); amir@qau.edu.pk (A.W.)

Abstract: This study used a novel combination of cellulose nanocrystals (CNCs) and calcium oxide (CaO) nanocomposite (CaO/CNCs) for the production of biodiesel from waste cooking oil. The filter paper was used as a raw cellulose source to produce the CNCs from the acid hydrolysis of cellulose with sulfuric acid. The as-synthesized CaO/CNC nanocomposite is recyclable and environmentally friendly and was characterized using Fourier transform infrared spectroscopy, energy dispersive X-ray, scanning electron microscopy, and X-ray diffraction. The optimum process parameters investigated are a 20:1 methanol-to-oil molar ratio, 3-weight percent catalyst concentration, 60 °C temperature, and 90 min of reaction time. Under the optimum conditions, a biodiesel yield of 84% was obtained. The CaO/CNC nanocomposite achieved five times reusability, indicating its effectiveness and reusability in the transesterification reaction. The synthesized biodiesel chemical composition was examined using FTIR, GCMS, ^{1}H-NMR, and ^{13}C-NMR, and its properties, including specific gravity, color, flash point, cloud point, pour point, viscosity, sulfur content, sediments, water content, total acid number, cetane number, and corrosion test, were ascertained using ASTM standard practices. The outcomes were determined to fulfill global biodiesel standards (ASTM 951, 6751). Five successive transesterification processes were used to test the regeneration of the catalyst; the first three showed no distinct change, while the fifth cycle showed a reduction of up to 79%. The innovative composite CaO/CNC and used cooking oil are stable, affordable, and extremely successful for long-term biodiesel generation.

Keywords: biofuel; cellulose nanocrystals; CaO nanoparticles; environmental sustainability; transesterification; waste cooking oil

Citation: Khosa, S.; Rani, M.; Saeed, M.; Ali, S.D.; Alhodaib, A.; Waseem, A. A Green Nanocatalyst for Fatty Acid Methyl Ester Conversion from Waste Cooking Oil. *Catalysts* **2024**, *14*, 244. https://doi.org/10.3390/catal14040244

Academic Editors: José María Encinar Martín and Sergio Nogales Delgado

Received: 18 March 2024
Revised: 4 April 2024
Accepted: 4 April 2024
Published: 6 April 2024

1. Introduction

Generally, renewable energy serves an important role in dealing with climate change and energy security challenges at the local, national, and worldwide levels. Most prosperous countries have made substantial strides toward reducing their reliance on fossil fuels through the excessive use of alternate and renewable energy sources. Despite these tremendous efforts, there is still an intense preference for energy derived from fossil fuels. However, despite the rapid development of technology, the transportation sector has consistently seen an increase in demand for oil [1]. Fossil fuels comprise 85% of the world's energy sources in today's world and are predicted to remain so throughout the foreseeable future as well [1]. It was pointed out that climate change has the greatest effect on the production of biomass, being a renewable energy source [2]. To solve these issues, suitable replacements for environmentally friendly fuels must be developed and put into practice. To replace fossil fuels like oil, coal, and diesel oil, appropriate, clean, and renewable energy sources have been the subject of much research, both in the past and present [2,3].

Fatty acid methyl esters (FAMEs), another name for biodiesel, are typically made by catalytically transesterifying vegetable or animal fats with methanol. Esterification and transesterification are the primary processes used in the manufacturing of biodiesel [4]. Researchers have centered on problems including (1) inexpensive raw materials, (2) reusable catalysts, and (3) effective reactors to carry out transesterification reactions to accomplish sustainable goals in the production of biodiesel [5–7]. In addition to that, the conversion of biomass to biofuels and life cycle assessment is also one of the most important aspects for policy decisions as biofuels should be based on evidence that biofuels are produced in a sustainable manner [8]. Thus, to produce biodiesel, it must be feasible to recognize widely accessible and unused feedstocks. Animal fats, used cooking oil, and non-edible seed oils have all expanded their availability as both environmentally and economically acceptable feedstocks for the generation of biofuels in recent years [3,9–12]. Among the many energy sources, biodiesel is biodegradable, sustainable, and non-toxic [5,13,14]. Both edible and nonedible vegetable oils as well as animal fats are possible sources of biodiesel feedstock. Yet, edible oils are not favored for the synthesis of biodiesel because of their high price, limited stock, and probable general consequences [15]. The general cost of manufacturing has decreased as a result of the widespread use of waste cooking oil (WCO) as a feedstock for biodiesel production. Moreover, using WCO as a feedstock can aid in addressing issues with environmental degradation [16].

Calcium oxide (CaO) is inexpensive and is one of several potential catalytic compositions; however, it is very relevant because of its high basicity, low toxicity, accessibility, and affordability. Numerous studies have shown that CaO alone leaves calcium ions in methanol if left in contact for a long time; therefore, CaO composite formation is recommended to avoid the use of decalcifying agents at an additional cost to ensure high purity in the end products [17,18]. However, the use of pristine CaO is limited by its long reaction time, and much interest has been shown in modifying pristine CaO to enhance reaction rates and basic active sites [19–21]. The emergence of cellulose nanocrystals (CNCs) as a novel class of nanomaterials can be attributed to their renewable, environmentally harmless, naturally occurring, biodegradable, and biocompatible properties [22]. Furthermore, CNCs have a large number of hydroxyl groups on their surface; and the surface sulfate ester groups in CNCs are produced by hydrolysis with sulfuric acid, which helps to stabilize metal nanoparticles (mNPs). Because of their exceptional qualities, researchers have focused a lot of attention on the use of CNCs in the production of mNP catalysts, which act as excellent support for heterogeneous catalysts [23]. Using waste material as a source for catalyst synthesis with active metal modification can thus be viewed as a viable strategy for producing cost-effective product biodiesel. For example, one study reported composites of CaO utilizing solid coconut waste as oil and eggshells as a CaO catalyst supported with polyvinyl alcohol (PVA) in a packed bed reactor. The biodiesel yield of 95% was obtained at a reaction temperature of 61 °C and reaction time of 3 h with a catalyst loading (CaO/PVA) of 2.29 wt.% [21]. In a similar study, a PVA-supported CaO/CNC/PVA catalyst was synthesized and utilized for biodiesel production from WCO in a packed bed reactor. Chicken bone and coconut residue were used as sources for CaO and CNC, and the supported catalyst (0.5 wt.%) was able to yield 98.40% biodiesel at 65 °C with a sufficiently large reaction time of 4 h [24]. Similarly, CaO (calcined from eggshells) supported on silica, which is obtained from treated waste glassware and further modified by impregnating cerium oxide to increase activity, acts as a bifunctional catalyst for the conversion of beef fat to biodiesel [20].

In this research, a calcium oxide and cellulose nanocrystal (CaO/CNC) nanocomposite was synthesized through the hydrothermal method without using support material. The cellulose source used was filter paper as a reference for CNC (cellulose nanocrystal) synthesis through acid hydrolysis; however, it can be replaced in the future by sustainable cellulose sources coming from wastes or residual biomass. The CaO/CNC nanocomposite, being a heterogeneous catalyst, offers a lack of sensitivity to free fatty acids, is non-corrosive in nature, and can be regenerated, reused, and separated easily from the product. Common laboratory filter paper, which is commercially available as a source of CNCs, is combined with CaO nanoparticles and employed as catalysts to assess their catalytic performance in the transesterification process for biodiesel production. The CaO/CNC catalyst underwent thorough characterization via techniques, i.e., SEM, EDX, XRD, and FTIR. The nanocomposite was utilized to produce biodiesel from waste cooking oil.

2. Results and Discussion

2.1. Catalyst Characterizations

2.1.1. X-ray Diffraction

The X-ray diffraction patterns of CaO, CNC, and CaO/CNC composites were examined. In close agreement with reference card no JCPDS 00-02-1088 (Joint Committee on Powder Diffraction Standards) using the software X'Pert HighScore®, version 2.1.1, the XRD pattern of CaO revealed distinctive peaks at 2θ values of 32.34°, 37°, 54.23°, and 63.47°, which, respectively, correspond to lattice planes (111), (200), (220), and (311) (Figure 1a,d).

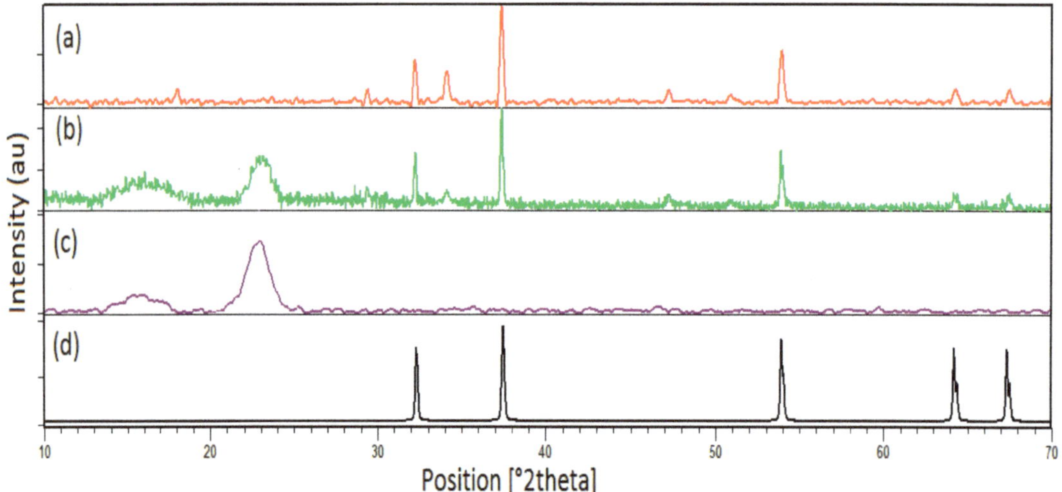

Figure 1. XRD patterns of (**a**) CaO, (**b**) CNC/CaO, and (**c**) CNC and (**d**) the standard reference pattern of CaO (JCPDS 00-02-1088).

In the case of CNCs, the XRD examination revealed two broad diffraction peaks at 15.1–17.5 and 22.7, which, as previously reported [25], correspond to the crystal planes (101), (101), and (002) of cellulose type I. (Figure 1c). Similar peaks to those of CaO and CNC were visible in the CaO/CNC composite; however, some peak intensities decreased as the composite formed (Figure 1b). The XRD pattern corroborated the calculated average crystalline size of 42 nm, which indicates the successful synthesis of the composite.

2.1.2. FTIR

The vibrational bands contained in CaO nanoparticles, CNC, and the CaO/CNC composite in the region of 4000–400 cm^{-1} were examined using the FTIR technique (Figure 2). The peak in the range of 500–450 cm^{-1}, due to the M–O bond present in CaO nanoparticles, as reported earlier, and the peak at 874 and 1418 cm^{-1} corresponding to the C–O bond, suggesting the carbonation of calcium oxide, were both seen in the spectra of CaO [26,27]. Peaks in CNC include the following: –OH stretch (3600–3200 cm^{-1}), C–H stretch of cellulose polysaccharides (2910 cm^{-1}), –OH bending vibration (1640 cm^{-1}), CH_2 bending (1430 cm^{-1}), and 1372 cm^{-1} (CH_3 bending), and –CO stretching (1060). Each of the positive detected peaks is in conformity with the published literature [28]. The peaks at 874 and 1418 cm^{-1} in the composite CaO/CNC FTIR spectrum are caused by CaO nanoparticles, which are not present in the CNC spectra. The peak of calcium oxide's carbonation coincided with the CH$_3$ and CH$_2$ bending vibrations in CaO/CNC, which are in the range of 1450–1300 cm^{-1}. These all provide confirmation that the CaO/CNC composite was successfully formed.

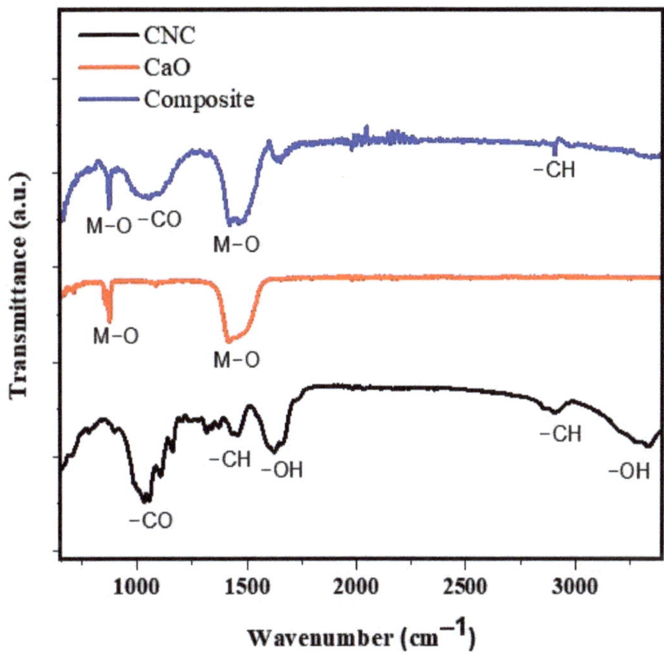

Figure 2. FT-IR spectra of CaO, CNCs, and composite CaO/CNC.

2.1.3. SEM and EDX

Scanning Electron Microscopy (SEM) and EDX for elemental analysis were used to examine the surface morphology of the produced materials including CaO, CNC, and CaO/CNC, as shown in Figure 3. Because of their polarity and electrostatic attraction, CaO nanoparticles have a smooth, regular, and flower-like appearance in their micrographs (Figure 3a,b). The CNC micrographs exhibit agglomeration and spherical particles (Figure 3c,d). Because of the presence of CaO and CNC, the composite displays porous nanoflakes and agglomerated particles of various sizes (Figure 3e,f). CaO, CNC, and CaO/CNC composite elements underwent EDX investigation. The purity of CaO NPs and the existence of C and O in cellulose are both confirmed by the presence of Ca and O in CaO nanoparticles. The presence of Ca, O, and C in the EDX spectrum effectively verifies the production of the composite CaO/CNC (Figure 4).

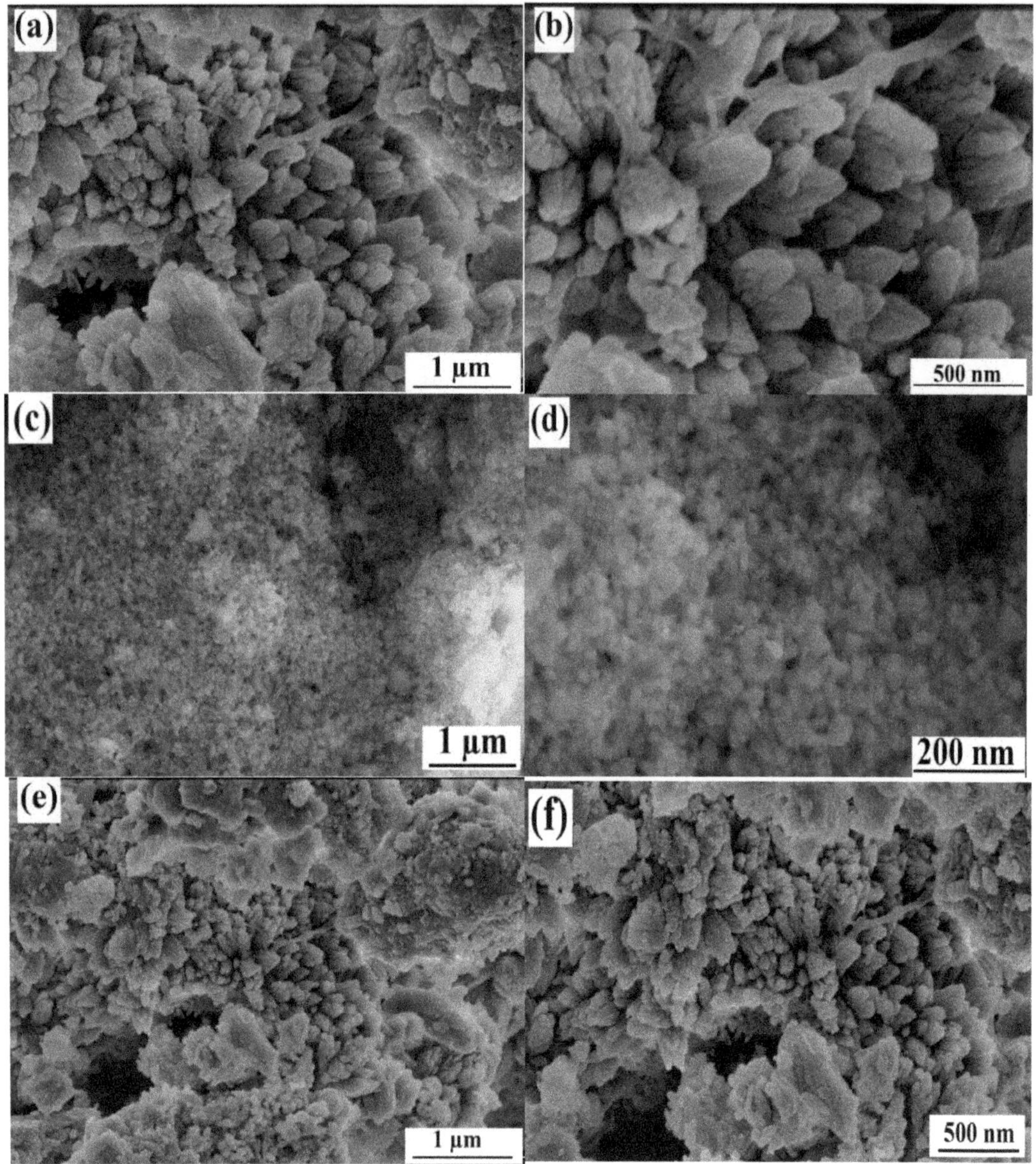

Figure 3. SEM images of (**a**,**b**) CaO, (**c**,**d**) CNCs, and (**e**,**f**) composite CaO/CNC at different resolutions.

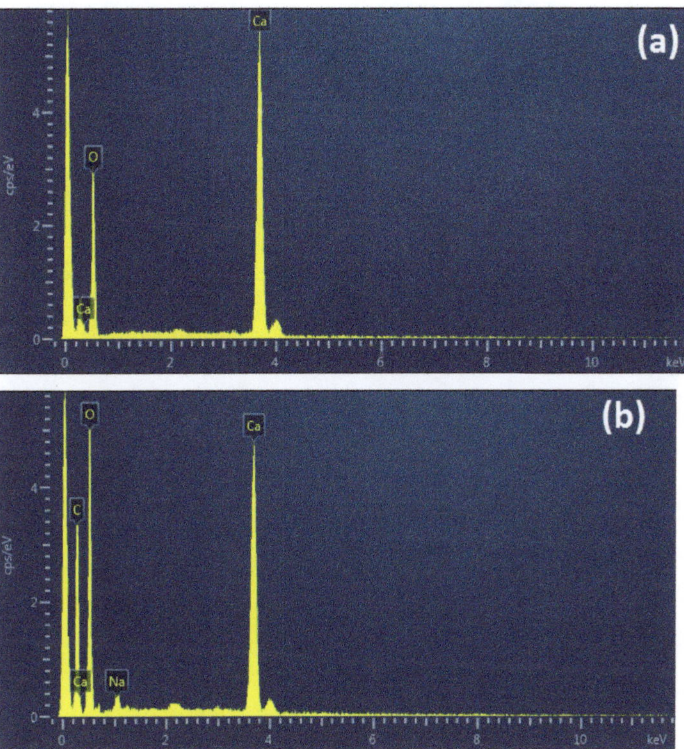

Figure 4. Scanning electron microscopy and energy dispersive X-ray spectroscopy of (**a**) CaO and the (**b**) CaO/CNC composite.

2.2. Effective Parameters of WCO Conversion to Biodiesel

To determine the impact of different effective parameters on the biodiesel yield of the used canola cooking oil, experiments were carried out three times. Temperature (50–100 °C), oil-to-methanol ratios (1:5–1:25), reaction time (30–150 min), and catalyst loading (1–5%) were the variables. The data were graphed to examine the effects of the various parameters on biodiesel yield.

2.2.1. Impact of Reaction Temperature

Temperature is an important factor in determining the conversion of triglycerides into biodiesel and affects the rate of the transesterification reaction. In Figure 5a, the impact of temperature on the catalyst's effectiveness is shown at temperatures varying between 40 and 80 °C with a MeOH: oil of 20:1, CaO/CNC amount of 3 wt.%, and the time of 120 min on biodiesel yield during the transesterification reaction of used fry cooking oil, which was investigated under reflux circumstances. The reaction temperature has a substantial impact on the biodiesel yield, as shown in Figure 5a. The initial biodiesel output was just 53% at 40 °C and 75% at 50 °C, but it grew further with just a 20 °C temperature rise and peaked at 84% at 60 °C. The biodiesel yield was lower at low temperatures because the oil viscosity was not reduced enough to allow for molecular collisions and a faster reaction rate. Hence, raising the reaction temperature is necessary for producing the maximum amount of biodiesel. A rise in temperature can accelerate the reaction rate and increase molecular collisions, which can reduce the activation energy [29]. Higher temperatures up to 80 °C are also used for WCO conversion to FAME [30]. The fatty acids acting as a nucleophile and the increased interaction between the catalyst and protons caused by the alcohol and mixed oxide catalyst with the alkyl group of triglycerides facilitated a higher conversion of oils to biodiesel at the reaction site. However, the biodiesel yield dropped to 75% at temperatures above 70 °C, possibly as a result of fast methanol evaporation. This pattern is consistent with our earlier observations [4,11,21]. Hence, 60 °C was determined to be the ideal temperature for subsequent tests in this inquiry.

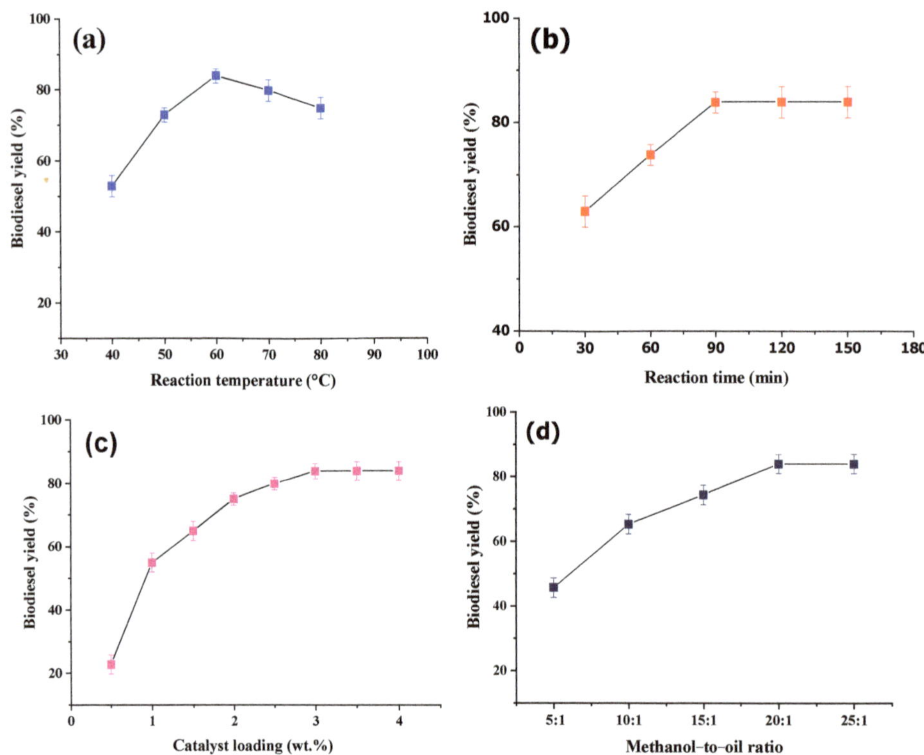

Figure 5. Effects of various transesterification reaction parameters on biodiesel yields: (**a**) reaction temperature, (**b**) reaction time, (**c**) catalyst loadings, and (**d**) methanol-to-oil molar ratio.

2.2.2. Impact of Reaction Time

The effect of transesterification reaction time on the synthesis of biodiesel from used cooking oil was investigated in this study. The 20:1 methanol-to-oil ratio, 3-weight percent catalyst loading, and reaction temperature of 60 °C were all held constant while the reaction period was changed from 30 to 150 min. The results showed that the production of biodiesel was only 63% after 30 min of reaction time (Figure 5b). The biodiesel output, however, rose as the reaction time was extended and peaked at 90 min and 84% yield. This shows that the reaction required additional time for the reactants to interact with each other and the catalyst, giving rise to higher biodiesel yield, which did not reach equilibrium after one hour. According to the present study, 90 min is the ideal amount of time for the transesterification reaction to occur. Several studies have also demonstrated that reaction time significantly affects biodiesel yield. For example, the highest biodiesel yield of 95% was attained when coconut waste oil was used as the feedstock for biodiesel synthesis by in situ transesterification using polymer-supported CaO/PVA of 2.29 wt.% and a time duration of 180 min [21]. According to earlier studies, the output of biodiesel increased as the reaction time was increased from 1 to 4 h [9,21,31]. In previous studies on the conversion of biodiesel, large reaction times were also reported ranging from 5 to 10 h [10,30,32,33].

2.2.3. Impact of Catalyst Loading

For transesterification processes, the involvement of the catalyst dose is crucial to produce biodiesel with a sufficient yield and quality. This study assessed the impact of catalyst loadings (1–5 wt.%) on the biodiesel yield from used cooking oil. As shown in Figure 5c, when the methanol-to-oil ratio was held at 20:1 and the reaction temperature was maintained at 60 °C for 90 min, the results showed that the lowest catalyst concentration (1%) only produced 55% biodiesel, but increasing the catalyst loading (3 wt.%) resulted in an increase in the yield by up to 84%. It was claimed earlier that at a lower catalyst concentration (ca. 0.5%), the active sites necessary to successfully convert triglycerides into biodiesel might not be present. It was discovered that using a 3 wt.% catalyst was best for producing biodiesel because of a higher number of available active sites [10]. Previous studies conducted by [31] showed that a catalyst amount of 5 wt.% could be used to convert marine fish waste oil to biodiesel. The maximum biodiesel yield of 86.5 wt.% was obtained with a methanol-to-oil molar ratio of 25:1 and a reaction time of 107 min [31]. Previously reported studies show the use of a large amount of catalyst loading, for example, a ZrO_2-supported heterogeneous catalyst of 12 wt.% showed a microwave-assisted conversion of soybean oil into biodiesel yield of up to 92.75% [34]. In another study, calcined salmon fish bone waste was used as a catalyst support for biodiesel production using sunflower oil. A catalyst loading of 10 wt.% was used for 3 h to obtain a biodiesel yield of 99.13% [35]. A pyrolytic rice straw ash supported with CaO (25–35 wt.%) using a catalyst loading of 4.87 wt.% and 175 min of reaction time resulted in a biodiesel yield of 96.49% [36]. It should be noted that these amounts are considerably higher than the optimal 3 wt.% catalyst loading obtained in this research. After careful consideration, we determined that the optimal conditions for further investigation were a catalyst concentration of 3 wt.% and a MeOH: oil ratio of 20:1.

2.2.4. Impact of the Methanol-to-Oil Ratio

The methanol-to-oil molar ratio also affects the transesterification of used cooking oil into biodiesel using a CaO/CNC catalyst. Various methanol-to-oil ratios of 5:1, 10:1, 20:1, and 25:1 were used. We maintained a univariate approach throughout the procedure, including a catalyst loading of 3 wt.%, a reaction temperature of 60 °C, and a reaction time of 90 min. A higher methanol-to-oil ratio was used earlier in many studies, for example, a maximum biodiesel yield of 86.5 wt.% was obtained with a methanol-to-oil molar ratio of 25:1 and reaction time of 107 min [31]. Excess methanol, however, enhances the transesterification procedure on the product side, leading to a full conversion of the reactants (used cooking oil) into products (FAMEs). Figure 5d shows the effects of various methanol-to-oil molar ratios on biodiesel production. Biodiesel yields climbed from 44% to a maximum of 84% as the methanol-to-oil ratio rose from 5:1 to 20:1. This is because higher methanol-to-oil ratios make the reactants more soluble, which raises the likelihood that methanol will attack the carbonyl/carboxylic acid functional groups in fatty acids and triglycerides through a nucleophilic attack. However, when the MeOH: oil ratio was raised to 20:1 and 25:1, the biodiesel yield remained the same. As a result, for further research, we determined that the optimal ratio is MeOH-to-oil at 20:1.

2.3. Biodiesel Characterization

2.3.1. FTIR Analysis

The identification of chemical bonds and functional groups in oil and biodiesel is carried out using this excellent method of analysis. Peaks are attained in the 400 to 4000 cm^{-1} range. The chemical structures of the FTIR spectra of the waste cooking oil and waste cooking oil biodiesel results are similar (Figure 6). However, small differences are observed in the intensities and frequencies of the absorption bands [4,9]. The presence of fatty acid methyl esters is responsible for the following peaks in waste cooking oil: 3006.9 cm^{-1} (sp_2 C–H stretch), 2921.7 cm^{-1} (sp_3 C–H stretch), 1377.1 cm^{-1} (CH_3 bending), 1463.7 cm^{-1} (CH_2 bending), 1098.6 cm^{-1} (C–C stretch), and 1743.6 cm^{-1} (C=O stretch).

With some small variations, similar peaks were seen in the produced biodiesel. It is notable that the methyl ester deformation peak, which appears at 1435.5 cm^{-1} in the spectrum of biodiesel, is absent from the oil spectrum, which is the methyl ester group with its deformation vibration. In addition, the biodiesel spectrum's fingerprint area peak at 1157.8 cm^{-1} was divided into two peaks at 1170.3 and 1195.7 cm^{-1}. The averaging of the energy over the triple ester group of the triglycerides is gone, which strongly supported the manufacture of biodiesel. These vibrational bands show that the heterogeneous catalytic transesterification event, which turns oil into biodiesel, has taken place.

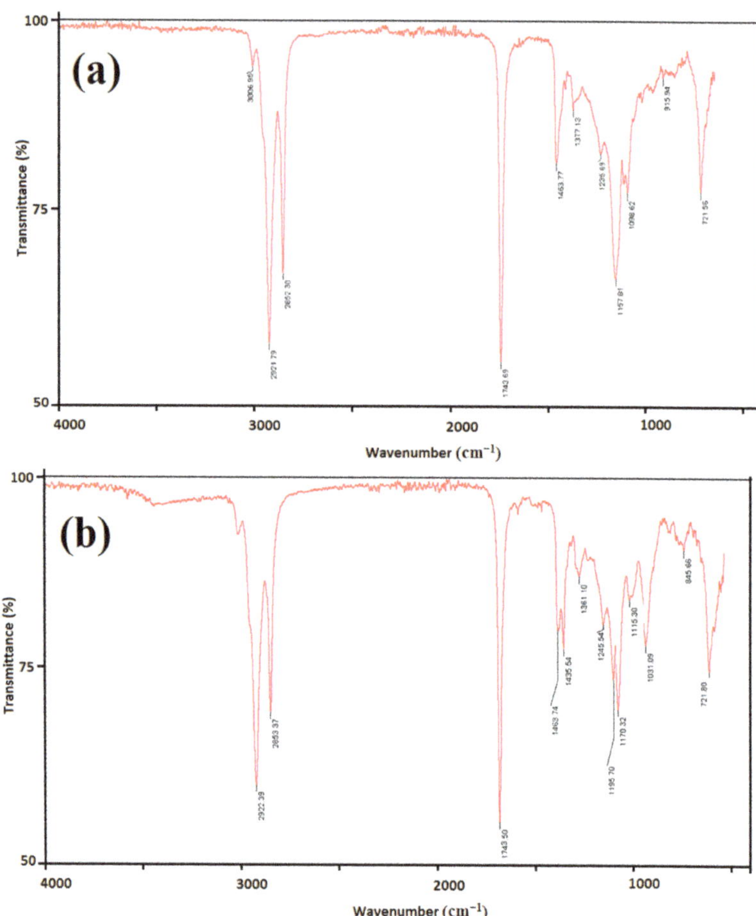

Figure 6. FTIR spectra of waste cooking oil (**a**) and biodiesel from waste cooking oil (**b**).

2.3.2. Nuclear Magnetic Resonance Analysis

^{1}H NMR was employed to determine the molecular composition of biodiesel, which is also used to demonstrate that methyl ester was produced during the transesterification procedure. The Figure 7a depicts the biodiesel ^{1}H NMR spectrum and its associated signals. The signal that appeared at 0.89 ppm is attributed to terminal methyl protons (C–CH$_3$), at 1.32–1.95 ppm to methylene groups (α–CH$_2$, β–CH$_2$), which suggested the existence of hydrogen atoms on the third carbon in an aliphatic fatty chain, and at 2.27 ppm to methylene protons (C–CH$_3$). The signal at 0.87 ppm is attributed to terminal methyl protons, while the peak of olefinic hydrogen (–CH=CH–) appears at 5.37 ppm, signifying the double bond integrated for two hydrogen atoms for each double bond group. The spectra peak at 2.32 ppm is for allylic hydrogen (–CH$_2$), which has two hydrogen atoms per non-conjugated group. The appearance of a peculiar single peak at 3.6 ppm, which is attributed to methoxy protons (–OCH$_3$), is evidence of the production of methyl esters. Figure 7b illustrates the ^{13}C NMR spectra, demonstrating distinctive signals at 31.91 ppm and 174.25 ppm that are related to (–CH$_2$) n and (–COO–), respectively. The unsaturated position in biodiesel (CH=CH) methyl ester was detected at 129.7 ppm for outer non-conjugated carbon. The conversion of oil into biodiesel was also confirmed by employing the following formula, as reported elsewhere [13,37,38]:

$$C = 100 \times \frac{2A\ (Me)}{3A\ (CH_2)}$$

The above equation, where C, A (Me), and A (CH$_2$) are the percentage conversion, integration values of methoxy protons, and alpha methylene proton, respectively, was used to calculate the biodiesel yield from waste cooking oil. The conversion of WCO to biodiesel was obtained as 82.8% by using the above formula, which is quite close to the practically obtained yield of 84%.

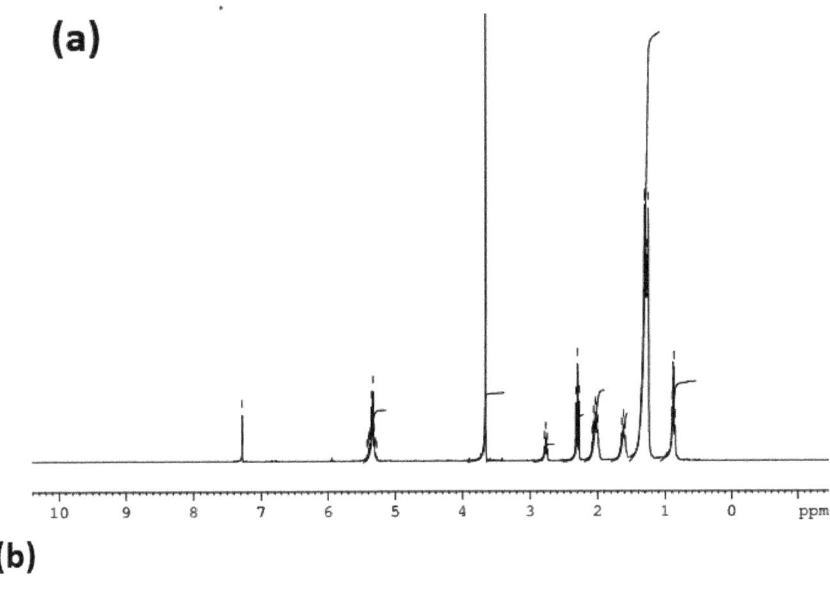

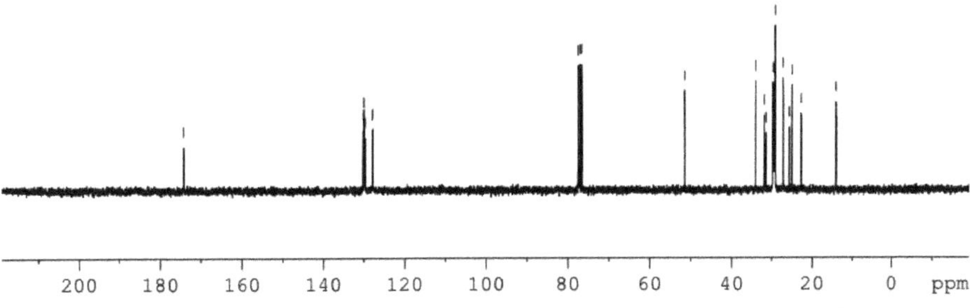

Figure 7. NMR spectra of (**a**) ^{1}H and (**b**) ^{13}C for biodiesel produced from waste cooking oil.

2.3.3. GC-MS Analysis

The produced biodiesel from the waste cooking oil chemical composition was analyzed further by GC-MS (Figure 8). The analysis of the biodiesel samples using library software (NO. NIST02) allowed for the identification of various FAMEs (fatty acid methyl esters). The mass spectra of docosanoic methyl ester (C_{22}:0 m/z 354) appeared at a retention time of 15,611 min and octadecanoic acid methyl ester (C_{18}:0 m/z 298) at a 12,177 min retention time, which is saturated fatty acid methyl ester. The mass spectrum shows important fragment ions characteristic of monosaturated fatty acid methyl esters such as tetracosenoic acid methyl ester (C_{25}:0, m/z 382) and other saturated fatty acid methyl esters including eicosanoic acid methyl ester (C_{21}:0, m/z 326), hexadecanoic methyl ester (C_{17}:0, m/z 270), and methyl tetradecanoate (C_{15}:0, m/z 242). Moreover, other monounsaturated fatty acid methyl esters were also observed, such as 10-octadecenoic acid methyl ester (C_{10}:1) and 11-eicosenic acid methyl ester (C_{21}:1) at m/z 324 and m/z 296. Furthermore, the current peaks exhibited saturated and unsaturated McLafferty rearrangements (m/z = 74 and 29). The fatty acid methyl ester profile is the key factor in determining feedstock's suitability in synthesizing biodiesel. However, it should be noted that the presence of a higher degree of unsaturated fatty acids can limit the suitability of (FAMEs) fatty acid methyl esters as fuel because of issues such as peroxidation and polymerization. Peroxidation occurs at an accelerated rate in combustion engines at higher temperatures, which can lead to engine clogging. As a result, feedstocks with a higher proportion of polyunsaturated fatty acids are not ideal for biodiesel production. Conversely, monounsaturated acids have a lower affinity for oxygen, minimizing the risk of peroxidation [39,40].

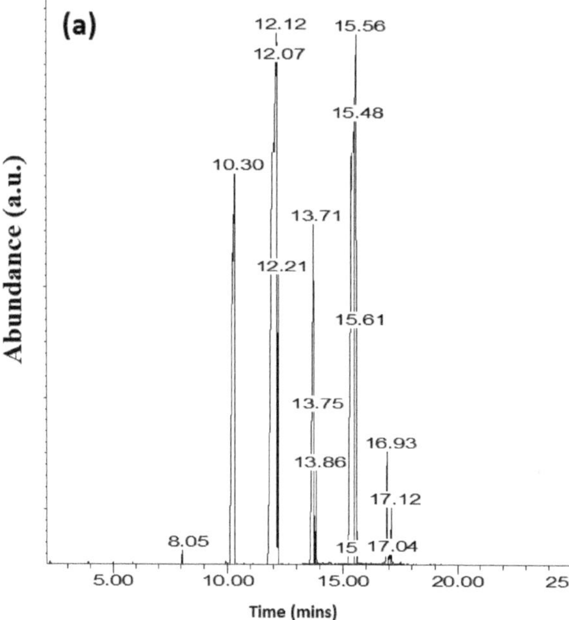

Figure 8. *Cont.*

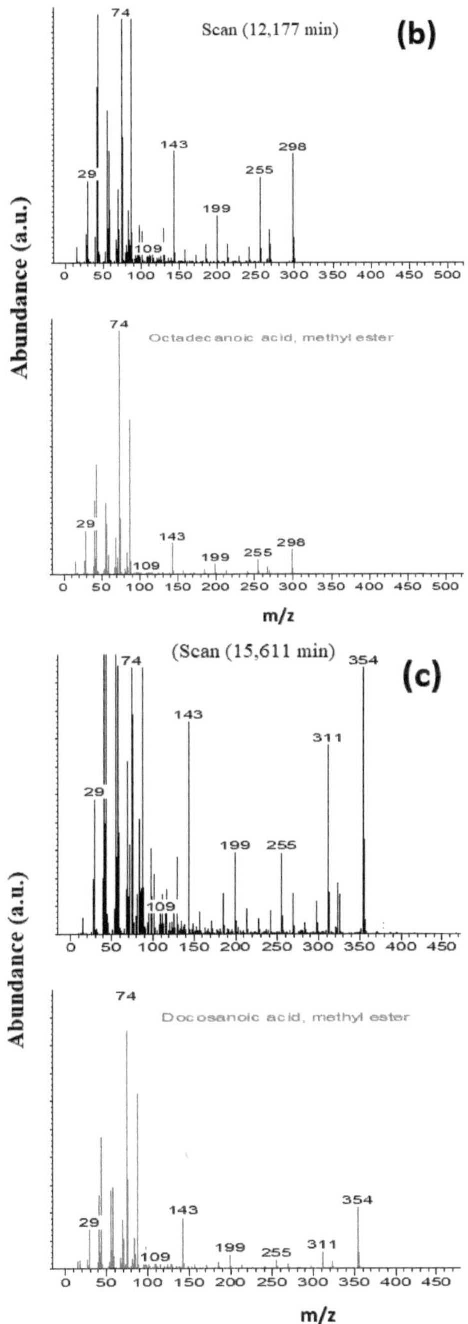

Figure 8. GC-MS chromatograms. (**a**) Total ion chromatogram of biodiesel, (**b**), Library match of peak at 12,177 min (octadecanoic acid methyl ester (C_{18}:0)), and (**c**) Library match of peak at 15,611 (docosanoic methyl ester (C_{22}:0)).

2.4. Comparison of Biodiesel Fuel Properties with International Standards

As recommended by ASTM, the fuel qualities of the synthesized biodiesel were evaluated. The test findings are consistent with international standards, including those indicated in Table 1 such as ASTM-951, 6751, Chinese international standards (GB/T 20828), and European Union standards (EU-14214). To establish the coherence of the synthesized biodiesel's properties with slight alterations, it was compared with earlier research. Higher acid numbers result in engine damage and poorer engine efficiency, and there is a correlation between the acid number and the level of free fatty acids (FFAs) in the fuel. Nonetheless, the synthetic biodiesel is within the permitted range for international standards with an acid value of 0.31 mg KOH/g. High-density fuels can contribute to engine viscosity issues and have an impact on atomization during combustion. For FAMEs, the density should be between 860 and 900 kg/m^3, as mandated by American (ASTM) and European (EU) requirements, which is somewhat higher than that of petro-diesel density (827.2 kg/m^3) [41,42]. This research discovered that the produced biodiesel's density was 0.8312 kg/m^3, which is within the bounds established by international regulations (Table 1). This shows that switching to biodiesel as a diesel fuel substitute will not have any negative consequences on the engine or the environment. We also examined kinematic viscosity, a fundamental property of biodiesel that is impacted by fuel spray, mix formation, and combustion. It is crucial to keep the kinematic viscosity as low as possible because high viscosity can cause deposits to build up and poor combustion in the engine. It was discovered that the kinematic viscosity of the biodiesel made from used cooking oil was 5.88 centistokes (cSt) (Table 1), which is slightly higher than the values discovered in prior research utilizing seed oils from the rubber raphnus L. and jatropha (5.65, 5.65, and 4.36 cSt, respectively) [10]. The biodiesel used in the current experiment has a flash point of 77 °C, which is within the permissible range for fuels with flash points higher than 66 °C, as defined by American (ASTM), Chinese (PRC), and European Standards (EU). That is, however, still below the targeted cap. The term "flash point" describes the temperature at which fuel begins to vaporize and catch fire when exposed to a spark [10].

Table 1. Comparison of fuel properties of the waste cooking oil biodiesel with international biodiesel standards and biodiesel from other studies.

Parameters	Unit	Test Method	Typical Values of Diesel (Market)	Observed Values
Specific gravity (15.6 °C)	No unit	ASTMD1298	<0.860	0.904
Color	No unit	ASTMD1500	Min. 3	4
Flash point	°C	ASTM D93	Min. 54	77
Cloud point	°C	ASTMD2500	Max. 9	−3
Pour point	°C	ASTM D97	Max. 9	−5
Viscosity (kinematic at 50 °C)	cSt or mm^2/s	ASTM D445	Max. 12	5.88
Sulfur	wt.%	ASTMD4294	Max. 1.8	0.0001
Sediments	wt.%	ASTM D473	Max. 0.05	0.014
Water content	wt.%	ASTM D95	Max. 0.25	0.046
Total acid number	mg KOH/gm	ASTM D664	Max. 0.5	0.31
Cetane number	No unit	ASTM D976	Min. 35	29
Copper strip (3 h corrosion at 100 °C)	No unit	ASTM D130	Max. 1	1a

2.5. Reusability of Catalyst

The reusability of the CaO/CNC catalyst's efficacy was assessed to evaluate the stability of nanocomposite in the transesterification of biodiesel production under optimal reaction conditions. The optimized transesterification reaction conditions, which included a 3-weight percent catalyst loading, a 20:1 methanol-to-oil ratio, a reaction temperature of 60 °C, and a 90-min reaction period, were employed to test the catalytic activity of the recycled catalyst. Up to five times reusability was possible for the catalyst after it was removed from the reaction mixture. The separation and cleaning procedure reduced the catalyst activity of the regenerated catalyst. After each process, the used catalyst was collected using filter paper, cleaned with methanol to remove any contaminants, and then utilized in tests on catalyst leaching. The catalyst activity in the following five reactions was assessed after the catalyst surface had completely dried at 100 °C for at least three hours. It was noted that throughout the three recycling procedures, the biodiesel yields fell somewhat from 84% to 82%. The biodiesel efficiency declined because of the leaching effect of active species in the reaction mixture after frequent use. In addition, CaO nanoparticles contained in the reaction can be solubilized in the glycerol attained by way of a by-product, i.e., calcium diglyceroxide species, which is the active phase in biodiesel manufacture [43]. The deactivation of catalytic active sites resulted in a further decline in biodiesel yields after each run, which explained why they reached 79% in the fifth cycle (Figure 9). However, the potentiality of the prepared nanocatalyst to be reused in biodiesel production can reduce the raw material cost of the overall production [44]. In a similar study [21], the yield of biodiesel was reduced from 94.55% to 67.07% after five cycles, and the pattern in terms of catalyst reusability was similar to the current study, where the yield of biodiesel was decreased slightly at the early cycle. The comparison between the CaO/CNC composite and pristine CaO was made using the same reaction conditions (Figure 9), and it was observed that the composite shows a better reaction rate and stability compared with pristine CaO. The reaction kinetics of pristine CaO is quite slow and takes time, which limits its suitability to be used as a catalyst. For this reason, researchers used modifications/doping or composite formation to enhance the reaction rates of CaO [20]. For example, Tang et al. modified CaO by treating it with bromooctane and obtained a lower reaction time of 3 h (99.2% conversion) compared with CaO (35.4% conversion) [19].

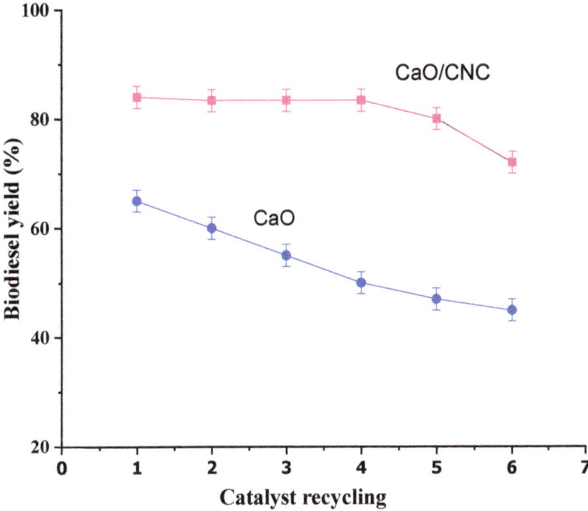

Figure 9. Recycling of CaO/CNC nanocomposite and CaO during transesterification reactions.

3. Materials and Methodology

3.1. Materials

The chemicals, which were used at analytical grade, came from Sigma Aldrich. These included sodium hydroxide (NaOH), calcium nitrate hexahydrate ($Ca(NO_3)_2 \cdot 6H_2O$), hydrochloric acid (HCl), methanol (CH_3OH), sulfuric acid (H_2SO_4), ethanol (C_2H_5OH), anhydrous magnesium sulfate ($MgSO_4$), and filter paper (No 41). Vegetable oil, canola waste cooking oil (WCO), was obtained from a nearby local restaurant.

3.2. Cellulose Nanocrystal Synthesis (CNC)

The world's most prevalent and renewable biopolymer is cellulose, which is derived from renewable resources like biomass [45]. Nanocrystalline cellulose (CNC) has a strong crystalline structure, is densely packed, inexpensive, and lightweight, and has distinct morphologies with nanoscale dimensions [46]. Acid hydrolysis was carried out under regulated conditions to separate the CNCs, ensuring that only the amorphous segments of the cellulose were acted on and the crystalline segments were left unaltered. Sulfuric acid and hydrochloric acid are the most often used acids for the acid hydrolysis of cellulose because of their considerable strength, which is necessary to hydrolyze the glycosidic link present in cellulose [47]. In fact, exposing cellulose fibers to acid hydrolysis can produce defect-free, rod-like crystalline residues [47]. The structure of cellulose is supramolecular, and acid hydrolysis causes a certain degree of cellulose structure breakage, resulting in lower molecular weight constituents and molecular weight distribution with smaller molecular fractions [48,49]. CNC was fabricated using acid hydrolysis of filter paper No 41, which is a cheap and easily available source of cellulose, requiring less effort and purification steps than other raw sources of cellulose. A homogenous suspension of 2 g of filter paper was prepared by cutting, pulverizing, and dissolving it in a 62% H_2SO_4 acid solution. The mixture was then placed in a water bath at 45 °C and continuously stirred for an hour. The addition of adequate DI water stopped hydrolysis (20-fold), and then the acid was neutralized to stop the hydrolysis process. The excess water was removed, and sediments were collected using centrifuging at 8000× g rpm for 15 min. Thereafter, the suspension was cleaned several times with deionized water (DI) water. The total solid content of CNC was determined by drying at 100 °C for 1 h.

3.3. CaO/CNC Nanocomposite

In the current experiment, pure calcium oxide (CaO) nanoparticles were produced by the direct precipitation method, which is one of the most useful, budget-effective, easy, and traditional methods, following the steps reported previously with some modifications [50]. $Ca(NO_3)_2 \cdot 6HO_2$ was properly weighed out and dissolved in (100 mL) distilled water at 60 °C under constant stirring for 1 h using a hotplate magnetic stirrer. A solution of NaOH (1 M) prepared in distilled water was added dropwise into the $Ca(NO_3)_2 \cdot 6H_2O$ solution (40 mL). The reaction process was carried out for 2 h at a temperature of 60 °C and pH = 10, which resulted in white-colored precipitates. These precipitates were filtered and washed with ethanol and distilled water. The end product of this chemical reaction was a precursor of CaO, that is, $Ca(OH)_2$, and dried in an oven for 6 h at 60 °C, which was finally calcinated at 900 °C for 5 h.

CaO/CNC nanocomposites were synthesized by the hydrothermal method. Separately, 1 g of CNC and 1 g of calcium oxide nanoparticles were added to 40 mL of 96% ethanol, which was then sonicated for 30 min separately to attain good dispersion. The solutions were slowly combined and sonicated for five minutes and then transferred to an autoclave Teflon for 12 h at 120 °C. The resulting product was centrifuged, dried at 60 °C overnight, and used for transesterification reaction.

3.4. Free Fatty Acid Content (FFA) Determination of WCO

It is essential to know the FFA content of the oil sample before transesterification. The acid–base titration method was used to determine the level of FFA in the used cooking oil.

Two grams of WCO was added in a 250 mL titration flask, and 100 mL of solution was prepared by adding ethanol–diethyl ether (1:1 *v/v*) by using a few drops of phenolphthalein added in the sample solution as an indicator and titrated with standardized potassium hydroxide 0.025 M solution. Using the following equation, the FFA content of the used cooking oil was determined [51]:

$$\%\text{FFA} = \frac{V\;(mL)\;C \times M \times 100}{1000 \times m\;(gram)}$$

where V is the volume of oil/solvent solution, C is the concentration of KOH in molarity, M is the molecular weight of oleic acid (most commonly used fatty acid), and m is the mass of oil used. A free fatty acid content of 3.5% was found in the waste cooking oil.

3.5. Process of Biodiesel Production

The presence of FFAs in biodiesel reduces its commercial value. As per international biodiesel standards, an acid number of less than 0.5% (mg KOH/g) is recommended. Therefore, after determining the acid value and FFA content of the WCO, acid esterification via pre-treatment was conducted to reduce its FFA content. In this study, the pre-treatment was performed using the conventional oil bath method. A three-necked round-bottom flask was utilized for the pre-treatment process. A reflux condenser was connected to the top neck of the flask to reduce the evaporative loss of alcohol, while the left neck was connected to a thermometer to measure the temperature. The right neck was closed with a stopper to reduce the evaporative loss of alcohol [52]. This was accomplished by an esterification process using cooking oil that was first filtered using a filter strainer to remove impurities. The methanol and oil were then mixed in a 2:1 molar ratio with 5 wt.% of HCl and 60 mL of WCO added. This mixture was then stirred for one hour at 60 °C on a reflux condenser. Distilled water was then added (50–100 mL) and repeatedly rinsed to remove unreacted acid and any other side products. The remaining oil phase was treated with $MgSO_4$ for 30 min until all the water in the oil was dried.

After the esterification of FFAs in the waste cooking oil was reduced to less than 0.5%, the CaO/CNC catalyst was used to produce biodiesel from esterified waste cooking oil. The transesterification reaction was performed using 30 mL of esterified oil for 90 min at 60 °C with a methanol-to-oil molar ratio of 20:1 and 3% CaO/CNC catalyst. Biodiesel and glycerol were produced during the transesterification reaction. Excess methanol was evaporated using a rotary evaporator, and the glycerol and biodiesel layers were separated using a separating funnel. The yield of biodiesel was estimated using the following equation, as per previously reported studies [11,13,30]:

$$Percentage\;yield = \frac{Grams\;of\;Biodiesel\;Produced}{Grams\;of\;Oil\;Used} \times 100$$

3.6. Instrumentation

Powder X-ray diffraction (PXRD) is a crucial technique for identifying the crystalline phases present in materials and for determining the structural properties. PXRD patterns of CaO, CNCs, and CaO/CNCs composite were obtained using a powder diffractometer (Model X'Pert powder X-ray diffractometer from PANalytical) equipped with a Cu-Kα radiation detector with a wavelength of 0.154 nm and a scanning rate of 2°/min in the 2θ range of 10–70°. The Scherrer equation, which relates the average crystallite size (D) to the diffraction width (β), was used to determine the crystallite size based on the broadening of the relevant X-ray spectral peaks. The Scherrer equation and the full-width half maximum (FWHM) of the sharpest peak were used to determine the crystallite sizes.

$$D = K \cdot \lambda / (\beta \cdot \cos\theta)$$

where D = Diameter of crystallite;

K = Shape factor, having a value of 0.9;

λ = Wavelength of the incident X-ray, having a value of 1.542 Å;

β = Full-width half maxima of the corresponding diffraction peak;

θ = Bragg's angle.

Functional groups and the bond formation of materials were studied by FT-IR (BRUKER-TENSOR-27) in the range of 4000 cm^{-1} to 400 cm^{-1} with a of resolution rate 1 cm^{-1} and 15 scans. Scanning electron microscopic study by NOVA Nano exposed the surface morphology (shape and size) and elemental analysis via energy-dispersive X-ray spectroscopy (EDX).

4. Conclusions

In this study, an efficient heterogeneous catalyst CaO/CNC nanocomposite is used for the transesterification of waste cooking oil. Filter paper is used as a cheap source for the development of CNCs, which typically possesses a high surface area and porosity. The filter paper can be replaced in the future with a cheap source of lignocellulosic biomass available abundantly for CNC preparation. The CaO/CNC nanocomposite possesses more active sites, which enhance the efficiency of the transesterification. The hydrothermally synthesized CaO/CNC composite is employed as a catalyst to synthesize biodiesel through transesterification, which is confirmed by different characterizations including XRD, SEM, and EDX. The impact of different operating conditions on the production of biodiesel was investigated thoroughly, such as a molar ratio of methanol-to-oil of 20:1, catalyst concentration (3 wt.%), temperature (60 °C), and time (90 min). Under these optimum conditions, a biodiesel yield of 84% was obtained. The CaO/CNC catalyst can be recycled up to five times without significantly lowering conversion efficiency and shows good stability compared with pristine CaO. To confirm the successful transesterification process, various analytical techniques such as ^{1}H-NMR, ^{13}C-NMR, GC-MS, and FT-IR spectroscopy were employed. The physicochemical characteristics evaluated in this study unequivocally demonstrate that waste cooking oil has the potential to serve as a valuable non-edible feedstock for the biodiesel industry. Its economic feasibility, local availability, environmental friendliness, and adaptability to diverse environmental conditions make it a promising candidate for sustainable biodiesel production. The properties of the biodiesel, such as cetane number, kinematic viscosity, flash point, copper strip corrosion, and water content, align with international standards. These biodiesel properties also meet the standards set by China, the European Union, and the United States. Consequently, biodiesel can be considered a compelling substitute for petroleum diesel. This research offered a cheap and environmentally beneficial technique for synthesizing biodiesel with high yield.

Author Contributions: Data curation, M.R., M.S. and S.D.A.; formal analysis, S.K.; funding acquisition, A.A.; investigation, S.K.; project administration, A.W.; resources, M.S. and A.A.; software, M.R. and S.D.A.; supervision, A.W.; writing—original draft, S.K.; writing—review and editing, A.W. All authors have read and agreed to the published version of the manuscript.

Funding: The researchers would like to thank the Deanship of Scientific Research, Qassim University, Saudi Arabia, for funding the publication of this project.

Data Availability Statement: The authors confirm that the data supporting the findings of this study are available within the article.

Conflicts of Interest: The authors declare that they have no known competing financial interests or personal relationships that could have appeared to influence the work reported in this paper.

References

1. Moradi, P.; Saidi, M. Biodiesel production from Chlorella Vulgaris microalgal-derived oil via electrochemical and thermal processes. *Fuel Process. Technol.* **2022**, *228*, 107158. [CrossRef]
2. Osman, A.I.; Chen, L.; Yang, M.; Msigwa, G.; Farghali, M.; Fawzy, S.; Rooney, D.W.; Yap, P.-S. Cost, environmental impact, and resilience of renewable energy under a changing climate: A review. *Environ. Chem. Lett.* **2023**, *21*, 741–764. [CrossRef]
3. Osman, A.I.; Elgarahy, A.M.; Eltaweil, A.S.; Abd El-Monaem, E.M.; El-Aqapa, H.G.; Park, Y.; Hwang, Y.; Ayati, A.; Farghali, M.; Ihara, I.; et al. Biofuel production, hydrogen production and water remediation by photocatalysis, biocatalysis and electrocatalysis. *Environ. Chem. Lett.* **2023**, *21*, 1315–1379. [CrossRef]
4. Jabeen, M.; Munir, M.; Abbas, M.M.; Ahmad, M.; Waseem, A.; Saeed, M.; Kalam, M.A.; Zafar, M.; Sultana, S.; Mohamed, A.; et al. Sustainable Production of Biodiesel from Novel and Non-Edible Ailanthus altissima (Mill.) Seed Oil from Green and Recyclable Potassium Hydroxide Activated Ailanthus Cake and Cadmium Sulfide Catalyst. *Sustainability* **2022**, *14*, 10962. [CrossRef]
5. Munir, M.; Saeed, M.; Ahmad, M.; Waseem, A.; Alsaady, M.; Asif, S.; Ahmed, A.; Shariq Khan, M.; Bokhari, A.; Mubashir, M.; et al. Cleaner production of biodiesel from novel non-edible seed oil (*Carthamus lanatus* L.) via highly reactive and recyclable green nano CoWO3@rGO composite in context of green energy adaptation. *Fuel* **2023**, *332*, 126265. [CrossRef]
6. Emmanouilidou, E.; Mitkidou, S.; Agapiou, A.; Kokkinos, N.C. Solid waste biomass as a potential feedstock for producing sustainable aviation fuel: A systematic review. *Renew. Energy* **2023**, *206*, 897–907. [CrossRef]
7. Emmanouilidou, E.; Lazaridou, A.; Mitkidou, S.; Kokkinos, N.C. A comparative study on biodiesel production from edible and non-edible biomasses. *J. Mol. Struct.* **2024**, *1306*, 137870. [CrossRef]
8. Osman, A.I.; Mehta, N.; Elgarahy, A.M.; Al-Hinai, A.; Al-Muhtaseb, A.A.H.; Rooney, D.W. Conversion of biomass to biofuels and life cycle assessment: A review. *Environ. Chem. Lett.* **2021**, *19*, 4075–4118. [CrossRef]
9. Munir, M.; Ahmad, M.; Mubashir, M.; Asif, S.; Waseem, A.; Mukhtar, A.; Saqib, S.; Siti Halimatul Munawaroh, H.; Lam, M.K.; Shiong Khoo, K.; et al. A practical approach for synthesis of biodiesel via non-edible seeds oils using trimetallic based montmorillonite nano-catalyst. *Bioresour. Technol.* **2021**, *328*, 124859. [CrossRef]
10. Munir, M.; Ahmad, M.; Rehan, M.; Saeed, M.; Lam, S.S.; Nizami, A.S.; Waseem, A.; Sultana, S.; Zafar, M. Production of high quality biodiesel from novel non-edible Raphnus raphanistrum L. seed oil using copper modified montmorillonite clay catalyst. *Environ. Res.* **2021**, *193*, 110398. [CrossRef]
11. Munir, M.; Ahmad, M.; Saeed, M.; Waseem, A.; Nizami, A.-S.; Sultana, S.; Zafar, M.; Rehan, M.; Srinivasan, G.R.; Ali, A.M. Biodiesel production from novel non-edible caper (*Capparis spinosa* L.) seeds oil employing Cu–Ni doped ZrO$_2$ catalyst. *Renew. Sustain. Energy Rev.* **2021**, *138*, 110558. [CrossRef]
12. Nasreen, S.; Liu, H.; Skala, D.; Waseem, A.; Wan, L. Preparation of biodiesel from soybean oil using La/Mn oxide catalyst. *Fuel Process. Technol.* **2015**, *131*, 290–296. [CrossRef]
13. Munir, M.; Ahmad, M.; Saeed, M.; Waseem, A.; Rehan, M.; Nizami, A.-S.; Zafar, M.; Arshad, M.; Sultana, S. Sustainable production of bioenergy from novel non-edible seed oil (*Prunus cerasoides*) using bimetallic impregnated montmorillonite clay catalyst. *Renew. Sustain. Energy Rev.* **2019**, *109*, 321–332. [CrossRef]
14. Munir, M.; Saeed, M.; Ahmad, M.; Waseem, A.; Sultana, S.; Zafar, M.; Srinivasan, G.R. Optimization of novel Lepidium perfoliatum Linn. Biodiesel using zirconium-modified montmorillonite clay catalyst. *Energy Sources Part A Recovery Util. Environ. Eff.* **2022**, *44*, 6632–6647. [CrossRef]
15. Foroutan, R.; Peighambardoust, S.J.; Mohammadi, R.; Peighambardoust, S.H.; Ramavandi, B. Application of walnut shell ash/ZnO/K2CO3 as a new composite catalyst for biodiesel generation from Moringa oleifera oil. *Fuel* **2022**, *311*, 122624. [CrossRef]
16. Yaakob, Z.; Mohammad, M.; Alherbawi, M.; Alam, Z.; Sopian, K. Overview of the production of biodiesel from Waste cooking oil. *Renew. Sustain. Energy Rev.* **2013**, *18*, 184–193. [CrossRef]
17. Atabani, A.E.; Silitonga, A.S.; Badruddin, I.A.; Mahlia, T.; Masjuki, H.; Mekhilef, S. A comprehensive review on biodiesel as an alternative energy resource and its characteristics. *Renew. Sustain. Energy Rev.* **2012**, *16*, 2070–2093. [CrossRef]
18. Huaping, Z.; Zongbin, W.; Yuanxiong, C.; Zhang, P.; Shijie, D.; Xiaohua, L.; Zongqiang, M. Preparation of biodiesel catalyzed by solid super base of calcium oxide and its refining process. *Chin. J. Catal.* **2006**, *27*, 391–396.
19. Tang, Y.; Xu, J.; Zhang, J.; Lu, Y. Biodiesel production from vegetable oil by using modified CaO as solid basic catalysts. *J. Clean. Prod.* **2013**, *42*, 198–203. [CrossRef]
20. Al-Muhtaseb, A.A.H.; Jamil, F.; Osman, A.I.; Tay Zar Myint, M.; Htet Kyaw, H.; Al-Hajri, R.; Hussain, M.; Ahmad, M.N.; Naushad, M. State-of-the-art novel catalyst synthesised from waste glassware and eggshells for cleaner fuel production. *Fuel* **2022**, *330*, 125526. [CrossRef]
21. Talha, N.S.; Sulaiman, S. In situ transesterification of solid coconut waste in a packed bed reactor with CaO/PVA catalyst. *Waste Manag.* **2018**, *78*, 929–937. [CrossRef]
22. Padalkar, S.; Capadona, J.R.; Rowan, S.J.; Weder, C.; Won, Y.-H.; Stanciu, L.A.; Moon, R.J. Natural biopolymers: Novel templates for the synthesis of nanostructures. *Langmuir* **2010**, *26*, 8497–8502. [CrossRef]
23. Rezayat, M.; Blundell, R.K.; Camp, J.E.; Walsh, D.A.; Thielemans, W. Green one-step synthesis of catalytically active palladium nanoparticles supported on cellulose nanocrystals. *ACS Sustain. Chem. Eng.* **2014**, *2*, 1241–1250. [CrossRef]
24. Zik, N.; Sulaiman, S.; Jamal, P. Biodiesel production from waste cooking oil using calcium oxide/nanocrystal cellulose/polyvinyl alcohol catalyst in a packed bed reactor. *Renew. Energy* **2020**, *155*, 267–277. [CrossRef]

25. Besbes, I.; Alila, S.; Boufi, S. Nanofibrillated cellulose from TEMPO-oxidized eucalyptus fibres: Effect of the carboxyl content. *Carbohydr. Polym.* **2011**, *84*, 975–983. [CrossRef]

26. Habte, L.; Shiferaw, N.; Mulatu, D.; Thenepalli, T.; Chilakala, R.; Ahn, J.W. Synthesis of nano-calcium oxide fromwaste eggshell by sol-gel method. *Sustainability* **2019**, *11*, 3196. [CrossRef]

27. Mirghiasi, Z.; Bakhtiari, F.; Darezereshki, E.; Esmaeilzadeh, E. Preparation and characterization of CaO nanoparticles from Ca(OH)2 by direct thermal decomposition method. *J. Ind. Eng. Chem.* **2014**, *20*, 113–117. [CrossRef]

28. Wang, Z.; Ding, Y.; Wang, J. Novel Polyvinyl Alcohol (PVA)/Cellulose Nanocrystal (CNC) Supramolecular Composite Hydrogels: Preparation and Application as Soil Conditioners. *Nanomaterials* **2019**, *9*, 1397. [CrossRef]

29. Brahma, S.; Basumatary, B.; Basumatary, S.F.; Das, B.; Brahma, S.; Rokhum, S.L.; Basumatary, S. Biodiesel production from quinary oil mixture using highly efficient Musa chinensis based heterogeneous catalyst. *Fuel* **2023**, *336*, 127150. [CrossRef]

30. Al-Muhtaseb, A.A.H.; Osman, A.I.; Jamil, F.; Al-Riyami, M.; Al-Haj, L.; Alothman, A.A.; Htet Kyaw, H.; Tay Zar Myint, M.; Abu-Jrai, A.; Ponnusamy, V.K. Facile technique towards clean fuel production by upgrading waste cooking oil in the presence of a heterogeneous catalyst. *J. King Saud Univ. Sci.* **2020**, *32*, 3410–3416. [CrossRef]

31. Karkal, S.; Kudre, T. Valorization of marine fish waste biomass and Gallus Gallus eggshells as feedstock and catalyst for biodiesel production. *Int. J. Environ. Sci. Technol.* **2023**, *20*, 7993–8016. [CrossRef]

32. Olutoye, M.A.; Wong, S.W.; Chin, L.H.; Amani, H.; Asif, M.; Hameed, B.H. Synthesis of fatty acid methyl esters via the transesterification of waste cooking oil by methanol with a barium-modified montmorillonite K10 catalyst. *Renew. Energy* **2016**, *86*, 392–398. [CrossRef]

33. Ma, L.; Wei, P.; Li, J.; Liang, L.; Li, G. Synthesis of nano-crystal PVMo2W9@[Cu$_6$O(TZI)$_3$(H$_2$O)$_6$]4·nH$_2$O for catalytically biodiesel preparation. *J. Solid State Chem.* **2024**, *329*, 124434. [CrossRef]

34. Fatimah, I.; Rubiyanto, D.; Taushiyah, A.; Najah, F.B.; Azmi, U.; Sim, Y.-L. Use of ZrO$_2$ supported on bamboo leaf ash as a heterogeneous catalyst in microwave-assisted biodiesel conversion. *Sustain. Chem. Pharm.* **2019**, *12*, 100129. [CrossRef]

35. Mohebolkhames, E.; Kazemeini, M.; Sadjadi, S. Utilization of Salmon fish bone wastes as a novel bio-based heterogeneous catalyst-support toward the production of biodiesel: Process optimizations and kinetics studies. *Mater. Chem. Phys.* **2024**, *311*, 128522. [CrossRef]

36. Saetiao, P.; Kongrit, N.; Cheng, C.K.; Jitjamnong, J.; Direksilp, C.; Khantikulanon, N. Catalytic conversion of palm oil into sustainable biodiesel using rice straw ash supported-calcium oxide as a heterogeneous catalyst: Process simulation and techno-economic analysis. *Case Stud. Chem. Environ. Eng.* **2023**, *8*, 100432. [CrossRef]

37. Gelbard, G.; Brès, O.; Vargas, R.M.; Vielfaure, F.; Schuchardt, U.F. 1H nuclear magnetic resonance determination of the yield of the transesterification of rapeseed oil with methanol. *J. Am. Oil Chem. Soc.* **1995**, *72*, 1239–1241. [CrossRef]

38. Meher, L.C.; Vidya Sagar, D.; Naik, S.N. Technical aspects of biodiesel production by transesterification—A review. *Renew. Sustain. Energy Rev.* **2006**, *10*, 248–268. [CrossRef]

39. Nomgboye, A.; Hansen, A. Prediction of cetane number of biodiesel fuel from the fatty acid methyl ester [FAME] composition. *Int. Agrophysics* **2008**, *22*, 21–29.

40. Sokoto, M.; Hassan, L.; Dangoggo, S.; Ahmad, H.; Uba, A. Influence of fatty acid methyl esters on fuel properties of biodiesel produced from the seeds oil of Curcubita pepo. *Niger. J. Basic Appl. Sci.* **2011**, *19*, 81–86. [CrossRef]

41. Amenaghawon, A.N.; Obahiagbon, K.; Isesele, V.; Usman, F. Optimized biodiesel production from waste cooking oil using a functionalized bio-based heterogeneous catalyst. *Clean. Eng. Technol.* **2022**, *8*, 100501. [CrossRef]

42. Patil, P.D.; Deng, S. Optimization of biodiesel production from edible and non-edible vegetable oils. *Fuel* **2009**, *88*, 1302–1306. [CrossRef]

43. Marques Correia, L.; Cecilia, J.A.; Rodríguez-Castellón, E.; Cavalcante, C.L.; Vieira, R.S. Relevance of the physicochemical properties of calcined quail eggshell (CaO) as a catalyst for biodiesel production. *J. Chem.* **2017**, *2017*, 5679512. [CrossRef]

44. Kirubakaran, M. Eggshell as heterogeneous catalyst for synthesis of biodiesel from high free fatty acid chicken fat and its working characteristics on a CI engine. *J. Environ. Chem. Eng.* **2018**, *6*, 4490–4503.

45. Pirani, S.; Hashaikeh, R. Nanocrystalline cellulose extraction process and utilization of the byproduct for biofuels production. *Carbohydr. Polym.* **2013**, *93*, 357–363. [CrossRef]

46. Habibi, Y.; Lucia, L.A.; Rojas, O.J. Cellulose nanocrystals: Chemistry, self-assembly, and applications. *Chem. Rev.* **2010**, *110*, 3479–3500. [CrossRef]

47. Siqueira, G.; Bras, J.; Dufresne, A. Luffa cylindrica as a lignocellulosic source of fiber, microfibrillated cellulose, and cellulose nanocrystals. *BioResources* **2010**, *5*, 727–740. [CrossRef]

48. Iranmahboob, J.; Nadim, F.; Monemi, S. Optimizing acid-hydrolysis: A critical step for production of ethanol from mixed wood chips. *Biomass Bioenergy* **2002**, *22*, 401–404. [CrossRef]

49. Duran, N.; Lemes, A.P.; Duran, M.; Freer, J.; Baeza, J. A minireview of cellulose nanocrystals and its potential integration as co-product in bioethanol production. *J. Chil. Chem. Soc.* **2011**, *56*, 672–677. [CrossRef]

50. Khine, E.E.; Koncz-Horvath, D.; Kristaly, F.; Ferenczi, T.; Karacs, G.; Baumli, P.; Kaptay, G. Synthesis and characterization of calcium oxide nanoparticles for CO$_2$ capture. *J. Nanopart. Res.* **2022**, *24*, 139. [CrossRef]

51. Varona, E.; Tres, A.; Rafecas, M.; Vichi, S.; Barroeta, A.C.; Guardiola, F. Methods to determine the quality of acid oils and fatty acid distillates used in animal feeding. *MethodsX* **2021**, *8*, 101334. [CrossRef]
52. Chai, M.; Tu, Q.; Lu, M.; Yang, Y.J. Esterification pretreatment of free fatty acid in biodiesel production, from laboratory to industry. *Fuel Process. Technol.* **2014**, *125*, 106–113. [CrossRef]

Review

A Review on Biolubricants Based on Vegetable Oils through Transesterification and the Role of Catalysts: Current Status and Future Trends

Sergio Nogales-Delgado [1],*, José María Encinar [2] and Juan Félix González [1]

[1] Department of Applied Physics, University of Extremadura, Avda. de Elvas s/n, 06006 Badajoz, Spain; jfelixgg@unex.es
[2] Department of Chemical Engineering and Physical-Chemistry, University of Extremadura, Avda. de Elvas s/n, 06006 Badajoz, Spain; jencinar@unex.es
* Correspondence: senogalesd@unex.es

Abstract: The use of biolubricants as an alternative to petroleum-based products has played an important role in the last decade. Due to the encouragement of global policies, which mainly support green chemistry and circular economy, there has been an increasing interest in bio-based products, including biolubricants, from scientific and industrial points of view. Their raw materials, production, and characteristics might vary, as biolubricants present different applications for a wide range of practical uses, making this field a continuously changing subject of study by researchers. The aim of this work was to study biolubricant production from vegetable oil crops from a bio-refinery perspective, paying attention to the main raw materials used, the corresponding production methods (with a special focus on double transesterification), the role of catalysts and some techno-economic studies. Thus, the main factors affecting quality parameters such as viscosity or oxidative stability have been covered, including catalyst addition, reaction temperature, or the use of raw materials, reagents, or additives were also analyzed. In conclusion, the search for suitable raw materials, the use of heterogeneous catalysts to improve the effectiveness and efficiency of the process, and the optimization of chemical conditions seem to be the most interesting research lines according to the literature.

Keywords: fatty acids; fatty acid methyl esters; transesterification; epoxidation; catalyst; viscosity; oxidation stability; acidity; biorefinery; sustainability; circular economy

Citation: Nogales-Delgado, S.; Encinar, J.M.; González, J.F. A Review on Biolubricants Based on Vegetable Oils through Transesterification and the Role of Catalysts: Current Status and Future Trends. *Catalysts* **2023**, *13*, 1299. https://doi.org/10.3390/catal13091299

Academic Editor: Jaehoon Kim

Received: 17 August 2023
Revised: 11 September 2023
Accepted: 14 September 2023
Published: 16 September 2023

1. Introduction

1.1. Global Energy and Materials Scenario: The Role of Biolubricants

Due to the global concern about environmental and sustainable subjects, the promotion of new concepts such as circular economy, green chemistry, or sustainability has become popular in recent years. Indeed, global agencies and organizations, as well as national or local governments (along with society's environmental awareness), have encouraged these kinds of practices, as in the case of Europe, has established a long-term goal to develop a competitive, resource-efficient, and low carbon economy by 2050, pointing out the important role of biomass in future bioeconomy policies [1].

On the other hand, the 2030 Horizon and Sustainable Development Goals (SDG) established by the United Nations (UN) are a reference for the launch of specific national or regional guidelines regarding sustainability, green practices, and circular economy, among others [2].

In this context, the role of bioproducts as a replacement for petroleum products can be strategic, fostering economic and sustainable growth of developing countries or regions (as well as developed countries whose green policies have changed their strategies in the medium and long term). For instance, the implementation of technologies devoted to biofuel production has been widely studied in the literature, including biodiesel or bioethanol, among others.

Regarding bioproducts, the specific case included in this review, that is, biolubricants, has been considered an interesting research field, as explained later. Biolubricants could act as the replacement for mineral lubricants, implying many advantages and challenges. This way, biolubricants are more biodegradable, presenting higher flash and combustion points compared to traditional lubricants. Also, biolubricant production usually presents a high atom efficiency. These properties imply a product with a reduced environmental impact, as well as increased safety during storage or transport. On the other hand, these biolubricants may also have disadvantages, like short oxidative stability values, which could worsen viscosity or acidity during storage, with subsequent problems for its direct use in machines, engines, or gears.

In general, apart from the abovementioned advantages, and depending on the raw material used and the kind of production, biolubricants could comply with some Sustainable Development Goals, like the ones included in Figure 1:

Figure 1. Main Sustainable Development Goals related to biolubricant production [2,3].

Therefore, as can be seen in this figure, biolubricant production, which can be considered in a biorefinery context, could comply directly with many of the Sustainable Development Goals (SDGs) established by the UN. For instance, as explained in the following sections, biolubricant production can present intermediate products, such as fatty acid methyl esters, which could contribute to affordable and clean energy.

Regarding decent work and economic growth, the use of bio-based products obtained from typical raw materials in a certain area can contribute to the energy and material independence of this region, developing sustainable economic growth. Indeed, local crops and the subsequent vegetable oils could be an interesting starting point for a biorefinery based on biodiesel and biolubricant production (usually through double transesterification), promoting sustainable industries, innovation, and responsible production of many components such as biodiesel, glycerol, biolubricants, methanol recovery, etc.

Finally, and since natural raw materials are used in a sustainable way, obtaining a biodegradable product, biolubricant production could help to improve climate action, protecting life on land and below water (as, for instance, a biolubricant spill would not be as harmful as its petrol-based equivalent). Furthermore, other SDGs could be positively influenced (such as "No poverty", "Good health and well-being", or "Reduced inequalities"), but in an indirect way.

These SDGs point out another interesting point that supports sustainable strategies, possibly being the real and definitive starting point for the implementation of these policies all over the world. Unfortunately (or not, if green policies are finally taken by national and international agencies at this point, after learning the hard way), the energy and material dependence of most countries (especially concerning natural gas or oil), along with the subsequent and unstable changes on many raw material prices (including oil and, as observed after the war between Russia and Ukraine, vegetable oils such as corn oil) has implied an unstable period where economic crisis (pressed by this energy instability) became the general tone, affecting almost every corner of the globe in a direct or indirect manner.

However, this is just a simple example of how international relations and wars in recent history influenced aspects such as energy, food, or transport [4,5]. This way, concepts such as renewable energy, circular economy, or green chemistry are nowadays as linked to geopolitics as ever (although this issue is not new), apart from the obvious economic and industrial concepts, in order to reduce the negative effect of international stress [6–9].

As observed in Figure 2, two significant trends are observed that point out the negative effects of energy and material dependency. First, according to Figure 2a, it is clear that oil price evolution has been a roller coaster since 2005, with increases of 100% in a few years, followed by abrupt decreases of over 50%. With this uncertainty related to oil prices, it is difficult for countries (especially developing ones) to carry out a correct industrial development plan, among other economic issues, with negative consequences for industries and, accordingly, citizens.

Another paradigmatic example is the Russo-Ukrainian war, whose consequences regarding fuel prices are included in Figure 2b. Considering that the Russian invasion of Ukraine took place in February 2022, its consequences were noticeable almost instantaneously, with a considerable increase in FAME (120%), gasoline (90%), and diesel (92%) prices compared to average prices in 2018–2019. As expected, these unstable prices (including some metals like gold or nickel) implied a global economic crisis, affecting developing countries to a greater extent [10]. Consequently, this event could be the "last straw" to definitively change the public and institutional support towards a green transition [11].

Another event to take into consideration was the COVID-19 pandemic, whose effect was also noticeable (apart from the obvious impact on health and economic issues) in energy consumption all over the world, especially in cities (where the different stringency responses taken by governments to mitigate the spread of COVID-19 had an influence on the reduction in urban energy consumption) [12], energy market (whose effect was persistent for an extended period) [13–15] or renewable energy, with an increasing interest in subjects such as green policies or clean energy investments (to revitalize the economy after the pandemic) in such a stable stage like the post-pandemic era.

Equally, energy efficiency research related to renewable energy has gained importance in recent years in order to compete with previous energy sources and promote the sustainable development of green processes [16,17]. As a result, many countries, such as those included in G20, have encouraged green policies through fiscal stimulus as a consequence of the coronavirus crisis [18]. Considering that similar events could take place in the medium to long term, it is no wonder that this trend can be enhanced in the future.

(a)

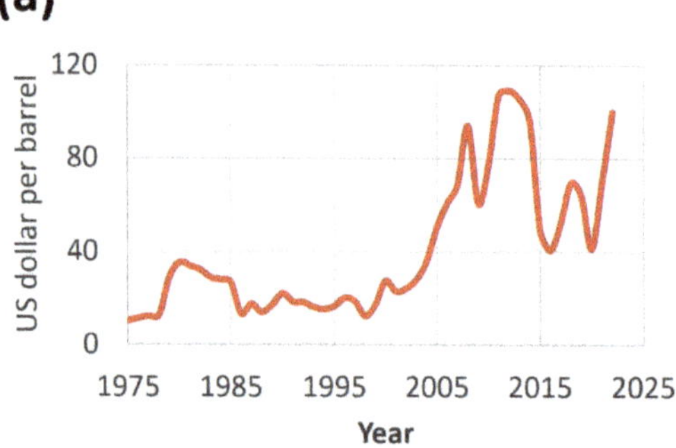

Figure 2. *Cont.*

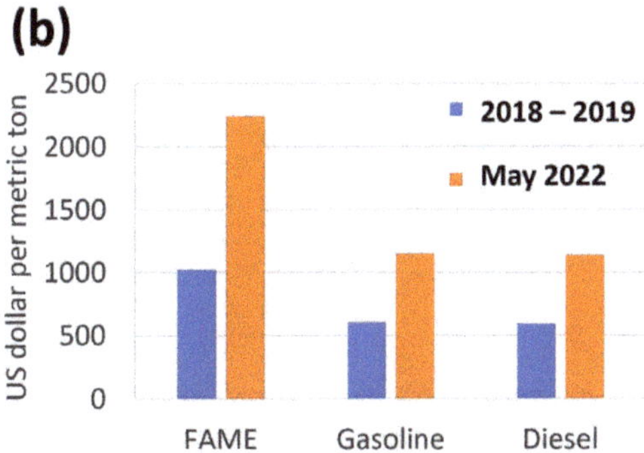

Figure 2. Different trends pointing out the negative effect of energy dependency: (**a**) Average annual OPEC crude oil price from 1975 to 2022; (**b**) Comparison of average wholesale motor fuel prices (FAME, gasoline, and diesel) in Europe in 2018/19 and May 2022 as a result of the Russia-Ukraine war, expressed in U.S. dollars per metric ton of oil equivalent. Sources: [19,20].

Consequently, green policies have become as important as ever, influencing or even changing the current international status when it comes to geopolitics during the energy transition [7,21,22].

Thus, energy and material independence (especially in developing countries, but also at a global level) is necessary, and the use of concepts such as green chemistry or circular economy could contribute to the sustainable development of industrial activities. In this case, as explained in further sections in this review work, the role of some raw materials such as vegetable oils (along with some wastes) could be vital, as they can contribute to the production of very interesting goods, serving as an energy source (for instance, through biodiesel production) or providing important products such as biolubricants (the perfect replacement for lubricants obtained from petroleum industry). But is the use of biolubricants really a great opportunity to contribute to the abovementioned purposes? It will depend on market opportunities and production efficiency, as will be discussed in the following sections.

1.2. Industrial Activity, Lubricant Demand, and the Subsequent Opportunity for Biolubricants

There is no doubt that industrial activity has been constantly increasing since the industrial revolution took place. Indeed, there are plenty of indicators that support this trend, not only in world powers such as China or the United States but also in other developing countries, especially in Africa. Africa's economic awakening is interesting in many ways, but as far as this work is concerned, there are three key points that should be considered, like the following:

- Its economic growth and industrial development are becoming more and more noticeable, with the subsequent risks if environmental measures are not suitably taken.
- There is a great opportunity for sustainable development according to the SDGs, with the subsequent decrease in economic dependence on traditional trade relations with some countries, especially in the case of former colonial powers.
- For this purpose, many sustainable processes can contribute to the industrial network of developing countries.

The role of vegetable oils could be crucial, as there are many oilseed crops (especially safflower, cardoon, or rapeseed) that can be easily adapted to extreme climate conditions and poor soils worldwide, which could encourage developing countries to implement or foster oilseed crops in non-arable areas, with the subsequent benefits related to the production of fuels and other bio-based products such as biolubricants. Indeed, the real implementation of biolubricant production will depend on the real demand for these products.

In that sense, there is a steady industrial development, which usually implies the use of industrial machinery and the subsequent demand for lubricants. Thus, considering the above, the role of biolubricants could be an interesting starting point for the implementation of sustainable processes, green chemistry, or circular economy. This fact can be supported by the demand for biolubricants, as shown in Figure 3. As observed, there was a steady increase in the global demand for biolubricants, forecasting for 2023 a demand of 37.4 million metric tons, that is, over an 8% increase compared to 2010. Consequently, biolubricant production presents a promising future, although it is still in an emerging stage.

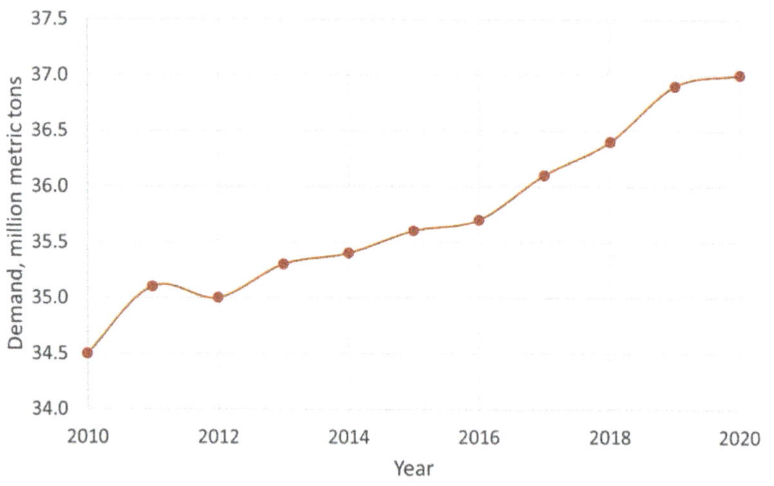

Figure 3. Global demand evolution for biolubricants in the last decade (source: [3]).

In parallel, lubricant production and demand can be broken down by region, as observed in Figure 4. It should be noted that the majority of lubricant demand (Figure 4a) and production (Figure 4b), exceeding 75%, is mainly concentrated in three regions that are, Asia-Pacific, Europe, and North America. On the contrary, lubricant production in Africa can be considered negligible for the time being. In that sense, the role of biolubricants can be doubly promising:

- On the one hand, considering the abovementioned concerns from international agencies and countries, the replacement for lubricants presents great potential in developed countries in Asia, Europe, and North America.
- On the other hand, considering the case of Africa, the use of biolubricants can meet the demands of developing countries, where the production of vegetable oils adapted to extreme climate conditions (such as cardoon or safflower) could contribute to an increase in industrial activity in a sustainable way, making these regions less dependent on energy or imported materials (like lubricants).

Thus, it is clear that there is great potential for biolubricant production based on vegetable oils. Even though there are some concerns related to energy crops (mainly based on oilseed species, both for edible and non-edible purposes) and their environmental impact or food competition, the sustainable use of poor soils, as well as the concept of crop rotation that could enhance soil alleviation have been widely studied as a sustainable way to take advantage of non-arable areas.

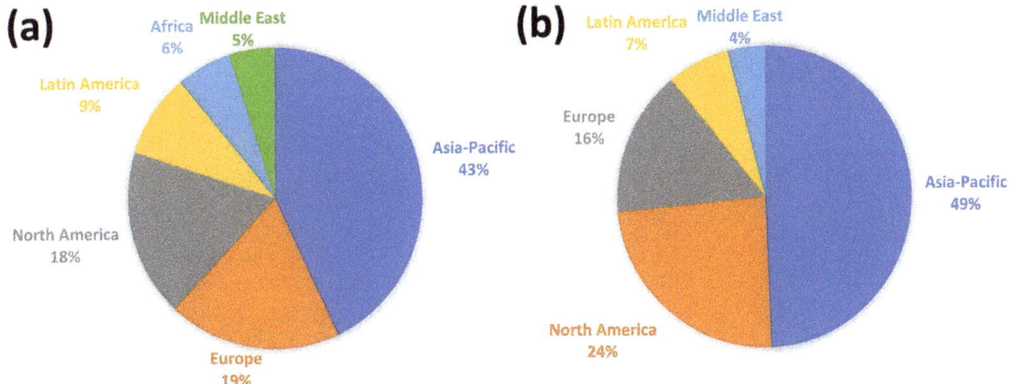

Figure 4. Lubricant demand (**a**) and production (**b**) shares by region in 2018 [23].

In that sense, some crops such as cardoon, rapeseed, or safflower [24–26] would be suitable for sustainable oil production, as they can be easily adapted to extreme climate conditions and poor soils that are not usually arable (which could be an advantage for developing areas where these conditions are usual) and present a long and deep tap root which can be used in rotation crops to recover superficial soils where typical crops such as corn, lettuce or garlic are used. Also, these crops are normally resistant to many plagues. In conclusion, if suitably managed, soils for oil production can be sustainable and would not imply any environmental concern.

Apart from that, there are some wastes related to the use of vegetable oils, especially after cooking and the subsequent frying oil generation, that should be properly managed to avoid environmental problems, especially related to soil and water pollution, where these wastes are especially pollutant. Thus, as commented in the following section, there is a real concern about frying oil management, which could be efficiently managed by its use as a biodiesel and biolubricant source. Equally, the use of these wastes to obtain biodiesel and biolubricants could also contribute to lower soil exploitation, implying a more sustainable production of biolubricants.

1.3. Waste Cooking Oil: A Real Challenge with Endless Opportunities (Such as Biolubricant Production)

Frying oils (FO), or waste cooking oils (WCO), which can be considered as used vegetable oils (UVO), is a waste that is increasingly generated in food and industrial sectors worldwide, which could imply a serious environmental problem if it is not suitably managed. Thus, about 200 million tons per year are produced, with 4 million tons in Europe (which constitutes around 2%, whereas the US, with 55%, and China, with 25%, are the majority of WCO generators) [27]. In any case, it should be noted that, in essence, WCO composition is relatively similar to the original vegetable oils used, presenting degradation compounds depending on many different factors such as frying cycles, temperatures used, etc., which could imply a relatively similar use of this waste compared to vegetable oils devoted, for instance, to biodiesel or biolubricant production (possibly adding some pre-treatments to remove solid residues or adjust pH).

Fortunately, there are plenty of ways to valorize this waste in order to obtain valuable products such as biofuels (like biodiesel), to produce energy, animal feeds, ecological solvents, composites materials, non-aqueous gas sorbent devices or, as in the case of the main subject of this review, biolubricants. For that purpose, new recycling processes have been developed, and the most traditional ones have been optimized to make the use of WCO a valuable chemical block (feasible and efficient), whereas other studies compared the environmental impact of multiple valorization options of waste cooking oil, showing low environmental impact and promising results for achieving circular economy goals [28–30]. Regarding biodiesel production (which is mainly carried out through a common chemical synthesis called transesterification, as will be explained in the following sections), it has proved the suitability of FO to be re-used in green chemistry or circular economy, whose market is expected to generate up to 8.88 billion dollars in 2026 [31].

As a result, WCO management is a matter that has attracted the attention of researchers. In that sense, our previous works have pointed out the possibility of using waste cooking oil as an interesting source for biodiesel and biolubricant production, with similar results to those obtained for equivalent vegetable oils such as corn, rapeseed, or safflower. Indeed, the only disadvantage of these productions was the short oxidative stability of final products, a challenge shared by the rest of the vegetable oils included in our experiences [32–34]. Apart from that, there are many examples of biolubricant production from WCO, offering interesting alternatives for the management of this waste [35]. Other authors have produced biolubricants from WCO through epoxidation, obtaining interesting products with high viscosity index values that could complement the standard ISO vegetable-grade lubricants in the market [36]. Moreover, from the same waste, some authors have synthesized an octylated branched biolubricant through hydrolysis, esterification, epoxidation, and a final reaction with octanoic acid, obtaining a product with better properties compared to mineral-based lubricants (for instance, improving friction coefficient and viscosity index) [37]. Furthermore, some experiences have assessed the possible implementation of double transesterification of WCO with methanol and trimethylolpropane to obtain biolubricants in a vertical pulsed column, carrying out an energy optimization through the surface response of the main operating parameters. As a result, an optimal reaction yield of around 83% was obtained, obtaining an interesting biolubricant with improved properties such as VI, flash point, and pour point [38].

But this interest has also been extended to global society and institutions, with the clear example of Europe, where there is a specific legislative framework about the management and the subsequent employment of this waste for the abovementioned industries, from disposal-collection to reconversion [28,31]. Thus, in the case of the United Kingdom, used cooking oil was the most resourceful feedstock for biofuel production, with 54% in 2020 [39]. But not only in Europe, there is a concern about WCO management. For instance, in 2021, there were more than 98% of biodiesel manufacturers in Japan whose main raw material was waste cooking oil, proving that the use of this waste is already a reality. Moreover, a high percentage of WCO comes from households (37.7%), proving the power of individuals to contribute to a change towards green and sustainable policies. That is, global change comes from local action [40].

Considering these facts, WCO "has earned the right" to be considered as valuable as any other vegetable oil for biolubricant production, being included throughout the discussion and reasoning of this work and playing a decisive role, as will be discussed later.

1.4. Scientific Interest in Biolubricants

In light of this evidence, it is quite clear that there is a real interest in the specific topic we are going to explain in this review work. In such a way, and according to Figure 5, there was a continuous interest in research about biolubricants, which implied an increasing number of published articles about this subject. Thus, especially from 2010, there was a considerable increase, reaching about 80 published articles per year about biolubricant production, application, or characterization. Among these publications, the use of vegetable oils through transesterification seemed to be popular, whereas epoxidation processes to obtain biolubricants were less researched.

As explained throughout this review, the higher interest in double transesterification could be due to the special characteristics of this chemical route, which could perfectly fit with the biorefinery concept, where biodegradable products and low quantities of wastes are obtained. In that sense, many studies about specific techno-economic aspects of biorefineries, as well as several patents about these subjects, have been developed (which will be further explored in the following sections), proving the increasing interest in double transesterification both at scientific and industrial level.

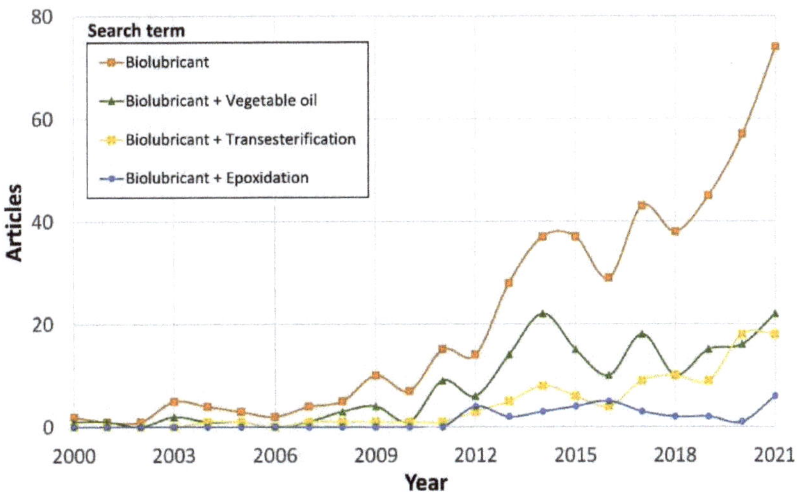

Figure 5. Current research trends on biolubricant (source: [41]). According to the following search criteria: "biolubricants"; "biolubricant" AND "vegetable oil"; "biolubricant" AND "transesterification"; "biolubricant" AND "epoxidation". Articles included research and review articles.

If published articles are broken down by country, the results obtained are included in Figure 6. As observed in this figure, Malaysia contributed the majority of articles about biolubricants based on vegetable oils (52), followed by India (46) and Spain (22). Interestingly enough, the two most important world players in research articles (the United States and China) were the fifth and seventh countries regarding article publication in this field, which is a relatively low position for such developed countries when it comes to research. This could be due to the fact that the former are countries with a long farming tradition, where some specific vegetable oils such as safflower or rapeseed, which are highly used as raw material for biodiesel or biolubricant production, are easily adapted to their climate. In other words, these oils are especially suitable for biolubricant production, and that is the reason why the research related to the main subject of this review is highly remarkable in the abovementioned countries at the expense of China or the United States.

Regarding the most cited articles about this subject (according to the search criterion "biolubricants" AND "vegetable" AND "oil"), the results are included in Table 1. These articles mainly dealt with the use of green chemical routes to obtain biolubricants, presenting their most representative characteristics (pointing out the sustainability of the process and their biodegradability) and focusing on tribological tests. In general, and even though there is room for improvement according to these works, the suitability of biolubricants has been proven, exploring the possibility of their production at the industrial level in a biorefinery context.

It should be noted the heterogeneous disciplines of the journals where these articles were published (from agriculture to tribology), which points out the interdisciplinary nature of the research teams that have carried out these research works. Also, these articles were recently published (many of them in the last 6 years), which proves the recent and great interest in this subject by the scientific community. Additionally, these recent works were highly cited (with almost 100 citations for one of the most recent articles, published in 2020, and up to 300 citations for the most popular work), proving the increasing interest in this subject by the scientific community.

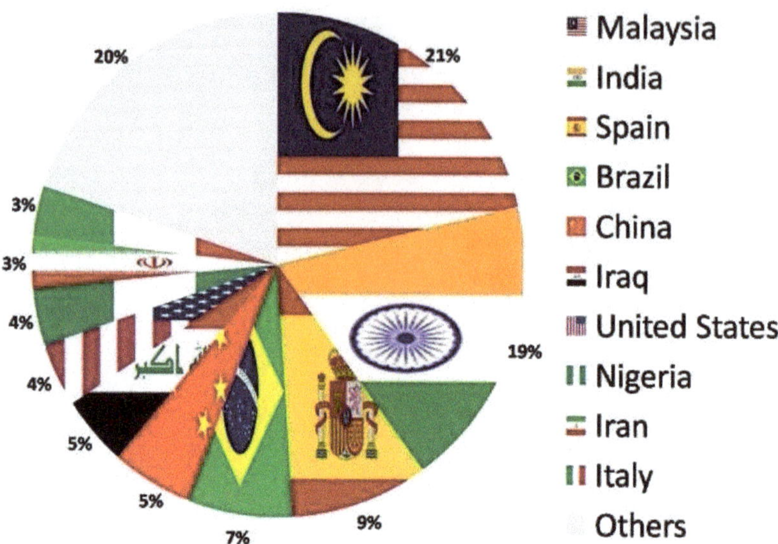

Figure 6. Articles about biolubricants based on vegetable oils by country [41]. According to the following search criterion: "biolubricant" AND "vegetable" AND "oil". Research and review articles were included.

Table 1. Top cited articles, and top cited authors dealing with scientific articles about biolubricants based on vegetable oils (source: [41]).

Article Title	Authors	Source	Year	Citations	Reference
The prospects of biolubricants as alternatives in automotive applications	Mobarak et al.	Renewable and Sustainable Energy Reviews	2014	304	[42]
Development of biolubricants from vegetable oils via chemical modification	McNutt et al.	Journal of Industrial and Engineering Chemistry	2016	218	[43]
Manufacturing of environment-friendly biolubricants from vegetable oils	Heikal et al.	Egyptian Journal of Petroleum	2017	171	[44]
Lubrication-enhanced mechanisms of titanium alloy grinding using lecithin biolubricant	Jia et al.	Tribology International	2022	148	[45]
Tribological behavior of biolubricant base stocks and additives	Chan et al.	Renewable and Sustainable Energy Reviews	2018	135	[46]
A review of biolubricants in drilling fluids: Recent research, performance, and applications	Kania et al.	Journal of Petroleum Science and Engineering	2015	129	[47]
Chemically modifying vegetable oils to prepare green lubricants	Karmakar et al.	Lubricants	2017	125	[48]
The physicochemical and tribological properties of oleic acid-based triester biolubricants	Salih et al.	Industrial Crops and Products	2011	122	[49]
Green synthesis of biolubricant base stock from canola oil	Madankar et al.	Industrial Crops and Products	2013	100	[50]
An overview of the biolubricant production process: Challenges and future perspectives	Cecilia et al.	Processes	2020	98	[51]

Table 1. *Cont.*

Article Title	Authors	Source	Year	Citations	Reference
Author Name	Published Articles about this Subject		Documents in Total		h-Index
Salimon, J.	11		207		35
Salih, N.	9		115		27
Yunus, R.	7		247		44
Yousif, E.	6		275		37
Freire, D. M. G.	6		284		52
Delgado, M. A.	6		50		22
Nogales-Delgado, S.	5		50		16
Habert, A. C.	5		63		20
Encinar, J. M.	5		80		36
Cavalcante, C. L.	5		164		41

Concerning the top publishing authors on this subject (also included in Table 1), it should be noted that their specific work about this research area was relatively short compared to their total published articles, pointing out that their careers are equally focused on other subjects and establishing, again, that researchers interested in this field come from multiple fields, focusing on biolubricant production from vegetable oils due to their scientific context and the great interest of this issue.

Also, it should be pointed out the great prestige and creativity of these top publishing authors, whose total publications (from 50 to 287 published works) and h-index (up to 52, which implies 52 papers with at least 52 citations each) are high, reflecting a great experience in research.

Regarding the subject area or field where these articles are focused on (according to the scientific journal where they were published), Figure 7 shows the distribution of the articles considered for this review:

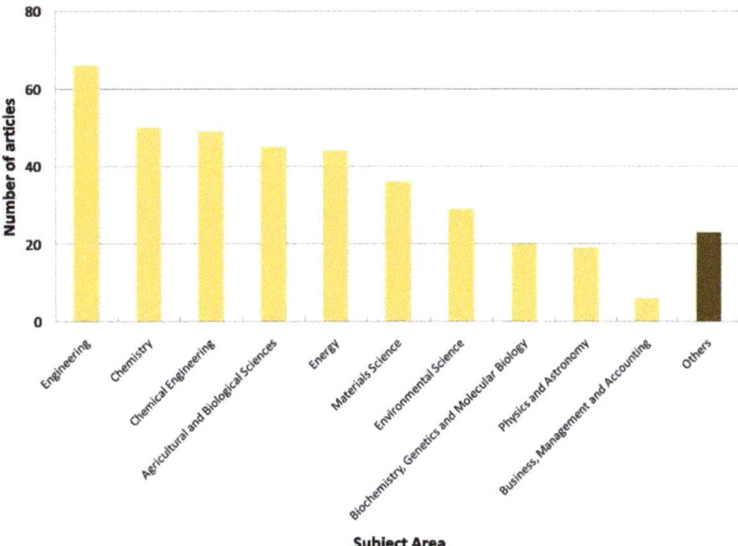

Figure 7. Main subject areas of the articles published about biolubricants obtained from vegetable oils [41]. According to the following search criterion: "biolubricant" AND "vegetable" AND "oil". Research and review articles were included.

As observed, most articles were focused on subject areas like engineering, chemistry, chemical engineering, agricultural and biological sciences, energy and materials, and environmental science. In that sense, the chemical routes for biolubricant production, as well as its industrial use and possible implementation in biorefineries (where products like biodiesel and glycerol can be used as fuels for different purposes), could explain the suitability of this research field for the abovementioned subject areas. As a result, biolubricants from vegetable oils offer a wide range of possibilities in research, implying the research work of multidisciplinary teams, with the subsequent synergy and enhancement of scientific production in this field.

Finally, concerning the role of catalysts during biolubricant production (which is essential to make the process efficient and competitive compared to traditional lubricant production), the evolution of published articles about this subject is included in Figure 8. As observed, the interest in catalytic conversion for biolubricant production arose in 2010, with a considerable increase in published works from then on. This could point out the fact that improvements in efficiency during this process are required, especially when the possible implementation at an industrial scale is considered. In that sense, the search for new catalysts (especially heterogeneous ones, in order to make the purification of biolubricant easier) could explain the increasing interest in this point. In addition, as observed in future sections, studies about techno-economic assessments and patents seem to show a special interest in the development of catalysts and the consequences on biolubricant production, especially when it comes to quality improvement of the final product and efficiency increase of the whole process mainly in a biorefinery context.

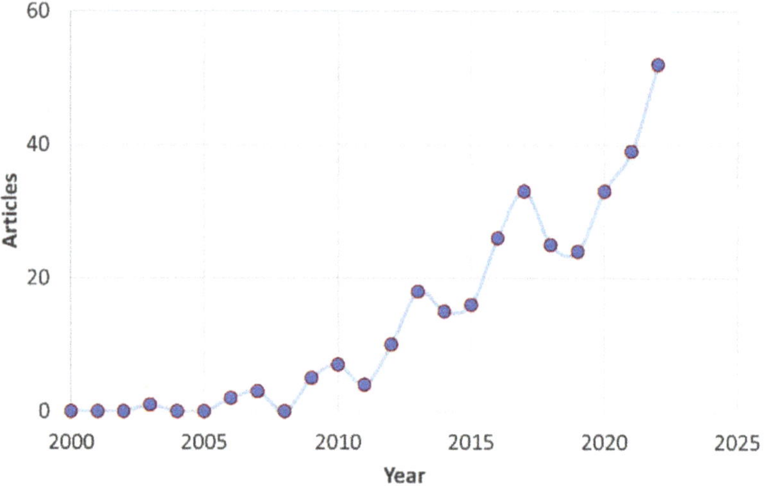

Figure 8. Articles published about catalytic biolubricant production (source: [41]). According to the following search criterion: "biolubricant" AND "catalysis". Research and review articles were included.

1.5. Aim of This Review

Considering the above, the aim of this review work was to carry out a state-of-the-art analysis of biolubricants based on vegetable oils, paying attention to the most recent developments in the following fields:

- Current use of vegetable oils in biolubricant production, especially concerning the possibilities related to the valorization of frying oil.
- Main biolubricant synthesis routes, paying special attention to transesterification processes, offer real potential for the implementation of biolubricant production in biorefineries based on vegetable oils.

- Use of catalysts, both homogeneous and heterogeneous, to make biolubricant production more competitive compared to the equivalent and traditional lubricant production.
- The main factors affecting quality during biolubricant production, with a special focus on viscosity and oxidative stability, are parameters that are vital to determining the use and service life of biolubricants. This way, the role of chemical conditions, as well as catalyst addition or additives such as antioxidants, will be especially enhanced.

For this purpose, a comparison of the most cited and interesting research works related to this field will be carried out in this review, trying to draw conclusions in this regard.

1.6. Scope and Bibliometric Analysis

To carry out this review, Scopus was investigated for all entries in the literature on the topics of biolubricant (including keywords such as vegetable oil, transesterification, epoxidation, and catalysts) for the last 20 years, with special attention to the last 5-year period (2018–2023), where there has been a considerable increase in published papers about this subject. The search, which was made from January to August 2023, returned 1560 results, from which up to 253 articles were considered for their inclusion in this work, including information about 152 published works (mainly research works and, to a lesser extent, proceeding papers and patents) in the final paper.

2. Biolubricants Based on Vegetable Oils: Main Sources and Characteristics

2.1. Biolubricants: Definition and Raw Materials

Biolubricants are a kind of lubricant based on plants (mainly from vegetable oils such as palm, safflower, or rapeseed oils), which makes them biodegradable and environmentally friendly. They mainly act as anti-friction media. Thus, the main purposes of this kind of product, as in the case of petroleum-based lubricants, are wear reduction by decreasing friction coefficient between two contacting surfaces, rust and oxidation prevention, and sealing effect against dirt, dust, or water [42,52].

In that sense, it is interesting to point out the versatility of biolubricants, as they can present different states of matter (solid, liquid, or semi-solid) obtained from different sources (from natural to synthetic oils), which make them suitable for multiple purposes such as automotive (engine or gearbox oils, transmission or brake fluids, etc.) or industrial oils (machine, hydraulic or compressor oils, for instance).

These products usually present some advantages compared to traditional lubricants based on petrol, like the following [52]:

- They are more sustainable and biodegradable, as raw materials are natural compared to oil. Also, there are no by-products with difficult management, as they can be used for other purposes or re-used in the same biolubricant production, as explained in further sections.
- Biolubricants usually have higher lubricity and viscosity index values (clearly exceeding 140–150, compared to the low values found for lubricants, at 90–100), which is important as it means that their viscosity is less dependent on temperature.
- They present, in general, higher flash and combustion points, which is a positive issue when it comes to safety during storage and shipping.

However, they also present some disadvantages or challenges, like the following:

- Due to the presence of saturated fatty acids, biolubricants might present a poor performance at low temperatures, which limits their worldwide marketability.
- Hydrolysis can take place in contact with moisture, increasing the possibility of corrosion in facilities by increasing free fatty acid levels.
- They usually have a short oxidative stability, which could imply a change in their properties during storage or oxidation, which is undesirable.

As mentioned above, there are plenty of oilseed crops that are easily adapted to extreme climate and soil conditions, and that can be suitable for biolubricant production, as the subsequent vegetable oil can be a perfect raw material for this purpose. These vegetable oils, along with the use of waste cooking oil, usually present homogeneous characteristics, although their fatty acid composition can present a considerable influence on the properties of these raw materials and, consequently, on the final biolubricant. Indeed, as many previous studies have pointed out, the role of fatty acids in some properties of biolubricants (for instance, viscosity or oxidative stability) is essential, as the presence of some functional groups such as hydroxyl groups (as in the case of ricinoleic acid) or even double bonds can alter these properties, as observed in Figure 9.

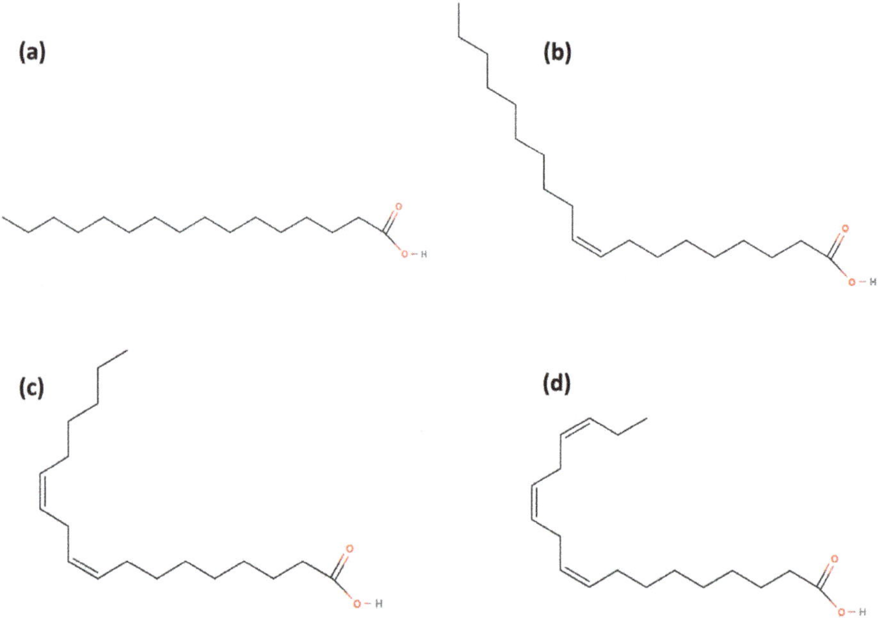

Figure 9. Molecular structure of some fatty acids: (**a**) palmitic acid, (**b**) oleic acid, (**c**) linoleic acid, (**d**) linolenic acid.

According to this figure, it can be observed how fatty acids basically differ from each other according to their molecular chain length and the presence of double bonds (which will be essential to understand the oxidative stability of biolubricants obtained through double transesterification). This way, a thorough knowledge of vegetable oils and their corresponding fatty acid composition is essential to assess, at least approximately, the main properties of the future biolubricant. It should be noted that the oil content in vegetable seeds might vary, according to the literature, from 20–36% for moringa to 45–70% for olive, with some representative crops such as rapeseed and palm presenting around 38–46% and 30–60%, respectively [42]. This fact should be considered in the final biolubricant yield as well as the agronomic performance of these crops.

Table 2 shows some examples of the main vegetable oils and their corresponding fatty acid composition, which will be an essential tool to understand some reasonings in this review work. As observed, some oils present a homogeneous fatty acid profile, whereas other crops (especially genetically modified) have a wide range of specific fatty acids, as in the case of safflower, whose oleic acid content can vary from less than 30% to up to 70%. In that sense, it is not a matter of "names" or "species", but a matter of fatty acid composition, as the fatty acid profile of some safflower oils could perfectly fit with the fatty acid profile of other vegetable oils like rapeseed.

Table 2. Potential raw materials for biolubricant production and their fatty acid composition (indicated in numerical symbol C:D, where C is the number of carbon atoms and D is the number of double bonds in their molecular structure). n.d. = not determined.

Vegetable Oil	16:0	16:1	18:0	18:1	18:2	18:3	Others	References
Castor	0.5–1.3	n.d.	0.5–1.2	4–5	2–8.4	0.4–1	83–88	[48,51,53–56]
Coconut	8–11	n.d.	1–3	5–8	0–1	n.d.	57–71	[48,51]
Corn	10.3–13	0.3	2–3	25–31	54–60	1	n.d.	[47,48,51]
Jatropha	13–16	n.d.	5–10.5	24–45	32–63	0.3–0.7	0.8–1.4	[48,51,57–59]
Olive	11.5–13.7	1.8	2.5	71–74	9.5–10	1.5	n.d.	[48,51,58]
Palm	37–47.9	0.4	3–8	37–45	1.9–10	0.3–0.5	1	[48,51,58,60]
Rapeseed	4–5	0.1–0.2	1–2	56–69.8	20–26	6.2–10	9.1	[48,51,54,61]
Safflower	5–7	0.08	1–4	10–21	73–79	n.d.	n.d.	[48,51,62]
Soybean	9.3–12	0.2–1.7	3–4.7	21–27.3	48.5–56.3	5.6–8	2.3	[48,51,56–58]
Sunflower	4.9–7	0.1–0.3	1.9–5	18.7–68	21–68.6	0.1–1.9	2.2	[48,51,55,57,63,64]
WCO	6.6–36.8	0.21–1.9	3–18.4	17.9–37.5	11.8–72.1	0.02–2.02	3.3	[33,57,65,66]

Thus, some interesting points can be inferred from this table:

- In general, palmitic, oleic, and linoleic acids are the majority of fatty acids found in most vegetable oils, which points out their vital role in some properties in the final biolubricant, as explained in further sections. In that sense, the presence of double bonds in their molecular structure (see Figure 9), which can be conserved in the molecular structure of the final biolubricant (especially in the case of double transesterification of fatty acids), could determine its oxidative stability. Thus, the knowledge of the ratio of some fatty acids (such as oleic/linoleic ratio) is usually interesting to understand oxidative stability in biodiesel or biolubricants.
- Nevertheless, there are some specific oils, such as castor oil, whose main fatty acid presents some special properties, as in the case of ricinoleic acid, with a hydroxyl group that can influence the properties of the final biolubricant regarding viscosity (as it promotes intermolecular interactions like hydrogen bonds, implying an increase in viscosity). In any case, there are other oils that present high quantities of other fatty acids, such as lauric and myristic acid in the case of coconut oil or icosenoic acid for rapeseed, which could imply changes in the properties of the corresponding biolubricants.
- The use of GM crops, as in the case of safflower, might vary the properties of the corresponding oil, with a considerable increase in oleic acid (exceeding 80%), improving some properties such as oxidative stability in the final biolubricant obtained [67]. Other studies point out the same aspect related to soybean oil, with a wide range of fatty acid contents depending on the gene technology used or the selection of soybean mutants [68]. Consequently, it is more interesting to consider FA profiles instead of kinds of vegetable oils, as it would give us a more exact idea about the raw material.
- The nature of WCO (and its subsequent fatty acid composition) might vary depending on the eating habits of the area where the research study was carried out. That is the reason why there was a wide range of oleic and linoleic acids in this table. In general, the main oils used for cooking are rapeseed, sunflower, soy, and olive oil, which can vary in the diet of some regions or areas. In any case, the fatty acid composition of this waste is relatively equivalent to the rest of vegetable oils, which supports the idea that its use as biodiesel and biolubricant source is feasible if a proper pre-treatment is carried out.

2.2. Main Characteristics of Biolubricants

It is clear that a biolubricant should present a series of characteristics that are essential for its use in lubrication processes. In that sense, there are plenty of requirements that should be accomplished to be a real alternative for lubricants [47,69]. In this review, we will focus on three main parameters such as viscosity, viscosity index, and oxidative stability.

- Viscosity: This is an essential parameter for a biolubricant, as it will determine its use for specific purposes. Thus, the resistance to the flow of a specific fluid (normally expressed in cSt) will influence many factors, such as the film thickness between the surfaces in contact or its permanence during lubrication (the higher the viscosity is, the thicker the film will be). Viscosity is influenced by factors such as temperature or pressure. It can be measured by dynamic or kinematic methods (by using Cannon-Fenske or Ostwald viscosimeters) at a specific temperature (normally 40 or 100 °C, which will be useful to determine VI) [70].

- Viscosity index (VI): This index indicates the changes in viscosity with temperature. Thus, a high viscosity index will imply a lower decrease in viscosity when temperature increases, which is a desirable effect as changes in temperature would not present a considerable influence on biolubricants [71,72]. High VI values will be obtained when the molecular structure of the final biolubricant is longer, and it will decrease with branching. That is the reason why the selection of a chemical route or specific reagents (like complex alcohols in double transesterification) will be vital to determine this parameter. It is calculated by measuring viscosity at 40 and 100 °C.

- Oxidative stability (OS): This parameter is related to the stability of biolubricants during oxidation processes, including storage. It is expressed in hours and determined through the induction point (IP) according to the Rancimat method [73,74]. This way, free radical generation (mainly due to the presence of double bonds in biolubricants) could start a chain reaction, where the propagation of free radicals could end up generating undesirable products such as free fatty acids (FFA) or polymers, among other decomposition compounds. The former could be related to the increase in acidity, which is an undesirable effect, especially when it comes to the maintenance of equipment and facilities. The latter is especially related to an increase in viscosity, as the polymerization of esters generates more complex molecules, which could imply an increase in molecular interactions such as Van der Waals or hydrogen bonds and, therefore, a higher resistance to the flow of biolubricants (that is, an increase in viscosity).

- Other parameters: in our opinion, the abovementioned properties are the most important ones to define the performance of a biolubricant, but there is a wide range of characteristics that should be considered, like pour point (the lowest temperature at which a biolubricant pours or flows, which is desirable to be as low as possible to be useful in cold climates), lubricity (that is, the reduction of friction between two surfaces in contact when the biolubricant is used), flash and combustion points (usually higher compared to traditional lubricants, which is a great advantage when it comes to safety during storage or shipping), hydrolytic stability (resistance of esters to hydrolyze, that is, to degrade in contact with water at high temperature) or biodegradability (which is considerably higher compared to petrol-based lubricants), among others.

Apart from the abovementioned properties, there are many other aspects that should be considered for the marketability of a biolubricant. For instance, acidity should be taken into account, as recently explained, whereas there are other parameters included in standards that should be considered. The changes in these parameters will be determined by different factors (such as raw materials used, the kind of chemical route selected for biolubricant production, or the chemical conditions that are chosen, including temperature, reaction time, reagents, etc.), as explained in further sections.

Thus, molecular factors such as functional groups and polarity (usually increase viscosity and tribofilm adhesion), numbers of branching (with a decrease in pour point and an improvement in oxidative stability), degree of unsaturation (with lower thermal and oxidative stabilities), or carbon chain length (improving viscosity and VI but with lower oxidative stability values) will determine the properties of a biolubricant, which are directly influenced by the abovementioned factors [46]

By way of example, Table 3 shows the most important characteristics of biolubricants mainly obtained through double transesterification (with methanol and the complex alcohol commented on in this table) from different sources.

Table 3. Some characteristics of biolubricants are obtained through double transesterification.

Biolubricant	Viscosity [1], cSt	VI	Pour Point, °C	Flash Point, °C	IP, h	References
Rapeseed-based 2-ethyl-1-hexyl-esters	7.97	n.d. [2]	n.d.	196	<6	[33]
Seed-based 2-ethyl-1-hexyl-esters	7.47	n.d.	n.d.	195	>3	[33]
WCO-based 2-ethyl-1-hexyl-ester	7.40	n.d.	n.d.	193	>3	[33,66]
WCO-based 2-ethyl-1-hexyl-ester	34.91	122	−1	216	n.d.	[66]
Mustard seed oil-based 2-ethyl-1-hexyl ester	8.6	n.d.	n.d.	n.d.	n.d.	[75]
Cardoon-based NPGE	18.85	184	n.d.	n.d.	<3	[76]
WCO-based NPGE	44.9	457	−1	238	n.d.	[66]
Palm-based PEE	68.4	188	−20	302	n.d.	[77]
High-oleic safflower-based PEE	77.7	155	n.d.	260	2.86	[67]
WCO-based PEE	68.5	144	n.d.	253	2.07	[34]
Palm-based TMPE	49.7	188	−1	n.d.	n.d.	[78]
High-oleic safflower-based TMPE	73.39	103	n.d.	216	6.7	[67]
High-oleic safflower-based TMPE	89.11	131	n.d.	220	>7	[79]
Rapeseed-based TMPE	75.5	128	n.d.	210	4.9	[80]
Jatropha-based TMPE	51.89	140	−3	n.d.	n.d.	[44]
Palm-based TMPE	38.25	171	5	240	n.d.	[44]
Sesame-based TMPE	35.55	193	−21	196	n.d.	[81]
WCO-based TMPE	30	179	n.d.	n.d.	n.d.	[82]
Palm kernel-based isoamyl ester	3-6	149	n.d.	n.d.	0.3	[83]

[1] at 40 °C. [2] not determined.

As inferred from this table, several factors should be pointed out, like the following:

- Viscosity values are mainly influenced by the alcohol used in the second transesterification, as it is the determining factor for the final molecular structure of the biolubricant. There are some exceptions where the raw material plays an important role, as in the case of castor oil, whose majority compound (ricinoleic acid) promotes a considerable increase in viscosity by itself. Thus, biolubricants obtained with complex alcohols usually present an increasing viscosity in that order: 2-ethyl-1-hexanol < NPG < TMP < PE.
- Another important factor is the conversion of the process. Thus, low conversions will imply a mixture with biodiesel (with a viscosity range between 3 and 6, in most cases). That is the reason why the role of catalysts is so important in order to obtain high conversion rates at mild reaction conditions.
- As previously explained, some properties of biolubricants are quite interesting, like high VI and flash points, which encourage the production of these compounds, as explained in the following section.

3. Biolubricant Production

As abovementioned in this review, there are different ways to produce biolubricants through specific chemical routes that will be explained in the following subsections. It should be noted that there are advantages and disadvantages for each route, mainly related to the properties of the final product or the technical/energy requirements.

3.1. Different Chemical Routes for Biolubricant Production

Among the main chemical routes to obtain biolubricants, there are some of them that have been widely studied, such as epoxidation, estolide formation, hydrolysis, hydrogenation (partial or complete), or transesterification/esterification [84].

- Epoxidation: This chemical route is carried out by using hydrogen peroxide proxy acids like formic or acetic acids, and the final product is a peracid that epoxidates double bonds included in fatty acids. This way, unsaturations are removed, which can increase the oxidative stability of biolubricants obtained by this chemical route.
- Estolide formation: These compounds are usually produced by using strong acids (sulphuric acid, methanesulfonic acid, or perchloric acids) as catalysts to activate alkene groups to produce estolides. Thus, they are generated by the bonding of a fatty acid's carboxylic acid functionality to an unsaturation of another fatty acid at relatively low temperatures (around 100 °C). This way, estolides usually have better cold flow properties and longer oxidative stability values compared to the original vegetable oil [43].
- Hydrolysis: It consists of the splitting of triglyceride molecules into fatty acids by using steam or water, implying an endothermic reaction. Some catalysts, such as metal oxides (ZnO), can be used to increase conversion and yield. Also, subcritical water is another interesting way to achieve hydrolysis of vegetable oils without catalyst addition [85].
- Hydrogenation: It normally implies the total or partial reaction with molecular hydrogen, taking place in an exothermic process. Partial hydrogenation can improve the oxidative and thermal stability of original vegetable oils or biolubricants, whereas other properties can be unaltered (such as low-temperature performance, viscosity, and VI, among others). It is usually carried out at relatively low temperatures (150–210 °C) and high pressures (21–35 bar), using some catalysts such as Ni, Pa, and Pt.
- Transesterification: This is the process generally used for biodiesel production from vegetable oils, removing glycerol from the triglyceride molecular structure. More details will be given later in this section.

To sum up, some of the advantages and disadvantages related to the most popular chemical routes to produce biolubricants are included in Table 4. As observed, every chemical route has its advantages and challenges, and in the case of double transesterification, the main problem is related to the possible low oxidative stability of the final product obtained if the raw material presents some specific characteristics like a fatty acid profile with a high percentage of linoleic or linolenic acids (at the expense of oleic or palmitic acids).

In any case, as explained in detail in further sections, these challenges can be easily overcome by the addition of low amounts of antioxidants, among other alternatives, and the possibility of implementing a biorefinery through double transesterification could offset these inconveniences in the long run, as the atom efficiency of these two processes is high, obtaining products that are highly biodegradable (such as fatty acid methyl esters and fatty acid esters). Moreover, other intermediate products obtained in the second transesterification, such as methanol, can be reused in the first transesterification, contributing to the abovementioned atom efficiency. Finally, according to Figure 5, where the publication trends were included, transesterification plays an important and increasing role in the scientific community, contributing to more published articles compared, for instance, with epoxidation. Regarding WCO, this double transesterification would be a suitable way to valorize this waste, as many valuable products are obtained, whose price will mainly depend on the raw material used. Thus, WCO, if collected properly, would be much cheaper than other raw materials studied in the literature (for instance, rapeseed oil), improving the abovementioned economic studies [80]. That is the reason why we will be focused on this chemical route in this review work.

Table 4. Main chemical routes for biolubricant production, including advantages and disadvantages [51,84].

Chemical Route	Details	Advantages	Disadvantages
Epoxidation	Double-bound removal and introduction of an epoxide functional group	Higher OS and lubricity	Viscosity, VO, and PP are usually low
Estolide formation	Estolide generation through different ways, reacting two acidic molecules	Low-temperature reaction, higher OS and lubricity, and better performance at low temperatures	Expensive
Esterification and Transesterification	Use of alcohols to transform fatty acids into fatty acid esters	High VI and flash point, improvement of low-temperature properties	High reaction temperatures. OS depends on the fatty acid profile of the raw material.
Hydrogenation	Reaction with molecular hydrogen	Better OS and lower unsaturation	Possible side reactions

Paying attention to the double transesterification process, Figures 10 and 11 present each transesterification route. Regarding the first transesterification, fatty acids react with methanol (or ethanol, but the former is preferred due to economic costs) to obtain fatty acid methyl esters and glycerol. It should be noted that the fatty acid composition might vary, with R_1, R_2, and R_3 representing the aliphatic chain of certain fatty acids such as palmitic, oleic, or linoleic acids. Thus, the presence of these chains might vary for each fatty acid depending on the nature of the original vegetable oil.

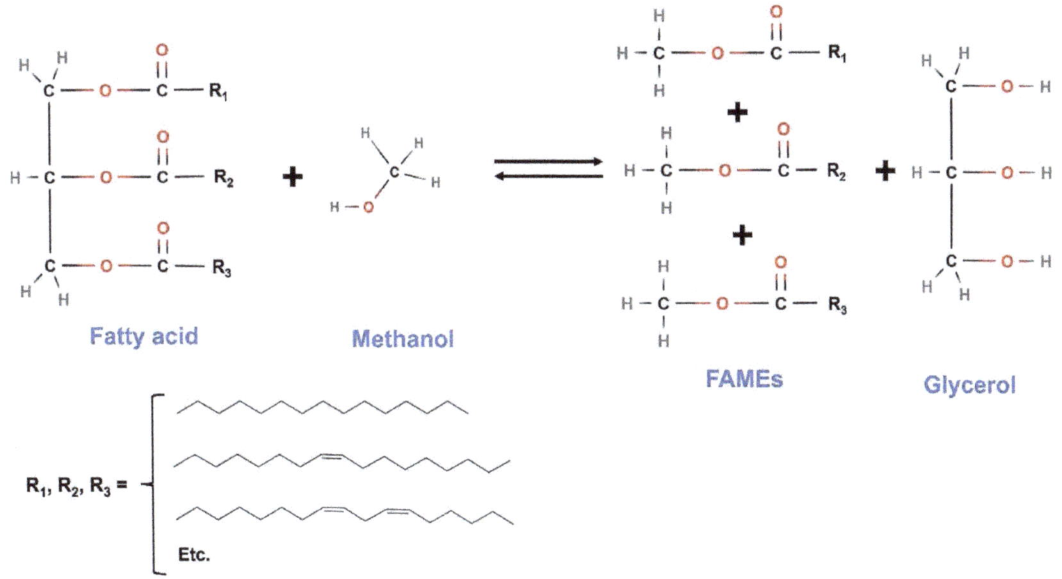

Figure 10. First, transesterification from fatty acids with methanol to obtain FAMEs as an intermediate step for biolubricant production.

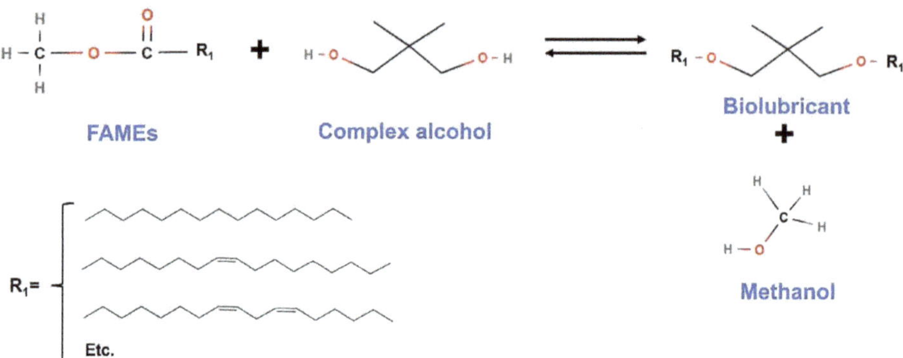

Figure 11. Example of the second transesterification from FAMEs to obtain biolubricants. The alcohol used was neopentyl glycol (2,2-Dimethyl-1,3-propanediol).

Regarding the second transesterification (Figure 11), it is similar to the abovementioned first stage, with the difference of using a more complex alcohol like those observed in Figure 12, which is widely used in the literature [86]. Thus, the final biolubricant is obtained, and methanol is released, which could be reused for the first transesterification process depending on the circumstances of this second transesterification and the recovery technique selected.

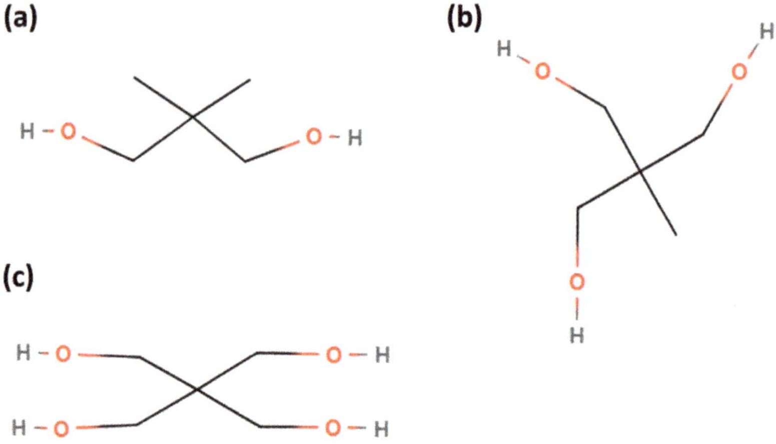

Figure 12. Molecular structure of main superior or complex alcohols used for biolubricant production during the second transesterification of fatty acid methyl esters: (**a**) neopentyl glycol; (**b**) trimethylolpropane; (**c**) pentaerythritol.

For this second transesterification, in order to promote the biolubricant generation, the use of a vacuum (that is, low working pressures) is recommended to remove methanol and shift the equilibrium towards product generation. However, as observed in Figure 12, some complex alcohols used for this purpose present low boiling points or a tendency to sublimation due to the spherical shape of their molecular structure. In that sense, the use of vacuum seems not to be suitable for NG, whereas TMP and PE seem to offer good results when working pressures below 300 mmHg are used.

In any case, the use of the alcohols included in Figure 12 (among others, as there is a wide range of products that can be used in this double transesterification) allow the production of biolubricants with endless opportunities, as it will be observed in further sections where the properties of biolubricants depending on the kind of alcohol (among other factors like conversion or the use of different raw materials) will be discussed, especially regarding viscosity and the subsequent use for industrial purposes of the biolubricant.

During the second transesterification, the complex alcohol must be dissolved in the reaction medium, that is, FAMEs, and due to dispersion phenomena and mixing, the reaction usually starts with a low reaction rate at the beginning. Once mixing is complete, the reaction rate will be increased, requiring the reaction a specific time to be carried out. If the reaction time is extended, it can proceed backward (that is, products can react to generate the reagents). That is the reason why factors such as temperature, reaction time, vacuum, etc., are essential to be optimized. Otherwise, low yields and efficiency can be found during the process.

As observed in Figures 10 and 11, these consecutive chemical routes can be perfectly coupled in a biorefinery context, obtaining several interesting products and reusing some intermediate by-products, as explained in the following section.

3.2. Double Transesterification as an Interesting Proposal for a Biorefinery Based on Vegetable Oils

As explained in the previous section, double transesterification presents a higher complexity compared to other simpler chemical routes. Nevertheless, apart from fatty acid esters that can be used as biolubricants, many different products are obtained during the whole process, which can be used for different purposes (see Figure 13):

- Fatty acid methyl esters: As a result of the first transesterification, FAMEs are obtained, which can be used as fuels in Diesel engines, being the perfect replacement for traditional fuels used for that purpose. In that sense, compared to the UNE-EN 14,214 standard [87], many of its requirements are clearly complied with by biodiesel, especially concerning key aspects such as viscosity or cold filter plugging point (CFPP), which are essential for the correct performance in a Diesel engine. However, in many cases, and due to the same factors affecting biolubricants, the oxidative stability of FAMEs is not high enough [88,89] (not reaching 8 h, which is the lower limit of the European standard, for instance), requiring the use of alternatives such as antioxidant addition (mainly TBHQ, PG or BHA) or genetically modified (GM) crops, among others [63,90,91]. In any case, it is a very important product obtained in this process, which can increase the valorization of the whole biorefinery [92].
- Glycerol: It has been one of the most abundant and versatile by-products obtained during biodiesel production. Comparing the similarities of biodiesel and biolubricant production (indeed, FAME production is the initial stage for further biolubricant synthesis), it is no wonder that glycerol could play an important role in this case. Depending on the degree of purity of glycerol, it can be used in different ways, such as an energy source (through dry or steam reforming) or the precursor of interesting products like acrolein, propanediols, or carboxylic acids, along with other products with a great interest in the pharmaceutical industry) obtained from routes such as hydrogenation, oxidation, or esterification [93–96]. If the recent trend in biodiesel generation, along with the possible incorporation of biolubricant production through double transesterification, glycerol production is expected to increase, with the subsequent opportunity for its valorization through the abovementioned chemical routes. This way, a feasible technology applied to glycerol could be synthesis gas generation (a mixture of hydrogen and carbon monoxide at different ratios), which could produce green fuels as glycerol is generally obtained from raw materials [97].
- Methanol: One of the byproducts generated during the second transesterification to produce biolubricants is equally interesting, as it can be reused in the first transesterification process, where it is one of the reagents used. Thus, the concept of circular economy is really connected to this process, which could make a biorefinery based on vegetable oils more efficient. However, some factors should be taken into account during this process, like the possibility of using a vacuum for a high biolubricant

production yield, which could make the recovery of methanol more difficult or less economically feasible.

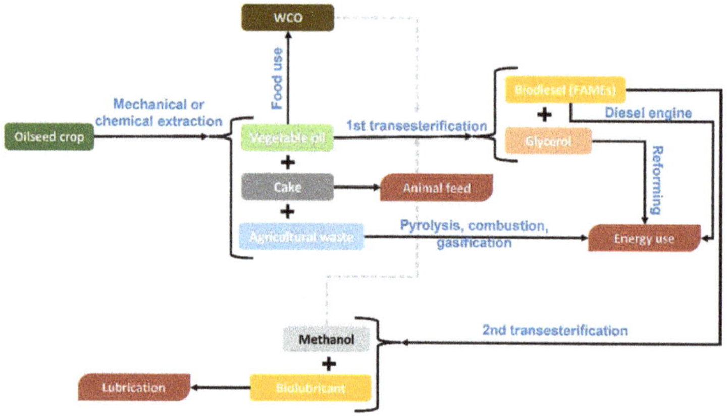

Figure 13. Biorefinery proposal according to the possibilities of oilseed crops. Dashed lines indicate the reusability of some products, whereas red boxes indicate the final use of each product taking part in this process. In blue are the main steps for each product.

Consequently, double transesterification could be a suitable solution for the implementation of a biorefinery from oilseed crops (or vegetable oils, VO), where plenty of products (apart from the abovementioned ones) can be valorized. In that sense, this process presents some similarities compared to other biorefineries, like the following [98]:

- Natural raw materials or wastes obtained in agricultural practices or food industries are normally used.
- Different products that are normally biodegradable are generated, as the raw materials are naturally obtained. The chemical transformations do not normally imply considerable changes in the molecular structure of the subsequent products (biodiesel and biolubricants), which makes these products (in case of spillage) easily assimilable by microorganisms.
- These products can be directly used for energy production or further synthesis, or they can be upgraded to different degrees. Thus, the versatility of this technology is a strong point to compete with traditional refineries.
- Some of the byproducts generated can be directly (or indirectly through purification processes) reused in the same process, which is one of the key points of the circular economy.
- With regard to these points, a biorefinery with these characteristics would perfectly fit the concept of green chemistry and circular economy, which is highly regarded by governments and society in general.

These particularities are essential to understand the role of biolubricants in sustainability and green policies, as it contributes to a circular economy production, with a high atom efficiency as most products can be directly used or reused in the process, with a considerable decrease in evolved products (or pollutants) to the environment. In that sense, some works have pointed out the potential of soy for its implementation in a biorefinery context, whereas food waste valorization (where WCO could be included) has also been assessed in previous studies [99–102]. It should be noted the possibility of implementation of this technology in previous biodiesel industries, as double transesterification could share equivalent processes and facilities that could be suitable for the industrial growth of traditional biodiesel plants, as previously considered by other authors [103].

One of the essential points of this review is the use of a catalyst to improve the efficiency of biolubricant production (in a biorefinery context) to compete with the equivalent and traditional processes to obtain lubricants. In that sense, high conversion of the two main reactions taking place in Figure 13 (that is, the first and second transesterification) at low reaction temperatures and times are required, which can be achieved by the use of catalysts. It should be noted, even though this is not the main purpose of this review, that catalysts are also present in other chemical routes included in this biorefinery, as in the case of pyrolysis, combustion, and gasification of agricultural waste or steam reforming of glycerol. In these cases, heterogeneous catalysts are generally used, with popular catalysts such as nickel catalysts supported on alumina with different promoters to increase their useful life and performance [104,105]. In the following section, a thorough study of this subject will be carried out, as catalysts play an essential role in the improvement of competitiveness of this technology compared to refineries based on petrol.

4. The Use of Catalysts in Biolubricant Production

As in any technology, production, or process, there are some aspects that should be considered to make them economically effective and efficient. In the case of green chemistry routes or circular economy procedures, effectiveness is especially important, as this will be the determining factor to make these processes competitive compared to previous or traditional technologies (for instance, based on petrol). That is the reason why economic and life-cycle assessments are so important in these cases. Consequently, every detail counts to make the process more efficient, and the role of catalysts in that sense is essential.

Biolubricant production through transesterification could take place in supercritical conditions, but the use of catalysts is generally accepted for this purpose. Thus, catalysts can contribute to a decrease in activation energy, which in turn could afford a decrease in chemical reaction conditions related to energy costs, mainly having to do with temperature. Regarding the main catalysts used in transesterification, there are three kinds of catalysts: homogeneous, heterogeneous, and biochemical (mainly lipases).

- Concerning homogeneous catalysts, they are usually classified as acidic and alkaline catalysts, which are normally more effective and provide faster reaction rates. Some examples of acidic catalysts are hydrochloric, sulfonic, sulfuric acids, etc., whereas examples of alkaline catalysts are sodium and potassium hydroxide and methoxide, among others. In the case of the latter, low moisture and FFA content in biodiesel is recommended to avoid a decrease in yield or side reactions. Figure 14 shows the main mechanism taking place when basic homogeneous catalysts are used. This way, the role of the catalyst was to generate the corresponding alkoxide (from neopentyl glycol, for instance) to carry out a nucleophilic substitution in the carboxyl group, with the final generation of the biolubricant and the release of methoxide and, subsequently, methanol.
- As far as heterogeneous catalysts are concerned, there are also acid and base catalysis, such as metal complexes, metal oxides, zeolites, membranes, or resins [106]. In this case, mass transfer phenomena, along with some important properties of the heterogeneous catalyst (such as porosimetry or reusability), should be taken into account in industrial design.
- Regarding lipases, their natural function is the hydrolysis of oils and fats to produce glycerol and free fatty acids, being one of the most resourceful enzymes in biocatalysis, as they are present in all organisms and their variety can offer different characteristics. Lipases can be used for different purposes and chemical routes, such as esterification, transesterification, acidolysis, or amidations, among others, being stable in different solvents, such as aqueous, organic, or ionic [107–109]. In that sense, as explained in previous studies, transesterification with methanol to produce biodiesel has been successfully carried out, offering satisfactory results [64,110,111]. This way, considering the equivalence between this chemical route and double transesterification to produce biolubricants, its possible use for this purpose seems to be feasible, as explained later on.

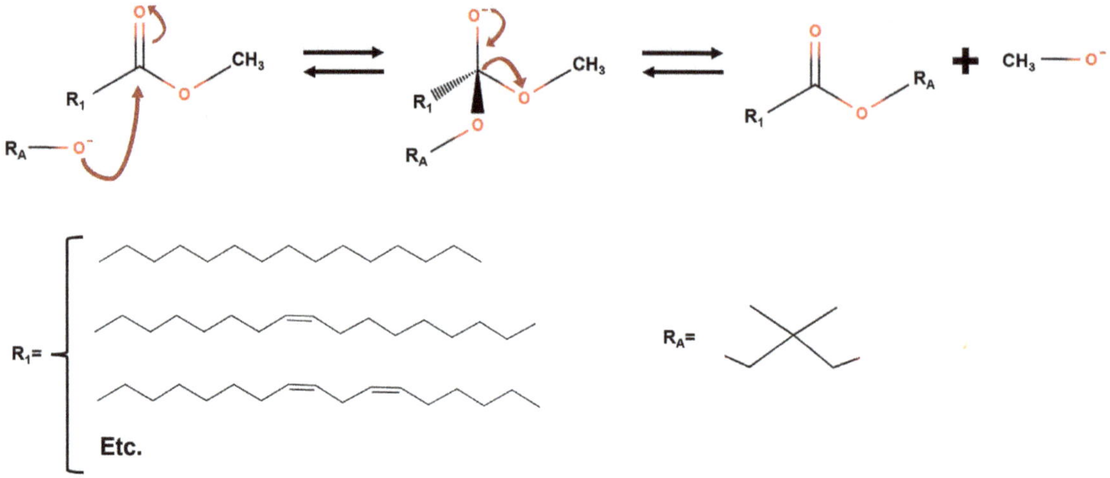

Figure 14. Mechanism of homogeneous basic catalysis for the second transesterification (FAME conversion to biolubricants).

This way, many research works carried out comparative studies with different kinds of catalysts or catalyst doses, proving the suitability of their use in biolubricant production and obtaining high production yields, as observed in Table 5 (homogeneous catalysis) and 6 (heterogeneous catalysis) for the case of double transesterification. As observed, different kinds of catalysts have been used for biolubricant production, but chemical reactions with heterogenous catalysts usually require longer reaction times and higher catalyst concentrations, not reaching, in some cases, high conversions compared to homogeneous catalysts. As previously explained, these catalysts were used for a wide range of raw materials, including WCO, which was successfully used in homogeneous and heterogeneous catalysis, proving the utility of this waste to be used to produce biolubricants.

One of the main issues related to the use of catalysts during biolubricant production is the subsequent purification process. It should be taken into account that, depending on the kind of catalyst (homogeneous or heterogeneous), this process could be drastically different:

- Regarding homogeneous catalysts, their removal from the final biolubricant might be complicated, as they are easily dissolved in the final product, increasing their level in some metal elements that could contribute to the acceleration of some degradation phenomena that can take place during storage (for instance, Na and K content due to catalyst addition could contribute to auto-oxidation acceleration). Thus, if a typical biodiesel purification (through successive washing steps) is considered an equivalent process to be carried out in this case, it should be noted that hydrolysis could take place, which is an undesirable effect. In that sense, some alternatives could be chosen, like the avoidance of catalyst removal by adding some additives to increase the oxidative stability of biolubricants or the use of more expensive alternatives, such as ion exchange columns.

- In the case of heterogeneous catalysts (see Table 6), the separation process is quite simple, mainly through filtration after biolubricant production. In that sense, porous catalysts could be an interesting starting point, as their interaction with raw materials is higher [60,112,113]. However, the use of these catalysts is usually related to some inconveniences, such as lower effectiveness compared to homogeneous catalysts or the low current reusability of these products, which presents considerable room for improvement in biodiesel and biolubricant synthesis.

Table 5. Studies related to catalytic conversion of vegetable oils to obtain biolubricants through double transesterification (using homogeneous catalysts).

Raw Material and Catalyst	Chemical Conditions [1]	Conversion, %	Reference
Rapeseed, corn, and sunflower mixture, and WCO with titanium isopropoxide	Transesterification with 2-ethyl-1-hexanol at 160 °C, 1.5% catalyst, and 1:1 molar ratio for 60 min	>96.5	[33]
Palm oil with H_2SO_4	Esterification of palm oil fatty acids with NPG at 138 °C, 1.12% catalyst, and 1:2.26 molar ratio for 4.79 h	87.6	[114]
Elaeis guineensis kernel oil with H_2SO_4	Transesterification with di-TMP at 150 °C, 1.7% catalyst, and 1.6:1 molar ratio for 4.6 h	79	[115]
Methyl oleate with K_2CO_3	Transesterification with TMP at 120 °C, 1.5% catalyst, and 4:1 molar ratio for 240 min	95.6	[116]
Babassu oil with sodium methoxide	Transesterification with TMP at 65 °C, 1.0% catalyst, and 3:1 molar ratio for 6 h at 700 mmHg	>90	[117]
Palm oil and sodium methoxide	Transesterification with pentaerythritol at 158 °C, 1.19% catalyst, and 4.5:1 molar ratio for 60 min	40.13	[77]
High-oleic safflower and sodium methoxide	Transesterification with pentaerythritol at 160 °C, 1.0% catalyst, and 1:1/3 molar ratio for 120 min (working pressure 400 mmHg)	>94	[67]
WCO and sodium methoxide	Transesterification with pentaerythritol at 160 °C, 1.0% catalyst, and 1:1/3 molar ratio for 120 min (working pressure 260 mmHg)	92.6	[34]
WCO and zinc acetate	Transesterification with different alcohols (1-heptanol, 2-ethyl hexanol, and neopentyl glycol) at 160 °C, 3.0% catalyst, and different molar ratio for 240 min	n.d.	[66]
Cardoon oil and sodium methoxide	Transesterification with NG at 130 °C, 1.5% catalyst, and 1:1 molar ratio for 120 min	>95	[76]
High-oleic safflower and sodium methoxide	Transesterification with TMP at 140 °C, 1.0% catalyst, and 1:1 molar ratio for 90 min (working pressure 400 mmHg)	>92	[118]
Jatropha oil and sodium methoxide	Transesterification with TMP at 200 °C, 1.0% catalyst, and 3.9:1 molar ratio for 3 h (working pressure 10 mbar)	47	[119]
Rapeseed and sodium methoxide	Transesterification with TMP at 120 °C, 1.5% catalyst, and 1:1 molar ratio for 90 min	>99	[80]
High-oleic safflower and sodium methoxide	Transesterification with TMP at 100 °C, 0.3% catalyst, and 1:1 molar ratio for 120 min and a working pressure of 210 mmHg	>94	[79]
Fish oil residue with sodium ethoxide	Transesterification with TMP at 100–140 °C under vacuum	84	[120]
Litsea cubeba kernel oil with KOH	Transesterification with TMP at 130 °C, 1/4:1 molar ratio for 60 min, and different working pressures	92	[121]
Cottonseed oil with sodium methoxide	Transesterification with TMP at 144 °C, 0.8% catalyst, and 1/4:1 molar ratio for 10 h at 25 mbar	>90	[122]
Jatropha oil with sodium hydroxide	Transesterification with ethylene glycol at 150 °C, 1.2% catalyst, and 2:1 molar ratio for 120 min and vacuum	98	[57]
Mustard seed oil with KOH	Transesterification with 2-ethyl-1-hexanol at 70 °C, 2% catalyst, 2:1 molar ratio for 65 min at 0.05 bar	93	[75]

[1] Otherwise explained, these transesterifications are with FAMEs. Alcohol/FAME ratios are expressed.

Table 6. Studies related to catalytic conversion of vegetable oils to obtain biolubricants through double transesterification (using heterogeneous/biochemical catalysts).

Raw Material and Catalyst	Chemical Conditions [1]	Conversion, %	Reference
Castor oil and lipase	Transesterification with TMP at 40 °C, 0.4% catalyst, and atmospheric pressure, using a pervaporation membrane to remove CH_4	59	[123]
Soybean oil and lipase	Esterification with NG and TMP at 45 °C, 4% catalyst, and 6 h	90	[124]
Palm oil and solid acid catalyst	Esterification with NG at 180 °C, 2% catalyst, 0.5:1 molar ratio and 4 h	85	[125]
Palm kernel oil with lipase	Transesterification with isoamyl alcohol at 45 °C, 4:1 molar ratio for 54 h	99	[83]
Palm oil and lipase	Esterification with TMP at 130 °C, 3% *w/w* catalyst, 3.45:1 molar ratio, 15.25 mbar for 48 h	82	[126]
WCO with CaO derived from waste cockle shell	Transesterification with TMP at 130 °C, 4% *w/w* catalyst, 3:1 molar ratio for 4 h	97	[82]
WCO with CaO	Transesterification with ethylene glycol at 130 °C, 1.2% catalyst, 3.5:1 molar ratio for 1.5 h	94	[127]
Palm oil with mixed oxides of Ca and Sr on CaO	Transesterification with TMP at 180 °C, 1% *w/w* mixed oxides of Ca and Sr catalyst with 5% *w/w* SrO on CaO, 2 mbar and 240 min	88	[128]
Soybean oil with Zn Al hydrotalcites	Transesterification with TMP at 140 °C, 5% catalyst, 4:1 molar ratio for 2 h	77	[129]
WCO with K_2CO_3-hydrotalcite	Transesterification with TMP at 160 °C, 2% catalyst, 3:1 molar ratio and 300 Pa for 2 h	80.6	[130]

[1] Otherwise explained, these transesterifications are with FAMEs. FAME/alcohol ratios are expressed.

It should be noted that there are other works dealing with the transesterification of vegetable oils (castor, coconut, and palm kernel oils) with methanol to obtain fatty acid methyl esters (that is, biodiesel), but for a different purpose, as thanks to the use of some additives biolubricants with different viscosity values were satisfactorily obtained [131]. In that sense, the biolubricant obtained from castor oil presented the highest viscosity value due to the high percentage of ricinoleic acid in the original oil, whose OH- group contributed to the increase in viscosity.

5. Influencing Factors on Quality Parameters of Biolubricants

As explained in previous sections, there are some quality parameters that are vital to understanding the right performance of biolubricants for industrial applications. Thus, a good biolubricant should present repeatable or stable properties once it is produced, stable properties during storage, and specific industrial use (unalterable due to circumstances such as auto-oxidation). Figure 15 shows the main influencing factors on biolubricant quality parameters, which are closely related to each other, usually presenting opposite effects depending on the circumstances. For instance, the use of a catalyst can contribute to a higher biolubricant yield, which is positive regarding a product with a stable composition, whereas high amounts of homogeneous catalysts can imply further auto-oxidation effects if they are not properly removed from the final product, with the subsequent decrease in oxidation stability. Additionally, some vegetable oils could present functional groups or molecular structures (such as branching), which could provide an interesting viscosity value for a specific industrial use, whereas the presence of unsaturations could worsen the oxidation stability of the final biolubricant.

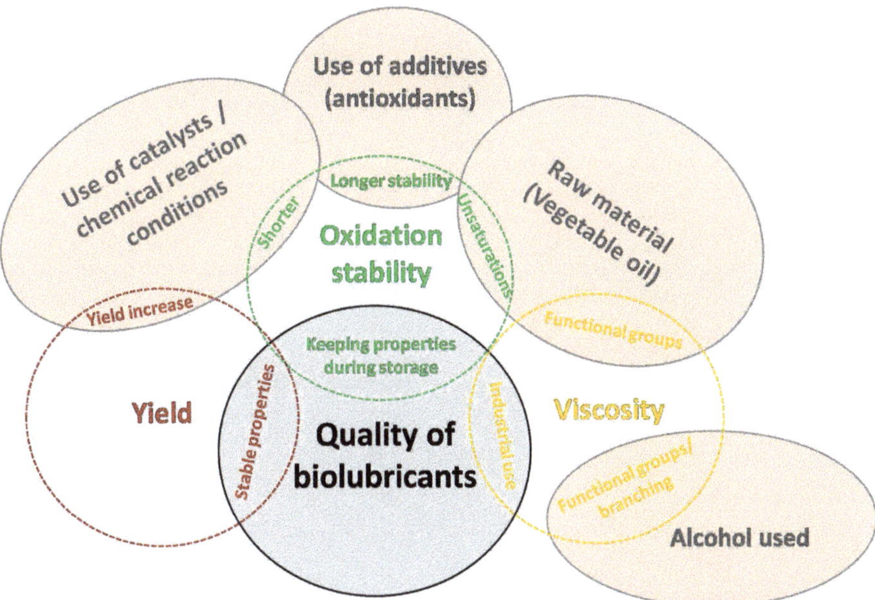

Figure 15. Main factors (in orange circles) affecting quality parameters of biolubricants obtained through transesterification.

A thorough explanation of Figure 15 is included in the following subsections. This way, the main quality parameters considered were yield or conversion, oxidation stability, and viscosity, which are vital to understanding the performance of a biolubricant:

5.1. Yield or Conversion

It is important to obtain a high conversion of biolubricant from vegetable oil or biodiesel, as its separation from the corresponding raw material is difficult, obtaining a mixture with combined properties like viscosity. Indeed, in previous studies, there was a good correlation between conversion and viscosity in the reaction medium for biolubricant production from some vegetable oils like high-oleic safflower biolubricant [118]. This way, high conversion of biodiesel or vegetable oil can provide more fixed viscosity values and, therefore, the characteristics of a biolubricant can be more predictable. On the other hand, the presence of biodiesel in the final biolubricant could imply a drastic decrease in viscosity in the medium, as the differences between biodiesel and biolubricant viscosities (much lower for the former, with 3.5 to 5.0 cSt) are considerable. That is the reason why high conversions are especially important in biolubricant production.

5.2. Oxidative Stability

As thoroughly explained, it is another important parameter that is vital to keep the main properties of biolubricant during storage, where auto-oxidation processes can take place. Thus, auto-oxidation can generate by-products like free fatty acids or polymers, which increase acidity (and the subsequent corrosion in containers or machines) and viscosity (with the subsequent change in its specific use), respectively. In that sense, as observed in the case of cardoon biolubricant (obtained through double transesterification with methanol and 2,2-dimethyl-1,3-propanediol) during storage, its viscosity increased up to 16% during storage for 12 months at room temperature, with a considerable decrease in viscosity index (which is also an undesirable effect) [76].

5.3. Viscosity

As previously stated, viscosity is highly related to the specific use of biolubricants in industry. Indeed, it can be considered the most important characteristic of a biolubricant, as viscosity can affect some functions, such as lubrication or sealing. Thus, depending on its viscosity, a biolubricant can be used in different areas, such as Diesel engines or machines in the textile industry.

Some of the key factors affecting these quality parameters are the chemical reaction conditions (where the use of catalysts plays a vital role), the use of additives, the raw material, and the alcohol selected. These factors can affect these quality parameters in different ways, or they can affect multiple quality parameters. For instance, the raw material can affect oxidation stability or viscosity, mainly depending on fatty acid composition.

5.4. Use of Catalysts/Chemical Reaction Conditions

As expected, the use of catalysts promotes the decrease in activation energy of a chemical reaction, which facilitates the completion of transesterification. Thus, on the one hand, the use of catalysts is positive as conversion increases, making the composition of the final biolubricant stable and predictable.

As explained in previous sections (see Tables 5 and 6), the use of catalysts might vary (from homogeneous to heterogeneous, with a wide range of compounds for each case), and in the case of esterification and transesterification, there seems to be a clear trend about the use of catalysts, with a special preference for homogeneous catalysts including sodium methoxide and ethoxide or sodium and potassium hydroxides, which are relatively cheap and easily obtained. These catalysts are usually added at low concentrations (from 0.3 to 2% w/w of total mass), which contributes to a low-cost impact in biolubricant production, obtaining high yields (at least 90% in most cases).

However, the use of homogeneous catalysts (usually containing Na and K), which are normally difficult to remove from the final biolubricant, can worsen the oxidative stability of biolubricants, as the presence of metal traces can act as catalysts for oxidation during storage. In that sense, the use of heterogeneous catalysts can present an additional advantage, like their easy removal from the chemical medium. Thus, the use of lipases, hydrotalcites, or CaO could avoid this problem, and even though their addition is relatively low (from 1% to 5% w/w, see Table 6), their costs are usually higher compared to homogeneous catalysts (which can explain the lower amount of research works about this matter). Nevertheless, they present an interesting advantage, like the total removal from the reaction medium, which would avoid the presence of catalysts in the final biolubricant and the abovementioned problems. However, the yields by using these catalysts are relatively low (up to 90% in many cases), and their reusability is limited, which is an interesting research line in the future, as their advantages could offset these inconveniences.

On the other hand, chemical reaction conditions (apart from catalyst concentration) can affect the final characteristics of the biolubricant. For instance, high temperatures during biolubricant production can affect its oxidative stability, as explained in previous studies. To avoid that, the use of a vacuum, when possible, is recommended to remove methanol released during the second transesterification, with the subsequent shift towards product generation at lower temperatures or catalyst concentrations [79].

5.5. Use of Additives (Antioxidants)

Apart from other additives that can improve the tribological performance of a biolubricant, as in the case of nanoparticles (for instance, Fe_3O_4 nanoparticles) to reduce the coefficient of friction between surfaces [132,133], we should pay attention to the use of antioxidants in final formulations. Thus, these antioxidants (both natural and synthetic) mainly affect oxidation stability in biodiesel and biolubricant samples. The use of products like propyl gallate (PG) or tert-butylhydroquinone (TBHQ) could increase oxidation stability, delaying some processes like free fatty acid generation or polymerization, which alter acidity and viscosity, respectively [67]. Thus, the use of TBHQ in WCO biolubricant showed a considerable increase in oxidative stability up to 10 h of induction point with very low quantities of antioxidants (2500 ppm) [34]. Figure 16 shows the molecular structure of some of the most popular antioxidants according to the literature in order to enhance the oxidative stability of biodiesel and biolubricants. As observed, all of them share one thing in common, that is, the aromatic ring that is responsible for free radical neutralization and the subsequent disruption (or delay) of the auto-oxidation process. It should be noted that once they react with free radicals, their molecular structure and effectiveness decrease, with the subsequent need to assess the amount of these compounds in biodiesel and biolubricant through analytical techniques. In that sense, the use of voltametric techniques seems to be effective in understanding the effect of antioxidant addition during oxidation processes [34,63,134].

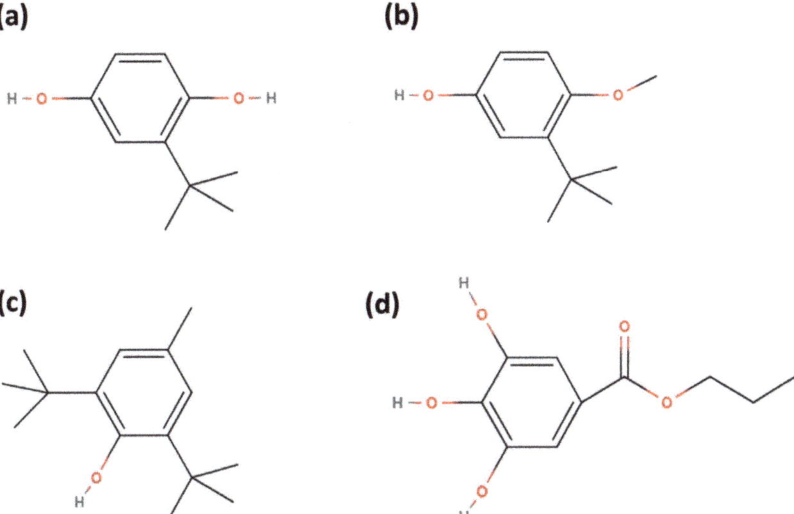

Figure 16. Molecular structure of some antioxidants used to extend the oxidative stability of biodiesel and biolubricants: (**a**) tert-butylhydroquinone (TBHQ); (**b**) butylated hydroxyanisole (BHA); (**c**) butylated hydroxytoluene (BHT); (**d**) propyl gallate.

In that sense, the use of antioxidants could offset the negative effect of trace metals included in biolubricants after the purification process, mainly due to homogeneous catalyst addition, although some removal techniques would be recommended to reduce Na and K content in biolubricants, such as the use of exchange-ion columns.

5.6. Raw Material (Vegetable Oil)

The raw material mainly affects, among other characteristics, the oxidation stability of the final product obtained. This is due to the fatty acid composition of vegetable oils, which are transformed into fatty acid methyl esters or fatty acid esters. Thus, the presence of double bonds can promote the generation of free radicals, the main starting point for auto-oxidation, decreasing oxidative stability. For instance, in the case of biolubricants obtained through double transesterification with methanol and 2-ethyl-1-hexanol from different vegetable oils (rapeseed, corn, and sunflower mix "seed oil" and WCO), different oxidative stability values were observed (with the following order from most to less stable: rapeseed, seed oil, and WCO) due to the different fatty acid content of the raw materials (rapeseed oil presented high oleic content, exceeding 60%, whereas seed oil and WCO had 31% and 27%, respectively [33]. Also, if fatty acids present some special functional groups like hydroxyl in the aliphatic chain (as in the case of ricinoleic acid), the final product obtained during the first transesterification could present a considerable viscosity, which can be used as a biolubricant for specific purposes [55].

5.7. Alcohols Used

Concerning transesterification, the superior alcohol used (usually trimethylolpropane and pentaerythritol, among others) plays an important role, as it is usually a polyol that, after transesterification with fatty acid methyl esters, generates a polyester that increases the viscosity of the final substance obtained. This way, the presence of more functional groups like hydroxyl in the polyol promotes the generation of more complex esters, increasing the possibility of higher intermolecular interactions and, therefore, a higher resistance to flow (in other words, a higher viscosity). Additionally, the stereochemistry of this second transesterification can make the reaction between hydroxyl groups in the superior alcohol and the corresponding fatty acid methyl ester require higher amounts of catalysts or temperature. In essence, the kind of alcohol equally affects the chemical conditions of the second transesterification and the yield of this chemical reaction.

As observed, these conditions are interrelated, and intermediate solutions might be advisable for sustainable and efficient biolubricant production. Indeed, in the following section, a brief exposition of works dealing with this subject (that is, techno-economic analyses and patents) is included to enhance the possibility of a real implementation of double transesterification at an industrial scale.

6. Techno-Economic Analyses Applied to Biolubricant Production in a Biorefinery Context and Patents

The role of a biorefinery can be attractive from an economic point of view if the efficiency of the process is comparable to equivalent industries based on petroleum. For that purpose, a correct design of a biorefinery should involve accurate data to estimate capital costs, using adequate indicators (especially for capital cost estimations) to carry out a suitable economic assessment [135].

Even though the industry of biobased products shows multiple economic and environmental benefits, there are still some challenges (especially in emerging industrial and developing countries) that should be addressed, like possible high production costs, lack of funds to invest in these kinds of technologies, inadequate policy support, fluctuating oil or feedstock prices, logistic performance, industrial competitiveness, etc. [136,137], considering several aspects such as logistics, existing infrastructure, feedstock supply chains, market opportunities, socio-economic issues and political context [137]. To overcome this inconvenience and concerning the feasibility of the implementation of biolubricant production based on vegetable oils, there are recent studies that assess the technical and economic issues that this green process must face to compete with equivalent processes based on petroleum.

As expected, the versatility and wide variety of biolubricant production processes implied the publication of dispersed works focused on specific aspects related to techno-economics. In any case, and even though there is a lack of these kinds of studies (which could imply an interesting research niche to contribute to the spread of knowledge), these research works share some points in common, like the following:

- The role of raw materials is essential to determine the marketability of biolubricants, as the costs related to agricultural practices (such as harvest), along with vegetable oil production and purification, usually imply a considerable percentage of fixed costs in a biorefinery based on vegetable oils. For instance, when it comes to estolide synthesis from oleic acid, the raw material implied 23.7% of total operating costs (around $6742 \cdot 10^3$ US$ to produce 1000 tons per year) [138], whereas, in the case of branched-chain glycerides production, this percentage was 18.2%, with total operating costs of $2247 \cdot 10^3$ US$ per year to produce 100 MT [139]. With this regard, again, the characteristics of WCO could be interesting, as many of the above-mentioned steps could be skipped, considerably reducing the costs related to the raw material.

- One of the main strong points of biolubricant production at an industrial scale is its integration in a biorefinery context, where multiple products can be obtained with the aim of adapting the performance of the biorefinery according to environmental policies or demand, among other factors.

- Another interesting aspect to consider is the reaction temperature (among other operating conditions), which usually takes a considerable percentage of energy consumption during biolubricant production and, therefore, is an important contribution to production costs. In that sense, a simple decrease of a few degrees could result in large savings. As proved in some abovementioned works (where the optimum reaction temperature was decreased at least 20 °C for biolubricant production from high-oleic safflower oil), reaction temperature can easily be reduced by using some improvements such as the vacuum (when this technique is possible, as in many cases it can remove some reagents in the reaction medium like complex alcohols that tend to sublimate due to their molecular structure, such as 2,2-dimethyl-1,3-propanediol for double transesterification) or the use of catalysts (both heterogeneous and, homogeneous, where a considerable decrease in reaction temperature was observed) [80].

- Indeed, the use of catalysts and their improvement related to their use in chemical routes to produce biolubricants has been widely studied in the literature. In that sense, their contribution to an increase in efficiency could be related to three aspects: First, the study of their optimum addition to avoiding extra costs; Second, the possibility of reuse (especially in the case of heterogeneous catalysts) or the purification process of biolubricants to remove homogeneous catalysts; Third, the effectiveness in reducing or improving chemical parameters that can contribute to energy or cost saving.

- Another interesting aspect to be considered is the possible real implementation of this technology by using pre-existing ones. Specifically, the use of double transesterification seems to be an interesting choice, as many of the facilities used for biodiesel production (that is, the first transesterification of fatty acids) could be perfectly adapted to the second transesterification process by adding some specific changes depending on the desired kind of biolubricant. Therefore, as similar facilities are required, the equipment acquisition would be easier and cheaper compared to other specific or customized facilities. Also, other chemical routes, such as epoxidation, could be easily adapted to this purpose, as explained to produce a biolubricant based on soybean oil [103].

- Finally, and even though some biolubricant productions could not be economically feasible at the industrial level, there is one interesting and favorable point to tilt the balance in favor of this green technology: the high environmental value and the favorable policies carried out by national and international agencies. Thus, the market value of these products could be higher compared to traditional lubricants, as there is an increasing demand for green products by customers in general. This fact could offset the initial and disadvantageous cost balance when these kinds of large-scale facilities are used.

Thus, Table 7 shows the main works related to techno-economic assessment applied to biolubricant production, even though it plays a secondary role within the biorefinery context of the corresponding vegetable oil.

As observed, these works point out the feasibility of biorefineries based on vegetable oils or agricultural wastes such as vegetable pulp, showing promising results that will encourage further studies in the near future.

Another remarkable point is the knowledge about the current patents about biolubricant production (see Table 8), specifically dealing with transesterification processes. In that sense, this is an interesting way to assess the practical application of a scientific subject or field, as patents are traditionally linked to the profitable exploitation of interesting findings, exploring their possible implementation at the industrial level.

In general, many of the chemical routes mentioned in this review are equally covered in patent registration (such as hydrolysis, epoxidation, and, especially, estolide formation, whose production was expensive, and these works are devoted to obtaining cheaper ways to exploit this concept), which points out the parallelisms with the scientific literature. Also, the use of multiple raw materials proves the versatility of this process. In this case, it should be noted the use of WCO, an interesting waste as explained in the introduction section, that is equally attractive to industrial processes (not only at the research level).

Table 7. Studies devoted to techno-economic assessment related to biolubricant production from vegetable oils.

Description	Details	References
Biolubricant production from rapeseed oil through double transesterification with methanol and TMP	High conversions were obtained for first (97%) and second (99%) transesterification and a reactor was designed (12 m^3) for industry level (production = 5550 tm·y^{-1})	[80]
Techno-economic analysis of biolubricants through different chemical routes	Non-edible oils (karanja, jatropha, WCO) through transesterification with TMP were proved as one sustainable way to obtain biolubricants for food lubrication and as automotive oils	[138]
Study of different biorefineries (based on first to fourth-generation raw materials, including edible and non-edible oils) for their design at different scale levels	The authors point out the importance of a suitable design of biorefineries depending on the purity and use of the final product obtained. Thus, high-quality products such as pharmaceutics are more adequate for small scales, whereas energy products could be useful at industrial scale	[140]
Biorefinery based on non-food agricultural feedstocks (vegetable pulp).	High lifecycle greenhouse gas savings can be obtained (up to 80%) if biofuels and biolubricant production are coupled in a biorefinery context	[141,142]
Biorefinery based on castor oil to produce biodiesel and multiple products, including a biolubricant	This biorefinery was mainly based on biodiesel production (exceeding 40% of total production) as well as other products, pointing out that multi-objective optimization is essential to obtain the optimal feedstock distribution and operating conditions to upgrade its performance	[143]

Table 8. Recent patents about biolubricant production through esterification or transesterification.

Description	Details	References
Preparation of heterogeneous catalyst for transesterification	It can be used for manufacturing commercial-grade biodiesel, biolubricant, and glycerol	[144]
Production of lubricating bio-oils	Use of soaps, WCO, and animal fats with initial hydrolysis to react with several alcohols and produce biolubricants	[145]
Production of biolubricants catalyzed by fermented solid	The reaction of methyl esters or free fatty acids with a polyhydroxylated alcohol is catalyzed by a fermented solid containing lipases	[146]
System for making biolubricants	A process for continuous preparation of biolubricants is described, including the use of acidic heterogeneous catalyst	[147]
Method for making biofuel and biolubricant	A process for producing biofuels and biolubricants from lipid material, pointing out the possibility of a biorefinery	[148]
Biolubricant production using fly ash as a catalyst	The reaction of fatty acids with different alcohols for the production of alkyl esters with C5 to C12 alcohols in the presence of fly ash as a catalyst	[149]
Method for producing neopentyl glycol diester as a biolubricant	Neopentyl glycol and vegetable fatty acids react using immobilized lipase	[150]

Thus, critical aspects such as high biolubricant conversion or catalyst durability are covered in these patents, which usually offer biolubricants for very specific uses.

7. Conclusions and Prospects

After carrying out a review of the most recent literature about the synthesis and characterization of biolubricants, the following remarks can be made:

- A definition of biolubricants, apart from the main chemical routes for their production (paying attention to double transesterification), has been shown. According to the literature, it has been proved the feasibility of biolubricant production from vegetable oils, obtaining high yields of this product regardless of the chemical route selected, and presenting some properties that are even better compared to traditional lubricants obtained from oil.

- Specifically, biolubricants through double transesterification present, in general, a higher viscosity index, which makes them more suitable to keep viscosity values under temperature changes.

- However, these products also present some challenges, like the short oxidative stability, which could imply a change in very characteristic properties such as viscosity during storage or auto-oxidation processes. Nevertheless, there are accessible alternatives to avoid this problem, like the use of antioxidants such as TBHQ, BHA, or PG, widely used in the scientific literature to increase the service life of this product.

- Regarding the possibility of implementation of this technology, it should be noted the potential of some wastes derived from VO, such as WCO, whose management would be difficult otherwise. As a matter of fact, the properties observed for biolubricants based on WCO are equivalent to those obtained from vegetable oils, with a slight decrease in oxidative stability in some cases, which could be easily solved by antioxidant addition, as explained above. This fact points out the versatility of biolubricant synthesis, being able to compete with traditional lubricant production.

- Considering the above, the possible integration of biolubricant production in biorefineries, or even the implementation of a biorefinery based on vegetable oils, could be an interesting starting point for this technology, especially in the case of double transesterification from fatty acids, as many bioproducts and byproducts that are easily reusable can be obtained, implying a green process that should replace traditional ones.

- In that sense, the role of catalysts is becoming more and more important in biolubricant production, as their use contributes to the higher efficiency of the process, which is an essential part of the real implementation of this technology at an industrial scale. This way, plenty of studies have been focused on the use of new and innovative catalysts, mainly heterogeneous and porous catalysts with the possibility of re-use in several cycles, to make the process more attractive when it comes to techno-economic analyses, especially by reducing the purification process to obtain the final biolubricant.

- Consequently, apart from the use of catalysts, the chemical conditions to carry out double transesterification usually have a strong influence on the final quality of biolubricants. Thus, "every detail counts", which should be pointed out especially in the case of temperature, catalyst addition, or reaction time, among other factors. Indeed, many studies addressed the use of milder reaction conditions, which will present a positive effect on techno-economic assessment, especially in biolubricant quality (especially concerning viscosity and oxidative stability). In other words, it would be the greatest exponent of sustainability, as it would be possible to obtain better products (in this case, biolubricants) under mild reaction conditions and the subsequent energy and material cost, implying a more attractive process for its implementation at an industrial scale.

- Finally, and according to the techno-economic assessments carried out by the scientific literature and recently published patents (where the role of catalysts is essential), the industrial scale implementation of biolubricant production is feasible, pointing out the high added-value product obtained, apart from other by-products equally interesting or reusable, and showing a promising starting point for the contribution to green chemistry and circular economy. In any case, due to the wide variety of processes, further studies would be advisable to cover this subject.

Author Contributions: Conceptualization, J.M.E. and J.F.G.; methodology, S.N.-D., J.M.E. and J.F.G.; validation, J.M.E. and J.F.G.; investigation, S.N.-D. and J.M.E.; resources, J.M.E. and J.F.G.; data curation, S.N.-D. and J.M.E.; writing—original draft preparation, S.N.-D.; writing—review and editing, S.N.-D. and J.M.E.; visualization, J.M.E. and J.F.G.; supervision, J.M.E. and J.F.G.; project administration, J.M.E.; funding acquisition, J.M.E. and J.F.G. All authors have read and agreed to the published version of the manuscript.

Funding: This research was funded by "Junta de Extremadura" and "FEDER" grant number IB18028 and GR18150.

Data Availability Statement: Not applicable.

Acknowledgments: Sergio Nogales-Delgado would like to thank the help and assistance of his project managers José María Encinar and Juan Félix González, whose support has been vital to fulfilling this project based on biolubricants. Also, Juan Félix González and Sergio Nogales-Delgado would like to wish José María Encinar a nice and well-earned retirement: "Thank you for your help and advice throughout all these years, we will miss you".

Conflicts of Interest: The authors declare no conflict of interest. The funders had no role in the design of the study; in the collection, analyses, or interpretation of data; in the writing of the manuscript; or in the decision to publish the results.

Abbreviations

Abbreviation	Term
BHA	Butylated hydroxyanisole
CFPP	Cold filter plugging point
FA	Fatty acid
FAE	Fatty acid esters
FAME	Fatty acid methyl ester
FFA	Free fatty acids
FO	Frying oil
GM	Genetically modified
IP	Induction point
NG	Neopentyl glycol
NGE	Neopentyl glycol ester
OPEC	Organization of the Petroleum Exporting Countries
OS	Oxidative stability
PE	Pentaerythritol
PEE	Pentaerythritol ester
PG	Propyl gallate
PP	Pour point
SDG	Sustainable development goal
TBHQ	Tert-Butylhydroquinone
TMP	Trimethylolpropane
TMPE	Trimethylolpropane ester
UN	United Nations
VI	Viscosity index
UVO	Used vegetable oil
VI	Viscosity index
VO	Vegetable oil
WCO	Waste cooking oil

References

1. Scarlat, N.; Dallemand, J.F.; Monforti-Ferrario, F.; Nita, V. The Role of Biomass and Bioenergy in a Future Bioeconomy: Policies and Facts. *Env. Dev.* **2015**, *15*, 3–34. [CrossRef]
2. UN. *UN Sustainable Development Goals*; UN: New York, NY, USA, 2019.
3. Statista Global Biolubricant Demand. Available online: https://www.statista.com/statistics/411616/lubricants-demand-worldwide/ (accessed on 16 March 2022).

4. Johnstone, P.; McLeish, C. World Wars and Sociotechnical Change in Energy, Food, and Transport: A Deep Transitions Perspective. *Technol. Forecast Soc. Chang.* **2022**, *174*, 121206. [CrossRef]

5. Johnstone, P.; McLeish, C. World Wars and the Age of Oil: Exploring Directionality in Deep Energy Transitions. *Energy Res. Soc. Sci.* **2020**, *69*, 101732. [CrossRef] [PubMed]

6. Scholten, D.; Bosman, R. The Geopolitics of Renewables; Exploring the Political Implications of Renewable Energy Systems. *Technol. Forecast. Soc. Chang.* **2016**, *103*, 273–283. [CrossRef]

7. Vakulchuk, R.; Overland, I.; Scholten, D. Renewable Energy and Geopolitics: A Review. *Renew. Sustain. Energy Rev.* **2020**, *122*, 109547. [CrossRef]

8. Palle, A. Bringing Geopolitics to Energy Transition Research. *Energy Res. Soc. Sci.* **2021**, *81*, 102233. [CrossRef]

9. Bricout, A.; Slade, R.; Staffell, I.; Halttunen, K. From the Geopolitics of Oil and Gas to the Geopolitics of the Energy Transition: Is There a Role for European Supermajors? *Energy Res. Soc. Sci.* **2022**, *88*, 102634. [CrossRef]

10. Umar, M.; Riaz, Y.; Yousaf, I. Impact of Russian-Ukraine War on Clean Energy, Conventional Energy, and Metal Markets: Evidence from Event Study Approach. *Resour. Policy* **2022**, *79*, 102966. [CrossRef]

11. Steffen, B.; Patt, A. A Historical Turning Point? Early Evidence on How the Russia-Ukraine War Changes Public Support for Clean Energy Policies. *Energy Res. Soc. Sci.* **2022**, *91*, 102758. [CrossRef]

12. Rowe, F.; Robinson, C.; Patias, N. Sensing Global Changes in Local Patterns of Energy Consumption in Cities during the Early Stages of the COVID-19 Pandemic. *Cities* **2022**, *129*, 103808. [CrossRef]

13. Ha, L.T. Storm after the Gloomy Days: Influences of COVID-19 Pandemic on Volatility of the Energy Market. *Resour. Policy* **2022**, *79*, 102921. [CrossRef] [PubMed]

14. Khan, K.; Su, C.W.; Zhu, M.N. Examining the Behaviour of Energy Prices to COVID-19 Uncertainty: A Quantile on Quantile Approach. *Energy* **2022**, *239*, 122430. [CrossRef] [PubMed]

15. Khan, K.; Su, C.W.; Khurshid, A.; Umar, M. COVID-19 Impact on Multifractality of Energy Prices: Asymmetric Multifractality Analysis. *Energy* **2022**, *256*, 124607. [CrossRef] [PubMed]

16. Wang, Q.; Huang, R.; Li, R. Towards Smart Energy Systems—A Survey about the Impact of COVID-19 Pandemic on Renewable Energy Research. *Energy Strategy Rev.* **2022**, *41*, 100845. [CrossRef]

17. Tian, J.; Yu, L.; Xue, R.; Zhuang, S.; Shan, Y. Global Low-Carbon Energy Transition in the Post-COVID-19 Era. *Appl. Energy* **2022**, *307*, 118205. [CrossRef]

18. Andrew, K.; Majerbi, B.; Rhodes, E. Slouching or Speeding toward Net Zero? Evidence from COVID-19 Energy-Related Stimulus Policies in the G20. *Ecol. Econ.* **2022**, *201*, 107586. [CrossRef]

19. Statista. *Comparison of Average Wholesale Motor Fuel Prices in Europe in 2018/19 and May 2022 as a Result of the Russia-Ukraine War, by Fuel Type (in U.S. Dollars per Metric Ton of Oil Equivalent)*; Statista Research Department: New York, NY, USA, 2023.

20. Statista. *Average Annual OPEC Crude Oil Price from 1960 to 2023 (in U.S. Dollars per Barrel)*; Statista Research Department: New York, NY, USA, 2023.

21. Overland, I.; Bazilian, M.; Ilimbek Uulu, T.; Vakulchuk, R.; Westphal, K. The GeGaLo Index: Geopolitical Gains and Losses after Energy Transition. *Energy Strategy Rev.* **2019**, *26*, 100406. [CrossRef]

22. Tiwari, A.K.; Boachie, M.K.; Suleman, M.T.; Gupta, R. Structure Dependence between Oil and Agricultural Commodities Returns: The Role of Geopolitical Risks. *Energy* **2021**, *219*, 119584. [CrossRef]

23. Statista Global Lubricant Demand. Available online: https://es.statista.com/ (accessed on 31 January 2023).

24. Anjani, K.; Yadav, P. High Yielding-High Oleic Non-Genetically Modified Indian Safflower Cultivars. *Ind. Crop. Prod.* **2017**, *104*, 7–12. [CrossRef]

25. Ciancolini, A.; Alignan, M.; Pagnotta, M.A.; Vilarem, G.; Crinò, P. Selection of Italian Cardoon Genotypes as Industrial Crop for Biomass and Polyphenol Production. *Ind. Crop. Prod.* **2013**, *51*, 145–151. [CrossRef]

26. Mihaela, P.; Josef, R.; Monica, N.; Rudolf, Z. Perspectives of Safflower Oil as Biodiesel Source for South Eastern Europe (Comparative Study: Safflower, Soybean and Rapeseed). *Fuel* **2013**, *111*, 114–119. [CrossRef]

27. Wan Azahar, W.N.A.; Bujang, M.; Jaya, R.P.; Hainin, M.R.; Mohamed, A.; Ngad, N.; Jayanti, D.S. The Potential of Waste Cooking Oil as Bio-Asphalt for Alternative Binder—An Overview. *J. Teknol.* **2016**, *78*, 111–116. [CrossRef]

28. Mannu, A.; Garroni, S.; Porras, J.I.; Mele, A. Available Technologies and Materials for Waste Cooking Oil Recycling. *Processes* **2020**, *8*, 366. [CrossRef]

29. Thushari, I.; Babel, S. Comparative Study of the Environmental Impacts of Used Cooking Oil Valorization Options in Thailand. *J. Env. Manag.* **2022**, *310*, 114810. [CrossRef]

30. Frota de Albuquerque Landi, F.; Fabiani, C.; Castellani, B.; Cotana, F.; Pisello, A.L. Environmental Assessment of Four Waste Cooking Oil Valorization Pathways. *Waste Manag.* **2022**, *138*, 219–233. [CrossRef]

31. Ibanez, J.; Martín, S.M.; Baldino, S.; Prandi, C.; Mannu, A. European Union Legislation Overview about Used Vegetable Oils Recycling: The Spanish and Italian Case Studies. *Processes* **2020**, *8*, 114810. [CrossRef]

32. Encinar, J.M.; Nogales, S.; González, J.F. Biorefinery Based on Different Vegetable Oils: Characterization of Biodiesel and Biolubricants. In Proceedings of the 3rd International Conference in Engineering Applications (ICEA), Sao Miguel, Portugal, 8–11 July 2019; Volume 2019.

33. Encinar, J.M.; Nogales, S.; González, J.F. Biodiesel and Biolubricant Production from Different Vegetable Oils through Transesterification. *Eng. Rep.* **2020**, *2*, e12190. [CrossRef]

34. Nogales-Delgado, S.; Cabanillas, A.G.; Romero, Á.G.; Encinar Martín, J.M. Monitoring Tert-Butylhydroquinone Content and Its Effect on a Biolubricant during Oxidation. *Molecules* **2022**, *27*, 8931. [CrossRef]

35. Perera, M.; Yan, J.; Xu, L.; Yang, M.; Yan, Y. Bioprocess Development for Biolubricant Production Using Non-Edible Oils, Agro-Industrial Byproducts and Wastes. *J. Clean. Prod.* **2022**, *357*, 131956. [CrossRef]

36. Paul, A.K.; Borugadda, V.B.; Goud, V.V. In-Situ Epoxidation of Waste Cooking Oil and Its Methyl Esters for Lubricant Applications: Characterization and Rheology. *Lubricants* **2021**, *9*, 27. [CrossRef]

37. Zhang, W.; Ji, H.; Song, Y.; Ma, S.; Xiong, W.; Chen, C.; Chen, B.; Zhang, X. Green Preparation of Branched Biolubricant by Chemically Modifying Waste Cooking Oil with Lipase and Ionic Liquid. *J. Clean. Prod.* **2020**, *274*, 122918. [CrossRef]

38. Dehghani Soufi, M.; Ghobadian, B.; Mousavi, S.M.; Najafi, G.; Aubin, J. Valorization of Waste Cooking Oil Based Biodiesel for Biolubricant Production in a Vertical Pulsed Column: Energy Efficient Process Approach. *Energy* **2019**, *189*, 116266. [CrossRef]

39. Department for Transport (UK); UK Petroleum Industry Association. *Biofuel Feedstock Consumption in the United Kingdom (UK) in 2020, by Type (in Million Liters)*; Department for Transport: London, UK, 2021.

40. Statista. *Number of Biodiesel Manufacturers in Japan in Fiscal Year 2021, by Type of Raw Material*; Statista Research Department: New York, NY, USA, 2023.

41. Scopus. Available online: https://www.scopus.com/home.uri (accessed on 29 August 2023).

42. Mobarak, H.M.; Niza Mohamad, E.; Masjuki, H.H.; Kalam, M.A.; Al Mahmud, K.A.H.; Habibullah, M.; Ashraful, A.M. The Prospects of Biolubricants as Alternatives in Automotive Applications. *Renew. Sustain. Energy Rev.* **2014**, *33*, 34–43. [CrossRef]

43. McNutt, J.; He, Q.S. Development of Biolubricants from Vegetable Oils via Chemical Modification. *J. Ind. Eng. Chem.* **2016**, *36*, 1–12. [CrossRef]

44. Heikal, E.K.; Elmelawy, M.S.; Khalil, S.A.; Elbasuny, N.M. Manufacturing of Environment Friendly Biolubricants from Vegetable Oils. *Egypt. J. Pet.* **2017**, *26*, 53–59. [CrossRef]

45. Jia, D.; Zhang, Y.; Li, C.; Yang, M.; Gao, T.; Said, Z.; Sharma, S. Lubrication-Enhanced Mechanisms of Titanium Alloy Grinding Using Lecithin Biolubricant. *Tribol. Int.* **2022**, *169*, 107461. [CrossRef]

46. Chan, C.H.; Tang, S.W.; Mohd, N.K.; Lim, W.H.; Yeong, S.K.; Idris, Z. Tribological Behavior of Biolubricant Base Stocks and Additives. *Renew. Sustain. Energy Rev.* **2018**, *93*, 145–157. [CrossRef]

47. Kania, D.; Yunus, R.; Omar, R.; Abdul Rashid, S.; Mohamad Jan, B. A Review of Biolubricants in Drilling Fluids: Recent Research, Performance, and Applications. *J. Pet. Sci. Eng.* **2015**, *135*, 177–184. [CrossRef]

48. Karmakar, G.; Ghosh, P.; Sharma, B.K. Chemically Modifying Vegetable Oils to Prepare Green Lubricants. *Lubricants* **2017**, *5*, 44. [CrossRef]

49. Salih, N.; Salimon, J.; Yousif, E. The Physicochemical and Tribological Properties of Oleic Acid Based Triester Biolubricants. *Ind. Crop. Prod.* **2011**, *34*, 1089–1096. [CrossRef]

50. Madankar, C.S.; Dalai, A.K.; Naik, S.N. Green Synthesis of Biolubricant Base Stock from Canola Oil. *Ind. Crop. Prod.* **2013**, *44*, 139–144. [CrossRef]

51. Cecilia, J.A.; Plata, D.B.; Saboya, R.M.A.; de Luna, F.M.T.; Cavalcante, C.L.; Rodríguez-Castellón, E. An Overview of the Biolubricant Production Process: Challenges and Future Perspectives. *Processes* **2020**, *8*, 257. [CrossRef]

52. Salimon, J.; Salih, N.; Yousif, E. Biolubricants: Raw Materials, Chemical Modifications and Environmental Benefits. *Eur. J. Lipid Sci. Technol.* **2010**, *112*, 519–530. [CrossRef]

53. Encinar, J.M.; González, J.F.; Pardal, A. Transesterification of Castor Oil under Ultrasonic Irradiation Conditions. Preliminary Results. *Fuel Process. Technol.* **2012**, *103*, 9–15. [CrossRef]

54. Encinar, J.M.; Nogales-Delgado, S.; Sánchez, N.; González, J.F. Biolubricants from Rapeseed and Castor Oil Transesterification by Using Titanium Isopropoxide as a Catalyst: Production and Characterization. *Catalysts* **2020**, *10*, 366. [CrossRef]

55. Sánchez, N.; Encinar, J.M.; Nogales, S.; González, J.F. Biodiesel Production from Castor Oil by Two-Step Catalytic Transesterification: Optimization of the Process and Economic Assessment. *Catalysts* **2019**, *9*, 864. [CrossRef]

56. Quinchia, L.A.; Delgado, M.A.; Reddyhoff, T.; Gallegos, C.; Spikes, H.A. Tribological Studies of Potential Vegetable Oil-Based Lubricants Containing Environmentally Friendly Viscosity Modifiers. *Tribol. Int.* **2014**, *69*, 110–117. [CrossRef]

57. Attia, N.K.; El-Mekkawi, S.A.; Elardy, O.A.; Abdelkader, E.A. Chemical and Rheological Assessment of Produced Biolubricants from Different Vegetable Oils. *Fuel* **2020**, *271*, 117578. [CrossRef]

58. Hájek, M.; Vávra, A.; De Paz Carmona, H.; Kocík, J. The Catalysed Transformation of Vegetable Oils or Animal Fats to Biofuels and Bio-Lubricants: A Review. *Catalyst* **2021**, *11*, 1118. [CrossRef]

59. Arianti, A.N.; Widayat, W. A Review of Bio-Lubricant Production from Vegetable Oils Using Esterification Transesterification Process. *MATEC Web Conf.* **2018**, *156*, 6007. [CrossRef]

60. Durango-Giraldo, G.; Zapata-Hernandez, C.; Santa, J.F.; Buitrago-Sierra, R. Palm Oil as a Biolubricant: Literature Review of Processing Parameters and Tribological Performance. *J. Ind. Eng. Chem.* **2022**, *107*, 31–44. [CrossRef]

61. Encinar, J.M.; Pardal, A.; Martínez, G. Transesterification of Rapeseed Oil in Subcritical Methanol Conditions. *Fuel Process. Technol.* **2012**, *94*, 40–46. [CrossRef]

62. Nogales-Delgado, S.; Encinar, J.M.; González, J.F. Safflower Biodiesel: Improvement of Its Oxidative Stability by Using BHA and TBHQ. *Energy* **2019**, *12*, 1940. [CrossRef]

63. Nogales-Delgado, S.; Encinar, J.M.; Guiberteau, A.; Márquez, S. The Effect of Antioxidants on Corn and Sunflower Biodiesel Properties under Extreme Oxidation Conditions. *JAOCS J. Am. Oil Chem. Soc.* **2019**, *97*, 201–212. [CrossRef]

64. Encinar, J.M.; González, J.F.; Sánchez, N.; Nogales-Delgado, S. Sunflower Oil Transesterification with Methanol Using Immobilized Lipase Enzymes. *Bioprocess Biosyst. Eng.* **2019**, *42*, 157–166. [CrossRef]
65. Bashiri, S.; Ghobadian, B.; Dehghani Soufi, M.; Gorjian, S. Chemical Modification of Sunflower Waste Cooking Oil for Biolubricant Production through Epoxidation Reaction. *Mater. Sci. Energy Technol.* **2021**, *4*, 119–127. [CrossRef]
66. Joshi, J.R.; Bhanderi, K.K.; Patel, J.V.; Karve, M. Chemical Modification of Waste Cooking Oil for the Biolubricant Production through Transesterification Process. *J. Indian Chem. Soc.* **2023**, *100*, 100909. [CrossRef]
67. Encinar, J.M.; Nogales-Delgado, S.; Álvez-Medina, C.M. High Oleic Safflower Biolubricant through Double Transesterification with Methanol and Pentaerythritol: Production, Characterization, and Antioxidant Addition. *Arab. J. Chem.* **2022**, *15*, 103796. [CrossRef]
68. Milazzo, M.F.; Spina, F.; Cavallaro, S.; Bart, J.C.J. Sustainable Soy Biodiesel. *Renew. Sustain. Energy Rev.* **2013**, *27*, 806–852. [CrossRef]
69. Kurre, S.K.; Yadav, J. A Review on Bio-Based Feedstock, Synthesis, and Chemical Modification to Enhance Tribological Properties of Biolubricants. *Ind. Crop. Prod* **2023**, *193*, 116122. [CrossRef]
70. *ISO 3104:1994*; UNE-EN ISO 3104/AC:1999 Petroleum Products. Transparent and Opaque Liquids. Determination of Kinematic Viscosity and Calculation of Dynamic Viscosity. ISO: Geneva, Switzerland, 1999.
71. Verdier, S.; Coutinho, J.A.P.; Silva, A.M.S.; Alkilde, O.F.; Hansen, J.A. A Critical Approach to Viscosity Index. *Fuel* **2009**, *88*, 2199–2206. [CrossRef]
72. *ASTM-D2270-10*; Standard Practice for Calculating Viscosity Index from Kinematic Viscosity at 40 °C and 100 °C. ASTM: West Conshohocken, PA, USA, 2016.
73. Focke, W.W.; Van Der Westhuizen, I.; Oosthuysen, X. Biodiesel Oxidative Stability from Rancimat Data. *Thermochim. Acta* **2016**, *633*, 116–121. [CrossRef]
74. *UNE-EN 14112*; Fat and Oil Derivatives—Fatty Acid Methyl Esters (FAME)—Determination of Oxidation Stability (Accelerated Oxidation Test). European Committee for Standardization: Brussels, Belgium, 2017.
75. Chen, J.; Bian, X.; Rapp, G.; Lang, J.; Montoya, A.; Trethowan, R.; Bouyssiere, B.; Portha, J.F.; Jaubert, J.N.; Pratt, P.; et al. From Ethyl Biodiesel to Biolubricants: Options for an Indian Mustard Integrated Biorefinery toward a Green and Circular Economy. *Ind. Crop. Prod* **2019**, *137*, 597–614. [CrossRef]
76. Nogales-Delgado, S.; Encinar Martín, J.M. Cardoon Biolubricant through Double Transesterification: Assessment of Its Oxidative, Thermal and Storage Stability. *Mater. Lett.* **2021**, *302*, 130454. [CrossRef]
77. Aziz, N.A.M.; Yunus, R.; Rashid, U.; Syam, A.M. Application of Response Surface Methodology (RSM) for Optimizing the Palm-Based Pentaerythritol Ester Synthesis. *Ind. Crop. Prod.* **2014**, *62*, 305–312. [CrossRef]
78. Yunus, R.; Fakhru'l-Razi, A.; Ooi, T.L.; Omar, R.; Idris, A. Synthesis of Palm Oil Based Trimethylolpropane Esters with Improved Pour Points. *Ind. Eng. Chem. Res.* **2005**, *44*, 8178–8183. [CrossRef]
79. Nogales-Delgado, S.; Encinar Martín, J.M.; Sánchez Ocaña, M. Use of Mild Reaction Conditions to Improve Quality Parameters and Sustainability during Biolubricant Production. *Biomass Bioenergy* **2022**, *161*, 106456. [CrossRef]
80. Encinar, J.M.; Nogales-Delgado, S.; Pinilla, A. Biolubricant Production through Double Transesterification: Reactor Design for the Implementation of a Biorefinery Based on Rapeseed. *Processes* **2021**, *9*, 1224. [CrossRef]
81. Ocholi, O.; Menkiti, M.; Auta, M.; Ezemagu, I. Optimization of the Operating Parameters for the Extractive Synthesis of Biolubricant from Sesame Seed Oil via Response Surface Methodology. *Egypt. J. Pet.* **2018**, *27*, 265–275. [CrossRef]
82. Ghafar, F.; Sapawe, N.; Dzazita Jemain, E.; Safwan Alikasturi, A.; Masripan, N. Study on the Potential of Waste Cockle Shell Derived Calcium Oxide for Biolubricant Production. *Mater. Today Proc.* **2019**, *19*, 1346–1353.
83. Cerón, A.A.; Vilas Boas, R.N.; Biaggio, F.C.; de Castro, H.F. Synthesis of Biolubricant by Transesterification of Palm Kernel Oil with Simulated Fusel Oil: Batch and Continuous Processes. *Biomass Bioenergy* **2018**, *119*, 166–172. [CrossRef]
84. Joshi, J.R.; Bhanderi, K.K.; Patel, J.V. A Review on Bio-Lubricants from Non-Edible Oils-Recent Advances, Chemical Modifications and Applications. *J. Indian Chem. Soc.* **2023**, *100*, 100849. [CrossRef]
85. Ho, C.K.; McAuley, K.B.; Peppley, B.A. Biolubricants through Renewable Hydrocarbons: A Perspective for New Opportunities. *Renew. Sustain. Energy Rev.* **2019**, *113*, 109261. [CrossRef]
86. Owuna, F.J.; Dabai, M.U.; Sokoto, M.A.; Dangoggo, S.M.; Bagudo, B.U.; Birnin-Yauri, U.A.; Hassan, L.G.; Sada, I.; Abubakar, A.L.; Jibrin, M.S. Chemical Modification of Vegetable Oils for the Production of Biolubricants Using Trimethylolpropane: A Review. *Egypt. J. Pet.* **2020**, *29*, 75–82. [CrossRef]
87. *UNE-EN 14214*; 2013 V2+A1:2018 Liquid Petroleum Products—Fatty Acid Methyl Esters (FAME) for Use in Diesel Engines and Heating Applications—Requirements and Test Methods. European Committee for Standardization: Brussels, Belgium, 2018.
88. Saluja, R.K.; Kumar, V.; Sham, R. Stability of Biodiesel—A Review. *Renew. Sustain. Energy Rev.* **2016**, *62*, 866–881. [CrossRef]
89. Sajjadi, B.; Raman, A.A.A.; Arandiyan, H. A Comprehensive Review on Properties of Edible and Non-Edible Vegetable Oil-Based Biodiesel: Composition, Specifications and Prediction Models. *Renew. Sustain. Energy Rev.* **2016**, *63*, 62–92. [CrossRef]
90. Tang, H.; De Guzman, R.C.; Salley, S.O.; Ng, S.K.Y. The Oxidative Stability of Biodiesel: Effects of FAME Composition and Antioxidant. *Lipid Technol.* **2008**, *20*, 249–252. [CrossRef]
91. Souza, A.G.; Medeiros, M.L.; Cordeiro, A.M.M.T.; Queiroz, N.; Soledade, L.E.B.; Souza, A.L. Efficient Antioxidant Formulations for Use in Biodiesel. *Energy Fuels* **2014**, *28*, 1074–1080. [CrossRef]
92. Knothe, G.; Razon, L.F. Biodiesel Fuels. *Prog. Energy Combust. Sci.* **2017**, *58*, 36–59. [CrossRef]

93. Checa, M.; Nogales-Delgado, S.; Montes, V.; Encinar, J.M. Recent Advances in Glycerol Catalytic Valorization: A Review. *Catalysts* **2020**, *10*, 1279. [CrossRef]

94. Hu, Y.; He, Q.; Xu, C. Catalytic Conversion of Glycerol into Hydrogen and Value-Added Chemicals: Recent Research Advances. *Catalysts* **2021**, *11*, 1455. [CrossRef]

95. Raza, M.; Inayat, A.; Abu-Jdayil, B. Crude Glycerol as a Potential Feedstock for Future Energy via Thermochemical Conversion Processes: A Review. *Sustainability* **2021**, *13*, 12813. [CrossRef]

96. Mota, C.; Pinto, B.P.; Lima, A.d. *Glycerol A Versatile Renewable Feedstock for the Chemical Industry*; Springer International Publishing: Berlin/Heidelberg, Germany, 2017; ISBN 978-3-319-59375-3.

97. Fasolini, A.; Cespi, D.; Tabanelli, T.; Cucciniello, R.; Cavani, F. Hydrogen from Renewables: A Case Study of Glycerol Reforming. *Catalysts* **2019**, *9*, 722. [CrossRef]

98. López-Bellido, L.; Wery, J.; López-Bellido, R.J. Energy Crops: Prospects in the Context of Sustainable Agriculture. *Eur. J. Agron.* **2014**, *60*, 1–12. [CrossRef]

99. Abdulkhani, A.; Alizadeh, P.; Hedjazi, S.; Hamzeh, Y. Potential of Soya as a Raw Material for a Whole Crop Biorefinery. *Renew. Sustain. Energy Rev.* **2017**, *75*, 1269–1280. [CrossRef]

100. Demichelis, F.; Fiore, S.; Pleissner, D.; Venus, J. Technical and Economic Assessment of Food Waste Valorization through a Biorefinery Chain. *Renew. Sustain. Energy Rev.* **2018**, *94*, 38–48. [CrossRef]

101. Esteban-Lustres, R.; Torres, M.D.; Piñeiro, B.; Enjamio, C.; Domínguez, H. Intensification and Biorefinery Approaches for the Valorization of Kitchen Wastes—A Review. *Bioresour. Technol.* **2022**, *360*, 127652. [CrossRef]

102. Kumar, V.; Sharma, N.; Umesh, M.; Selvaraj, M.; Al-Shehri, B.M.; Chakraborty, P.; Duhan, L.; Sharma, S.; Pasrija, R.; Awasthi, M.K.; et al. Emerging Challenges for the Agro-Industrial Food Waste Utilization: A Review on Food Waste Biorefinery. *Bioresour. Technol.* **2022**, *362*, 127790. [CrossRef]

103. Parente, E.J.; Marques, J.P.C.; Rios, I.C.; Cecilia, J.A.; Rodríguez-Castellón, E.; Luna, F.M.T.; Cavalcante, C.L. Production of Biolubricants from Soybean Oil: Studies for an Integrated Process with the Current Biodiesel Industry. *Chem. Eng. Res. Des.* **2021**, *165*, 456–466. [CrossRef]

104. Iriondo, A.; Barrio, V.L.; Cambra, J.F.; Arias, P.L.; Güemez, M.B.; Navarro, R.M.; Sánchez-Sánchez, M.C.; Fierro, J.L.G. Hydrogen Production from Glycerol Over Nickel Catalysts Supported on Al2O3 Modified by Mg, Zr, Ce or La. *Top. Catal.* **2008**, *49*, 46–58. [CrossRef]

105. Dahdah, E.; Estephane, J.; Gennequin, C.; Aboukaïs, A.; Abi-Aad, E.; Aouad, S. Zirconia Supported Nickel Catalysts for Glycerol Steam Reforming: Effect of Zirconia Structure on the Catalytic Performance. *Int. J. Hydrog. Energy* **2020**, *45*, 4457–4467. [CrossRef]

106. Ahmad, U.; Naqvi, S.R.; Ali, I.; Naqvi, M.; Asif, S.; Bokhari, A.; Juchelková, D.; Klemeš, J.J. A Review on Properties, Challenges and Commercial Aspects of Eco-Friendly Biolubricants Productions. *Chemosphere* **2022**, *309*, 136622. [CrossRef]

107. Monteiro, R.R.C.; Berenguer-Murcia, Á.; Rocha-Martin, J.; Vieira, R.S.; Fernandez-Lafuente, R. Biocatalytic Production of Biolubricants: Strategies, Problems and Future Trends. *Biotechnol. Adv.* **2023**, *68*, 108215. [CrossRef] [PubMed]

108. Brêda, G.C.; Aguieiras, E.C.G.; Cipolatti, E.P.; Greco-Duarte, J.; Collaço, A.C.D.A.; Costa Cavalcanti, E.D.; de Castro, A.M.; Freire, D.M.G. Current Approaches to Use Oil Crops By-Products for Biodiesel and Biolubricant Production: Focus on Biocatalysis. *Bioresour. Technol. Rep.* **2022**, *18*, 101030. [CrossRef]

109. De Sousa, I.G.; Mota, G.F.; Cavalcante, A.L.G.; Rocha, T.G.; Da Silva Sousa, P.; Holanda Alexandre, J.Y.N.; Da Silva Souza, J.E.; Neto, F.S.; Cavalcante, F.T.T.; Lopes, A.A.S.; et al. Renewable Processes of Synthesis of Biolubricants Catalyzed by Lipases. *J. Env. Chem. Eng.* **2023**, *11*, 109006. [CrossRef]

110. Dizge, N.; Keskinler, B. Enzymatic Production of Biodiesel from Canola Oil Using Immobilized Lipase. *Biomass Bioenergy* **2008**, *32*, 1274–1278. [CrossRef]

111. Kumar, D.; Das, T.; Giri, B.S.; Verma, B. Preparation and Characterization of Novel Hybrid Bio-Support Material Immobilized from Pseudomonas Cepacia Lipase and Its Application to Enhance Biodiesel Production. *Renew Energy* **2020**, *147*, 11–24. [CrossRef]

112. Masudi, A.; Muraza, O. Vegetable Oil to Biolubricants: Review on Advanced Porous Catalysts. *Energy Fuels* **2018**, *32*, 10295–10310. [CrossRef]

113. Kerman, C.O.; Gaber, Y.; Ghani, N.A.; Lämsä, M.; Hatti-Kaul, R. Clean Synthesis of Biolubricants for Low Temperature Applications Using Heterogeneous Catalysts. *J. Mol. Catal. B Enzym.* **2011**, *72*, 263–269. [CrossRef]

114. Nor, N.M.; Salih, N.; Salimon, J. Optimization and Lubrication Properties of Malaysian Crude Palm Oil Fatty Acids Based Neopentyl Glycol Diester Green Biolubricant. *Renew Energy* **2022**, *200*, 942–956. [CrossRef]

115. Bahadi, M.; Salimon, J.; Derawi, D. Synthesis of Di-Trimethylolpropane Tetraester-Based Biolubricant from Elaeis Guineensis Kernel Oil via Homogeneous Acid-Catalyzed Transesterification. *Renew Energy* **2021**, *171*, 981–993. [CrossRef]

116. Xie, Q.; Zhu, H.; Xu, P.; Xing, K.; Yu, S.; Liang, X.; Ji, W.; Nie, Y.; Ji, J. Transesterification of Methyl Oleate for Sustainable Production of Biolubricant: Process Optimization and Kinetic Study. *Ind. Crop. Prod.* **2022**, *182*, 114879. [CrossRef]

117. Bezerra, R.C.F.; Rodrigues, F.E.A.; Arruda, T.B.M.G.; Moreira, F.B.F.; Chaves, P.O.B.; Assunção, J.C.C.; Ricardo, N.M.P.S. Babassu-Oil-Based Biolubricant: Chemical Characterization and Physicochemical Behavior as Additive to Naphthenic Lubricant NH-10. *Ind. Crop. Prod.* **2020**, *154*, 112624. [CrossRef]

118. Nogales-Delgado, S.; Encinar, J.M.; González Cortés, Á. High Oleic Safflower Oil as a Feedstock for Stable Biodiesel and Biolubricant Production. *Ind. Crop. Prod.* **2021**, *170*, 113701. [CrossRef]

119. Gunam Resul, M.F.M.; Tinia, T.I.; Idris, A. Kinetic Study of Jatropha Biolubricant from Transesterification of Jatropha Curcas Oil with Trimethylolpropane: Effects of Temperature. *Ind. Crop. Prod.* **2012**, *38*, 87–92. [CrossRef]
120. Angulo, B.; Fraile, J.M.; Gil, L.; Herrerías, C.I. Bio-Lubricants Production from Fish Oil Residue by Transesterification with Trimethylolpropane. *J. Clean. Prod.* **2018**, *202*, 81–87. [CrossRef]
121. Cai, Z.; Zhuang, X.; Yang, X.; Huang, F.; Wang, Y.; Li, Y. Litsea Cubeba Kernel Oil as a Promising New Medium-Chain Saturated Fatty Acid Feedstock for Biolubricant Base Oil Synthesis. *Ind. Crop. Prod.* **2021**, *167*, 113564. [CrossRef]
122. Gul, M.; Zulkifli, N.W.M.; Masjuki, H.H.; Kalam, M.A.; Mujtaba, M.A.; Harith, M.H.; Syahir, A.Z.; Ahmed, W.; Bari Farooq, A. Effect of TMP-Based-Cottonseed Oil-Biolubricant Blends on Tribological Behavior of Cylinder Liner-Piston Ring Combinations. *Fuel* **2020**, *278*, 118242. [CrossRef]
123. Diaz, P.A.B.; Kronemberger, F.D.A.; Habert, A.C. A Pervaporation-Assisted Bioreactor to Enhance Efficiency in the Synthesis of a Novel Biolubricant Based on the Enzymatic Transesterification of a Castor Oil Based Biodiesel. *Fuel* **2017**, *204*, 98–105. [CrossRef]
124. Fernandes, K.V.; Cavalcanti, E.D.C.; Cipolatti, E.P.; Aguieiras, E.C.G.; Pinto, M.C.C.; Tavares, F.A.; da Silva, P.R.; Fernandez-Lafuente, R.; Arana-Peña, S.; Pinto, J.C.; et al. Enzymatic Synthesis of Biolubricants from By-Product of Soybean Oil Processing Catalyzed by Different Biocatalysts of Candida Rugosa Lipase. *Catal. Today* **2021**, *362*, 122–129. [CrossRef]
125. Ng, B.Y.S.; Ong, H.C.; Lau, H.L.N.; Ishak, N.S.; Elfasakhany, A.; Lee, H.V. Production of Sustainable Two-Stroke Engine Biolubricant Ester Base Oil from Palm Fatty Acid Distillate. *Ind. Crop. Prod.* **2022**, *175*, 114224. [CrossRef]
126. Abd Wafti, N.S.; Yunus, R.; Lau, H.L.N.; Choong, T.S.Y.; Abd-Aziz, S. Enzymatic Synthesis of Palm Oil-Based Trimethylolpropane Ester as Biolubricant Base Stock Catalyzed by Lipozyme 435. *Energy* **2022**, *260*, 125061. [CrossRef]
127. Hussein, R.Z.K.; Attia, N.K.; Fouad, M.K.; ElSheltawy, S.T. Experimental Investigation and Process Simulation of Biolubricant Production from Waste Cooking Oil. *Biomass Bioenergy* **2021**, *144*, 105850. [CrossRef]
128. Ivan-Tan, C.T.; Islam, A.; Yunus, R.; Taufiq-Yap, Y.H. Screening of Solid Base Catalysts on Palm Oil Based Biolubricant Synthesis. *J. Clean. Prod.* **2017**, *148*, 441–451. [CrossRef]
129. Shrivastava, S.; Prajapati, P.; Virendra; Srivastava, P.; Lodhi, A.P.S.; Kumar, D.; Sharma, V.; Srivastava, S.K.; Agarwal, D.D. Chemical Transesterification of Soybean Oil as a Feedstock for Stable Biodiesel and Biolubricant Production by Using Zn Al Hydrotalcites as a Catalyst and Perform Tribological Assessment. *Ind. Crop. Prod.* **2023**, *192*, 116002. [CrossRef]
130. Sun, G.; Li, Y.; Cai, Z.; Teng, Y.; Wang, Y.; Reaney, M.J.T. K2CO3-Loaded Hydrotalcite: A Promising Heterogeneous Solid Base Catalyst for Biolubricant Base Oil Production from Waste Cooking Oils. *Appl. Catal. B* **2017**, *209*, 118–127. [CrossRef]
131. Tulashie, S.K.; Kotoka, F. The Potential of Castor, Palm Kernel, and Coconut Oils as Biolubricant Base Oil via Chemical Modification and Formulation. *Therm. Sci. Eng. Prog.* **2020**, *16*, 100480. [CrossRef]
132. Sikdar, S.; Rahman, M.H.; Menezes, P.L. Synergistic Study of Solid Lubricant Nano-Additives Incorporated in Canola Oil for Enhancing Energy Efficiency and Sustainability. *Sustainability* **2022**, *14*, 290. [CrossRef]
133. Ahmad, U.; Raza Naqvi, S.; Ali, I.; Saleem, F.; Taqi Mehran, M.; Sikandar, U.; Juchelková, D. Biolubricant Production from Castor Oil Using Iron Oxide Nanoparticles as an Additive: Experimental, Modelling and Tribological Assessment. *Fuel* **2022**, *324*, 124565. [CrossRef]
134. Nogales-Delgado, S.; Guiberteau, A.; Encinar, J.M. Effect of Tert-Butylhydroquinone on Biodiesel Properties during Extreme Oxidation Conditions. *Fuel* **2022**, *310*, 122339. [CrossRef]
135. Solarte-Toro, J.C.; Rueda-Duran, C.A.; Ortiz-Sanchez, M.; Cardona Alzate, C.A. A Comprehensive Review on the Economic Assessment of Biorefineries: The First Step towards Sustainable Biomass Conversion. *Bioresour. Technol. Rep.* **2021**, *15*, 100776. [CrossRef]
136. Mousavi-Avval, S.H.; Sahoo, K.; Nepal, P.; Runge, T.; Bergman, R. Environmental Impacts and Techno-Economic Assessments of Biobased Products: A Review. *Renew. Sustain. Energy Rev.* **2023**, *180*, 113302. [CrossRef]
137. Solarte-Toro, J.C.; Laghezza, M.; Fiore, S.; Berruti, F.; Moustakas, K.; Cardona Alzate, C.A. Review of the Impact of Socio-Economic Conditions on the Development and Implementation of Biorefineries. *Fuel* **2022**, *328*, 125169. [CrossRef]
138. Khan, S.; Das, P.; Quadir, M.A.; Thaher, M.; Annamalai, S.N.; Mahata, C.; Hawari, A.H.; Al Jabri, H. A Comparative Physico-chemical Property Assessment and Techno-Economic Analysis of Biolubricants Produced Using Chemical Modification and Additive-Based Routes. *Sci. Total Environ.* **2022**, *847*, 157648. [CrossRef] [PubMed]
139. Ngo, H.; Latona, R.; Sarker, M.I.; Yee, W.; Hums, M.; Moreau, R.A. A Process to Convert Sunflower Oil into a Value Added Branched Chain Oil with Unique Properties. *Ind. Crop. Prod.* **2019**, *139*, 111457. [CrossRef]
140. Moncada, B.J.; Aristizábal, M.V.; Cardona, A.C.A. Design Strategies for Sustainable Biorefineries. *Biochem. Eng. J.* **2016**, *116*, 122–134. [CrossRef]
141. Budzianowski, W.M. High-Value Low-Volume Bioproducts Coupled to Bioenergies with Potential to Enhance Business Development of Sustainable Biorefineries. *Renew. Sustain. Energy Rev.* **2017**, *70*, 793–804. [CrossRef]
142. Balakrishnan, M.; Sacia, E.R.; Sreekumar, S.; Gunbas, G.; Gokhale, A.A.; Scown, C.D.; Toste, F.D.; Bell, A.T. Novel Pathways for Fuels and Lubricants from Biomass Optimized Using Life-Cycle Greenhouse Gas Assessment. *Proc. Natl. Acad. Sci. USA* **2015**, *112*, 7645–7649. [CrossRef] [PubMed]
143. Acevedo-García, B.; Santibañez-Aguilar, J.E.; Alvarez, A.J. Integrated Multiproduct Biorefinery from Ricinus Communis in Mexico: Conceptual Design, Evaluation, and Optimization, Based on Environmental and Economic Aspects. *Bioresour. Technol. Rep.* **2022**, *19*, 101201. [CrossRef]

144. Joshi, U.P.; Ham, P.G. Heterogeneous Catalyst for Transesterification and Method of Preparing Same. US Patent USRE49551 (E), 13 June 2023.
145. Di Serio, M.; Gallo, F. Process for the Production of Lubricating Biooils. International Patent WO2023126789 (A1), 6 July 2023.
146. Cavalcanti da Silva, J.A.; Silva, G.B.; Guimaraes Freire, D.M.; Gonçalves Aguieiras, E.C.; Cavalcanti Oliveira, E.D.; Greco Duarte, J.; Ignacio, K.L.; Soares, V.F.; Da Silva, P.R. Process for Producing Esters and Biolubricants, Catalysed By fermented Solid. US Patent US2021189442 (A1), 24 June 2021.
147. Summers, W.A.; Williams, R.; Gulledge, D.; Tripp, R.B. System and Methods for Making Bioproducts. US Patent US2018319733 (A1), 8 November 2018.
148. Salazar, J.A.; Joshi, M. Method for Making Biofuels and Biolubricants. Canadian Patent CA2816018 (A1), 10 May 2012.
149. Nagabhushana, K.; Mal, N.; Shinde, T.; Dapurkar, S.; Kumar, R. A Process for Production of Biolubricant Using Fly Ash as Acatalyst. International Patent WO2011007361 (A1), 20 January 2011.
150. Hwan, K.I.; Won, K.J. Method for Producing Neopentyl Glycol Diester as a Biolubricant Using Enzymaticreaction. Korean Patent KR20220101428 (A), 19 July 2022.

Article

Acid-Catalyzed Etherification of Glycerol with *Tert*-Butanol: Reaction Monitoring through a Complete Identification of the Produced Alkyl Ethers

Alfonso Cornejo [1,2], Inés Reyero [1,2], Idoia Campo [1], Gurutze Arzamendi [1,2] and Luis M. Gandía [1,2,*]

[1] Departamento de Ciencias, Universidad Pública de Navarra (UPNA), 31006 Pamplona, Spain; alfonso.cornejo@unavarra.es (A.C.); ines.reyero@unavarra.es (I.R.); icampoaran@educacion.navarra.es (I.C.); garzamendi@unavarra.es (G.A.)

[2] Institute for Advanced Materials and Mathematics (InaMat2), Universidad Pública de Navarra (UPNA), 31006 Pamplona, Spain

* Correspondence: lgandia@unavarra.es

Abstract: Higher *tert*-Butyl glycerol ethers (*t*BGEs) are interesting glycerol derivatives that can be produced from *tert*-butyl alcohol (TBA) and glycerol using an acid catalyst. Glycerol *tert*-butylation is a complex reaction that leads to the formation of five *t*BGEs (two monoethers, two diethers, and one triether). In order to gain insight into the reaction progress, the present work reports on the monitoring of glycerol etherification with TBA and *p*-toluensulfonic acid (PTSA) as homogeneous catalysts. Two analytical techniques were used: gas chromatography (GC), which constitutes the benchmark method, and ^{1}H nuclear magnetic resonance (^{1}H NMR), whose use for this purpose has not been reported to date. A method for the quantitative analysis of *t*BGEs and glycerol based on ^{1}H NMR is presented that greatly reduced the analysis time and relative error compared with GC-based methods. The combined use of both techniques allowed for a complete quantitative and qualitative description of the glycerol *tert*-butylation progress. The set of experimental results collected showed the influence of the catalyst concentration and TBA/glycerol ratio on the etherification reaction and evidenced the intrinsic difficulties of this process to achieve high selectivities and yields to the triether.

Keywords: etherification; glycerol; homogeneous acid catalyst; reaction monitoring; *tert*-butylation

Citation: Cornejo, A.; Reyero, I.; Campo, I.; Arzamendi, G.; Gandía, L.M. Acid-Catalyzed Etherification of Glycerol with *Tert*-Butanol: Reaction Monitoring through a Complete Identification of the Produced Alkyl Ethers. *Catalysts* **2023**, *13*, 1386. https://doi.org/10.3390/catal13101386

Academic Editors: José María Encinar Martín and Sergio Nogales Delgado

Received: 3 September 2023
Revised: 17 October 2023
Accepted: 18 October 2023
Published: 23 October 2023

1. Introduction

Glycerol is currently produced in large amounts as a byproduct of the biodiesel industry. According to ChemAnalyst the global glycerol market was about 1 million tons in 2021, and it is expected to grow at a compound annual rate of 4.5% until 2030. The personal care, cosmetic, and pharmaceutical industry sectors dominate this demand [1]. On the other hand, the production of glycerol associated to biodiesel is much higher. Indeed, according to the International Energy Agency [2], 45,712 million liters of biodiesel were produced in 2021, which allows estimating the biodiesel production at about 40.2 million tons (assuming a mean biodiesel density of 0.88 g/cm^3) and that of glycerol in crude (non-refined) form at 4.4 million tons. Therefore, there is great interest in developing new uses capable of absorbing the surplus in order to improve the economic balance of the biodiesel production processes and introducing that sustainable resource in the value chain, thus contributing to the circular economy.

Since the large-scale emergence of biodiesel as an alternative fuel, some 25 years ago, many review papers have appeared reporting on the progress made in the methodss for transforming glycerol into value-added products. Referring to some of the recent studies, Morais Lima et al. [3] described the production of propylene glycol, acrolein, epichloro-hydrin, dioxalane, dioxane, and glycerol carbonate through chemical routes and that of 1,3-propanediol, *n*-butanol, citric acid, ethanol, butanol, propionic acid, mono-, di-, and triacylglycerols, cynamoil esters, glycerol acetate, and benzoic acid by means of biochemical processes, mainly enzymatic. Checa et al. [4] discussed the rational formulation of the catalysts required depending on the chemistry of the transformation route according to reforming (steam and aqueous phase), hydrogenolysis, reduction, selective oxidation, and acetalization reactions. Direct uses of crude glycerol and recent valorization approaches such as the production of alkyl-aromatics and activated carbon were also highlighted. Other important conversion processes were dehydration, pyrolysis, gasification, selective trans-esterification, etherification, fermentation, oligomerization, and polymerization [5]. Kaur et al. [6] emphasized the environmental advantages of the biological conversion of crude glycerol and included among the valuable products polyglycerols, polyhydroxyalkanoates, solketal, trehalose, and various organic acids (lactic, glyceric, succinic, docosahexaenoic, and eicosapentaenoic). A recently proposed and very promising route of glycerol valorization is its catalytic deoxygenation for (bio)olefin (e.g., propylene) production [7]. The plethora of possible products that can be obtained from glycerol illustrates its frequent designation as a platform chemical.

There is also big interest in the applications of glycerol as a fuel (through combustion) and as a source of fuels (e.g., hydrogen, biogas, syngas, and ethanol) and fuel additives [5,8,9]. Fuel additives are commonly used in order to improve thermal engines performance, reduce their pollutant emissions, and modify specific physicochemical properties of commercial gasoline, diesel, and biodiesel. Oxygenated derivatives of glycerol such as ethers, acetals, and esters (acetates) have been reported as fuel additives [10,11]. The *tert*-butyl glycerol ethers (*t*BGEs) resulting from the reaction between isobutylene or *tert*-butanol and glycerol are precisely of particular interest as concerns the present work. Depending on the number of hydroxyl groups of the glycerol molecule that become alkylated, this reaction, also known as glycerol *tert*-butylation (see Figure 1), leads to the formation of two monoether isomers (*t*B1GE-a and *t*B1GE-b), two diether isomers (*t*B2GE-a and *t*B2GE-b), and a triether (*t*B3GE). Due to the limited solubility of the monoethers in the most common fuels, the products preferred as additives are the diethers and, especially, the triether. The *t*BGEs have been found to increase the octane number of gasoline and have been claimed as substitutes for methyl and ethyl *tert*-butyl ethers (MTBE and ETBE, respectively). When used as diesel and biodiesel additives, the main positive effects of *t*BGEs are the reduction in particulate matter and soot emissions [11]. On the other hand, alkylated glycerol monoethers have interesting surfactant and biological properties and find application as components of cosmetics and personal care and pharmaceutical products [12–14].

Glycerol etherification reactions have been thoroughly reviewed by Palanychamy et al. [15]. Glycerol *tert*-butylation involves three consecutive and reversible steps, leading to the successive formation of *tert*-butyl glycerol mono-, di-, and triethers (see Figure 1). The reaction is typically carried out in the presence of strong acid catalysts, and when *tert*-butanol (*tert*-butyl alcohol, TBA) is used as the alkylating agent, each step is accompanied by the liberation of a water molecule. Much of the early work on this field has been performed reacting glycerol and isobutylene (isobutene, 2-methylpropene, IB); a commercial process was developed based on this synthetic route [16]. IB requires operating the reactor under pressure (around 20 atm) in order to keep the olefin in the liquid state, although it is immiscible with glycerol, thus leading to a heterogeneous reaction system characterized by mass transport limitations [16,17]. These features, and the possibility of IB oligomerization to form diisobutylenes as side reaction, have been considered disadvantages that have encouraged the use of TBA instead of IB in more recent works. Nevertheless, the presence of coproduced water has been found to negatively affect the acid catalysts and introduce thermodynamic limitations that introduce thermodynamic limitations that make more complex reaching high yields of the higher (di-, and specially, tri-) *t*BGEs with TBA than with IB [18]. As for the catalysts required, homogeneous acids, i.e., those that are soluble in glycerol, such as *p*-toluensulfonic acid or the heteropoly acid $H_3PW_{12}O_{40}$, are much more active than the heterogeneous ones and allow obtaining significantly higher yields of di- and tri-*t*BGEs [16]. The interest in avoiding the use of the homogeneous acids due to corrosion, safety, and environmental issues has fostered the search for solid acid catalysts among which cation exchange resins with highly crosslinked structure, large pore zeolites, sulfonated mesostructured silicas and carbons, and supported tungstophosphoric acid have provided the best results with both IB [19–25] and TBA [19,26–32]. These materials require a fine-tuning of their acid and textural properties in order to develop suitable catalytic activity and selectivity toward higher *t*BGEs; at the same time, they are also very sensitive to water, which solvates the hydrophilic active sites, rendering them poorly active.

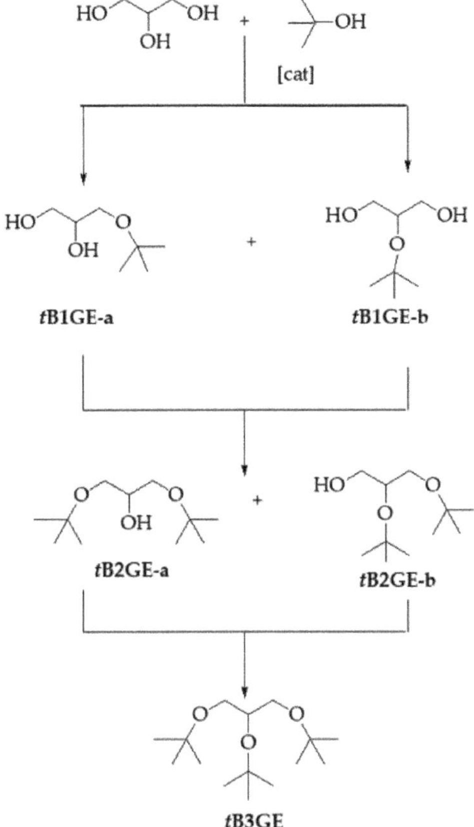

Figure 1. Products of the glycerol *tert*-butylation reaction with indication of the nomenclature used to refer to the several *tert*-butyl glycerol ethers.

In the vast majority of glycerol *tert*-butylation reports available, it is customary to lump the isomers as monoethers and diethers, and even the diethers and triether are sometimes lumped as higher ethers. In the present work, procedures for the identification and analysis of the different *tert*-butyl glycerol ethers are presented. The five *t*BGEs were obtained in our laboratories, isolated, and completely characterized by HRMS-ESI+, ATR-FT-IR, and NMR. A straightforward methodology is presented that allows for fast and reliable monitoring of the reaction between glycerol and *tert*-butanol (TBA) catalyzed with *p*-toluensulfonic acid (PTSA) combining ^{1}H NMR and conventional GC-FID analyses. It is expected this way to contribute to a complete characterization of the reacting system, as well as providing an improved description of the steps involved in the *tert*-butylation reaction.

Gas chromatography (GC) is the benchmark technique for the quantitative analysis of *t*BGEs; however, except made of *t*B1GE-a, the *t*BGEs are not easily available, which complicates the equipment calibration. Melero et al. [20,33] proposed to extrapolate the response factor obtained for *t*B1GE-a to the higher ethers. Other authors determined the response factors for all the individual ethers after column chromatography separation and purification from the reaction mixture [23,24,34], which is quite laborious.

Nuclear magnetic resonance (NMR) has become in recent years a high-throughput analytical technique for the characterization of complex mixtures. This is the case, for example, of the monitoring and/or quantitative analysis of reaction mixtures from the digestion of woody biomass [35–38], lignin depolymerization [39], or transesterification reaction for biodiesel production [40,41]. However, the purification of tertiary mono and di *tert*-butyl glycerol ethers is elusive, and to the best of our knowledge, their characterization has not been reported to date. As for the secondary mono and di-*tert*-butyl glycerol ethers and tri-*tert*-butyl glycerol ether, their NMR spectra have been reported [42–44]. Nevertheless, descriptions are, in some cases, imprecise when providing the chemical shifts for the ^{1}H NMR [44] or simplistic when explaining the spin systems of etherification products [42]. Indeed, the chemical shifts reported by Jamróz et al. [44], which describe the spin system, and those reported by González et al. [42] corresponding either to the glycerol skeleton hydrogen atoms or to the methyl groups in the *tert*-butyl moieties did not match at all between them.

2. Results and Discussion

2.1. Characterization of Etherification Products

The five *tert*-butyl ethers of glycerol were synthetized as indicated in the Materials and Methods Section 3.3. After their isolation and purification, the corresponding chemical structure and expected formula were confirmed by NMR and ESI$^+$ (see Figure A1). Glycerol and di- and tri-*tert*-butyl ethers presented the expected $[M + Na]^+$ as the major peak (*m/z* 227.1613 for *t*B2GE-a, *m/z* 227.1626 for *t*B2GE-b, and *m/z* 283.2267 for *t*B3GE). This peak was accompanied by $2M + Na^+$ in the case of *t*B2GE-a. More reactive monoethers presented the peak corresponding to $[M + Na]^+$ (*m/z* 171.0996). Condensation of the primary hydroxyl groups was observed under the ESI$^+$ analysis conditions for *t*B1GE. Thus, in the case of *t*B1GE-a, *m/z* 301.1988 was detected after the condensation of two molecules under the analysis conditions, producing $[2M\text{-}H_2O + Na]^+$. *t*B1GE-b presented an additional primary hydroxyl group, and in addition to $[M + Na]^+$ (*m/z* 171.0996), the major peak appeared at *m/z* 413.2647, which corresponds to $[3M\text{-}3H_2O + Na^+]$ as a result of the higher reactivity of these primary hydroxyl groups, due to their lower steric hindrance, to become a crown-ether like structure under the analysis conditions.

ATR-FT-IR spectra of the isolated compounds showed the gradual disappearance of the hydroxyl O–H stretching band at ca. 3400 cm^{-1} as the degree of etherification increased accompanied by the intensification of the aliphatic bands between 2974 and 2834 cm^{-1} that correspond to the C–H stretching mode (see Figure A2). As expected, no major differences were observed between *t*B2GE-a and *t*B2GE-b with this technique.

As for the NMR spectra, Table 1 gathers the chemical shifts and coupling constant for the hydrogen atoms on the glycerol skeleton that are identified in Figure 2. Given the symmetry of *t*B1GE-b, *t*B2GE-a, and *t*B3GE, the ^{1}H NMR signals were easier to assign (see Figures A5, A7 and A11). In contrast, the secondary C_2 carbon of *t*B1GE-a and *t*B2GE-b was asymmetric, and therefore, C_1 was diasterotopic, which complicated the interpretation of their ^{1}H NMR spectra (see Figures A3 and A9). The hydrogen atoms on the glycerol skeleton of *t*B1GE-a appeared as a set of three ^{1}H NMR signals in the range from 3.42 ppm to 3.81 ppm (see Figure 3a), whereas both hydroxyl hydrogen atoms appeared as broad shoulders at 2.40 ppm. As expected, the *tert*-butyl moiety appeared as a singlet at 1.19 ppm. The assignation of H and C signals was performed using ^{13}C APT and HMBC correlation. The ^{1}H signal at 1.19 ppm from –C–CH$_3$ presented long-range correlation with the quaternary C atom at 73.66 ppm (see Figure 3b). These signals also presented long-range correlation with the signal of H_1 centered at 3.43 ppm, which also correlated with C_3 at 63.92 ppm in the HMBC spectrum (see Figure 3b). C_2 is a chiral center; therefore, both H_1 and H_3 are diasterotopic. Hence, they appeared as a set of two signals each (H_{1a} and H_{1b}; H_{3a} and H_{3b}) with large geminal coupling constants, JH_{1a}-H_{1b} = 9.1 Hz and JH_{3a}-H_{3b} = 11.4 Hz. The coupling constants with H_2 were in the 3.9–5.8 Hz range. Because of the coupling with the non-equivalent H_1 and H_3 atoms, the ^{1}H NMR signals for H_2 were shown as multiplet centered at 3.78 ppm. The ^{1}H NMR spectra of tB1GE-a could be satisfactorily simulated using WINDNMR [45].

Table 1. Chemical shifts and coupling constant for the hydrogen atoms on the glycerol skeleton of the *t*BGEs as identified in Figure 2.

Ether	H_{1a} $(J_{1a,2})$	H_{1b} $(J_{1b,2})$	$(J_{1a,1b})$	H_2	H_{3a} $(J_{3a,2})$	H_{3b} $(J_{3b,2})$	$(J_{3a,3b})$	C_1–OC $[CH_3]_3$	C_2–OC $[CH_3]_3$	C_3–OC $[CH_3]_3$
*t*B1GE-a	3.43 (5.88)	3.49 (3.92)	(9.10)	3.78	3.65 (4.95)	3.70 (3.91)	(11.4)	1.20	-	-
*t*B1GE-b	3.65	-	-	3.71	3.65 [a]	-	-	1.24	-	-
*t*B2GE-a	3.37 (5.92)	3.42 (5.06)	(8.97)	3.78	3.37 [a] (5.92) [a]	3.42 [b] (5.06) [b]	(8.98)	1.19	-	1.19
*t*B2GE-b	3.34 (8.58)	3.41 (4.30)	(8.62)	3.65	3.61	3.61	-	1.21	1.19	-
*t*B3GE	3.27 (5.32)	3.37 (5.92)	(9.22)	3.60	3.27 [a] (5.32) [a]	3.37 [b] (5.92) [b]	(9.22) [a,b]	1.17	1.20	1.17

[a] Symmetry H-1a and H-3a. [b] Symmetry H-1b and H-3b.

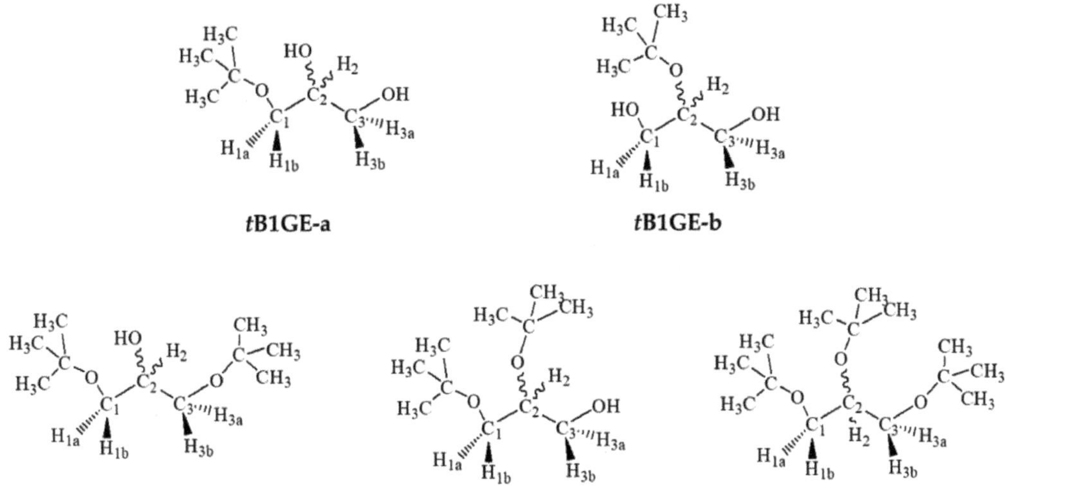

tB1GE-a

tB1GE-b

tB2GE-a

tB2GE-b

tB3GE

Figure 2. Identification of the hydrogen and carbon atoms on the glycerol skeleton of the *t*BGEs for ^{1}H NMR chemical shift assignment (see Table 1).

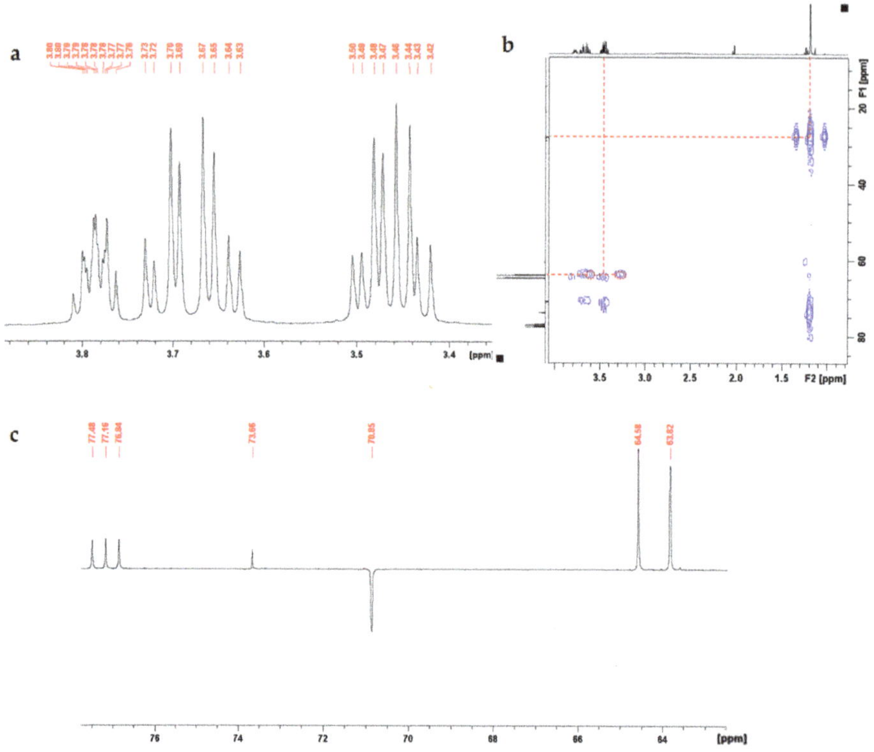

Figure 3. Details of *t*B1GE-a spectra: (**a**) ^{1}H NMR spectrum in the 3.35–3.85 ppm region, (**b**) HMBC experiment, and (**c**) ^{13}C APT spectrum.

Analysis of the spectra for symmetric *t*B2GE-a and tB3GE was much simpler. In the case of *t*B2GE-a, the ^{1}H NMR spectrum showed the signal of the hydrogen atom on the *tert*-butyl group as a singlet at 1.19 ppm, whereas the signal corresponding to the hydroxyl group appeared as a doublet at 2.54 ppm (J = 4.48 Hz). The signal for H_2 appeared as a hexuplet centered at 3.78 ppm showing an apparent coupling constant of 5.27 Hz. Because of the symmetry of the molecule, H_1 and H_3 were equivalent and turned out a unique signal at 3.39 ppm. However, as in the case of *t*B1GE-a, H_a and H_b presented slightly different chemical shifts because of their diasterotopic character. Indeed, nuclei $H_{1a,3a}$ and $H_{1b,3b}$ showed slightly different chemical shifts of 3.37 ppm and 3.42 ppm, respectively, with a large geminal coupling constant of J_{ab} = 8.97 Hz being J_{a2} = 5.92 Hz and J_{b2} = 5.06 Hz. Therefore, the signal for H_2 centered at 3.78 ppm actually corresponded to a triple triplet with J_{a2} = 5.92 Hz and J_{b2} = 5.06 Hz but could not be accurately resolved. The ^{13}C APT spectrum (Figure 3c) allowed the easy assignation of carbon atom signals, as indicated in Table 2.

The ^{1}H NMR spectrum of tB3GE presented a similar pattern to that of *t*B2GE-a. H_{1a} and H_{3a} appeared at 3.27 ppm, showing large coupling constants with H_{1b} and H_{3b}, J_{ab} = 9.22 Hz, and J_{a2} = 5.32 Hz, whereas H_{1b} and H_{3b} appeared at 3.37 ppm with J_{b2} = 5.92 Hz. The similarity for the J_2 coupling constants suggested that H_2 had the appearance of a well-defined quintuplet whose apparent coupling constant (J = 5.57 Hz) averaged J_{a2} and J_{b2}. The ^{13}C APT spectra for *t*B2GE-a and tB3GE were far simpler and allowed easier identification of the corresponding signals (see Appendix B). In the case of *t*B2GE-a, the signal at 27.52 ppm was attributed to the primary methyl groups on the *tert*-butyl moieties, the signal at 72.99 ppm to the quaternary carbon on the *tert*-butyl moieties, and that at 62.9 ppm to the secondary C_1 and C_3. Similarly, for *t*B3GE, the HMBC spectrum showed long-range correlations between the hydrogen signal at 1.17 ppm and the quaternary C atom at 72.56 ppm, as well as the hydrogen signal at 1.20 ppm and the quaternary C atom at 73.68 ppm, allowing the assignation of these quaternary C atoms.

Table 2. Chemical shifts for the carbon atoms of the *t*BGEs.

Ether	C_1	C_1–O–C–($\underline{C}H_3)_3$ [a]	C_1–O–$\underline{C}$–(CH$_3)_3$ [a]	C_2	C_2–O–C–($\underline{C}H_3)_3$ [a]	C_2–O–$\underline{C}$–(CH$_3)_3$ [a]	C_3
*t*B1GE-a	63.92	27.59	73.69	70.80	-	-	64.69
*t*B1GE-b	63.83	-	-	71.12	28.68	74.75	63.83
*t*B2GE-a	63.09	27.70	73.17	70.38	-	-	63.09
*t*B2GE-b	64.19	28.31	74.24	69.82	27.37	73.32	65.49
*t*B3GE	63.53	27.72	72.75	71.34	28.57	73.87	63.53

[a] Chemical shifts correspond to the carbon atom written in italics and underlined.

Elusive *t*B1GE-b and *t*B2GE-b have been recently identified by GC-MS [43], although their NMR characterization has not been reported. *t*B1GE-b, as in the case of *t*B2GE-a and *t*B3GE, shows a symmetry plane, so that a simple spectrum could be expected. Nevertheless, chemical shifts of H_2, H_a, and H_b were so close that signals corresponding to H_a and H_b were broad and appeared in the 3.60–3.68 ppm range, and the coupling constants could not be accurately determined. In the case of H_2, it appeared as an apparent quintuplet ($J = 4.67$ Hz) centered at 3.71 ppm. Concerning the *t*B1GE-b ^{13}C spectrum, it was recorded using an APT sequence that allowed the fast assignation of the signal at 28.68 ppm to the primary methyl carbon and the ones at 63.83 ppm and 71.12 ppm to the C_1 and C_3 secondary carbons and the C_2 tertiary carbon, respectively (see Appendix B). The long-range correlation of the signal at 1.24 ppm that corresponds to the CH$_3$ groups allowed the identification of a small signal at 74.75 ppm assigned to the quaternary carbon atom on the *tert*-butyl moiety.

Finally, the ^{1}H NMR spectrum for *t*B2GE-b showed four groups of signals in the glycerol skeleton and two singlets corresponding to the *tert*-butyl groups at 1.21 ppm and 1.19 ppm. The hydroxyl group appeared as a double doublet at 2.51 ppm ($J = 3.78$ Hz, $J = 7.94$ Hz) due to coupling with the diasterotopic H_3. This hydroxyl signal showed strong HSQC-TOCSY correlation with the carbon atom at 65.65 ppm that corresponded then to C_3 (Figure 4a). Once C_3 was assigned, the heteronuclear ^{1}H–^{13}C experiment combined with ^{13}C APT allowed easy assignation of C_1, C_2, H_1, and H_2 (Figure 4b). The hydrogen atom on the hydroxyl group presented a clear NOE effect with the hydrogen atoms on the *tert*-butyl group at 1.19 ppm (Figure 4c) that were hence assigned to the *tert*-butyl group on C_2. The long-range correlation in the HMBC spectrum (Figure 4d) allowed the identification of the quaternary carbon atoms.

The ^{1}H signals for H_2 and H_3 signals overlapped, making the resolution of the system difficult. As for H_1, two ^{1}H NMR signals were observed, the first centered at 3.41 ppm and the second at 3.34 ppm (Figure 5). The geminal coupling constant for H_{1a} and H_{1b} was 8.60 Hz, and they both coupled with H_2 with $J = 4.30$ Hz and $J = 8.58$ Hz, respectively, causing the triplet aspect of the signal at 3.34 ppm that indeed corresponded to a double doublet. The chemical shifts for H_2 and H_{3a} and H_{3b} were determined using the HSQC-TOCSY cross-signals with C_2 and C_3, respectively. The coupling constants for H_2 and H_3 needed to be determined using the spectrum simulation module in Topspin 3.6.2 and are gathered in Table 1.

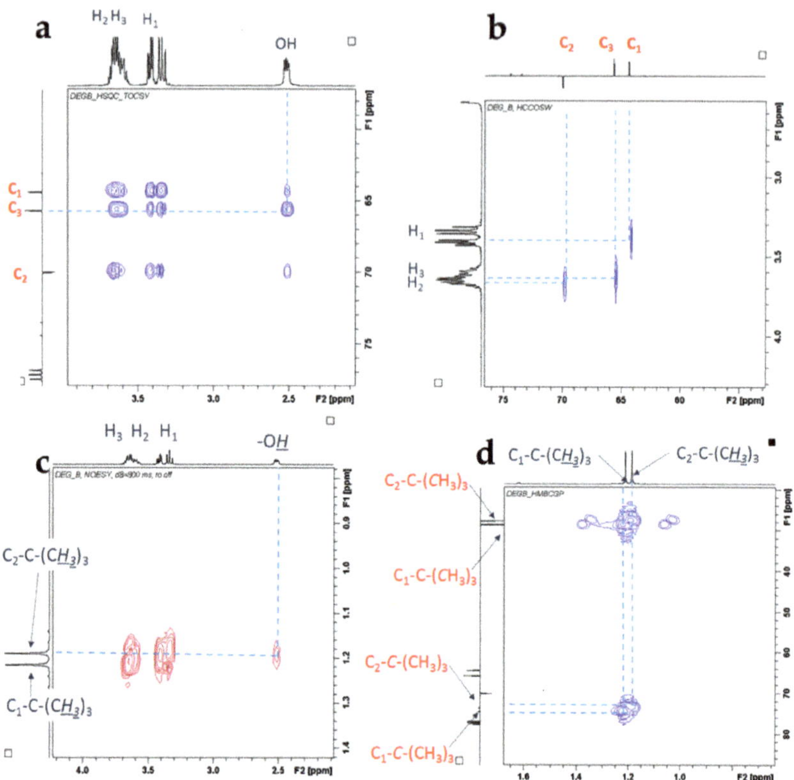

Figure 4. Details of the *t*B2GE-b spectra: (**a**) the HSQC-TOCSY spectrum, (**b**) the ^{1}H–^{13}C heteronuclear correlation, (**c**) the NOESY experiment, and (**d**) the HMBCGP experiment.

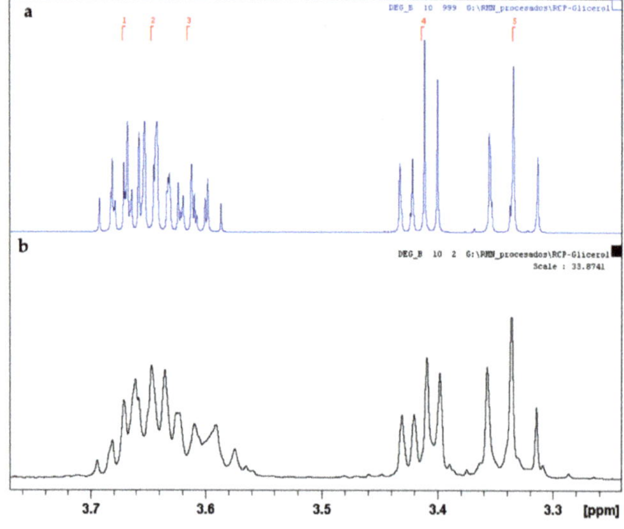

Figure 5. (**a**) Simulation of the *t*B2GE-b ^{1}H NMR spectrum using Topspin. (**b**) Detail of the *t*B2GE-b ^{1}H NMR spectrum.

2.2. Glycerol Tert-Butylation Monitoring through GC Analyses

Gas chromatography (GC) is the benchmark analytical technique for monitoring etherification reactions between glycerol and *tert*-butyl alcohol (TBA), although it presents several drawbacks. Polyols like glycerol have low vapor pressures that makes necessary the use of relatively harsh analysis conditions together with long analysis times that do not ensure the precise quantification of glycerol and, consequently, accurate mass balances. In addition, the complete derivatization by silylation of the reaction mixture is difficult and laborious and requires high quantities of specific reagents. Nevertheless, as shown in Figure 6, it allows the appropriate separation of the five *t*BGEs, glycerol, and the internal standard in ca. 20 min, although unreacted TBA cannot be quantified as it elutes with the solvent used to dilute the sample taken from the reaction mixture.

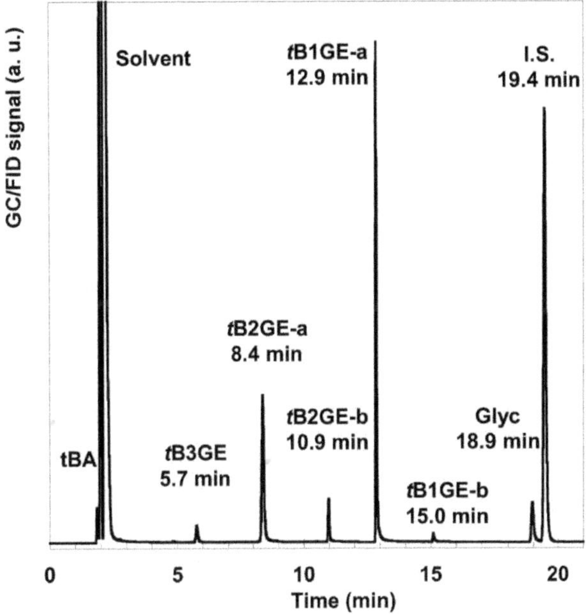

Figure 6. Typical GC-FID chromatogram of a *tert*-butylation mixture sample. Glyc stands for glycerol, and S.I. stands for internal standard.

Figure 7 shows the *t*BGE content of the reaction mixture expressed as molar fractions as a function of the glycerol conversion achieved after 24 h of *tert*-butylation reaction at temperatures between 70 and 110 °C, 8 wt.% PTSA, referred to the initial glycerol content and TBA/glycerol molar ratios within the 4:1–16:1 range. It can be observed that *t*B1GE-a and *t*B2GE-a were the main *tert*-butyl glycerol ethers produced. In addition, the *t*B1GE-a content was much higher than that of *t*B2GE-a, which only reached significant concentrations once the monoether was sufficiently abundant, in accordance to a reaction scheme in series. The formation of 1- and 1,3-ethers from the condensation of primary hydroxyl groups of glycerol to produce *t*B1GE-a and *t*B2GE-a was more probable against the formation of 2- and 1,2-ethers, *t*B1GE-b, and *t*B2GE-b [26]. Indeed, glycerol was a triol having double the number of primary than secondary hydroxyl groups. In addition, primary hydroxyls were preferred for *tert*-butylation due to steric effects because the *tert*-butyl group was a voluminous moiety. This explained in part the very low concentrations of the triether achieved, which was present in detectable amounts when the glycerol conversions reached values above ca. 0.75. Another reason was the thermodynamic limitations that appeared when TBA was used as the alkylating agent [18]. The *t*B1GE-b and *t*B2GE-b contents were also very low. The evolution of the molar fractions suggested that *t*B1GE-b disappeared to form *t*B2GE-b and that this diether reacted to form the triether *t*B3GE, as indicated by arrows in Figure 7.

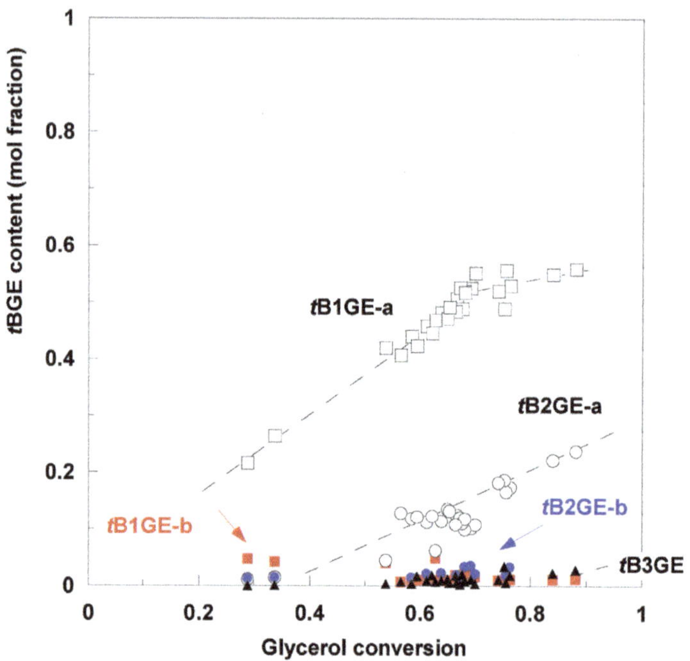

Figure 7. Molar fractions of the *t*BGEs according to GC-FID analyses of reaction mixtures after 24 h at 70–110 °C, 8 wt.% PTSA, referred to the initial glycerol content and TBA/glycerol molar ratios between 4:1 and 16:1.

2.3. Glycerol Tert-Butylation Monitoring through ^{1}H NMR Analyses

Given that the ^{1}H NMR signal was directly proportional to the amount of hydrogen atoms present in the sample, in principle, no calibration was required for the quantification of samples whose analysis required only 90 s. Initially, the ^{1}H NMR spectra of the isolated glycerol and the *t*B1GE-a, *t*B2GE-a, and *t*B3GE. *t*BGEs were superimposed (Figure 8a) in an attempt to find out a relation between the results of the integration of the different spectra regions and the molar fraction of each compound. However, the spectra of the reaction mixtures were slightly different from those corresponding to the isolated product superposition (Figure 8b). This was due to the differences in the solvent dielectric properties due to the presence of high amounts of TBA and glycerol in the reaction mixture. Hence, the spectra of the samples from the etherification reactions were much easier to interpret than those corresponding to the component superposition, which allowed the simplification of the quantitative analysis defining four integration regions (denoted as R_j, j = 1, 2, 3, 4) and assuming negligible the contributions of *t*B1GE-b and *t*B2GE-b. As for the rest of the compounds, their contributions to the several integration regions are gathered in Table 3. An obvious drawback of this procedure was that no distinction could be made between both monoethers and diethers.

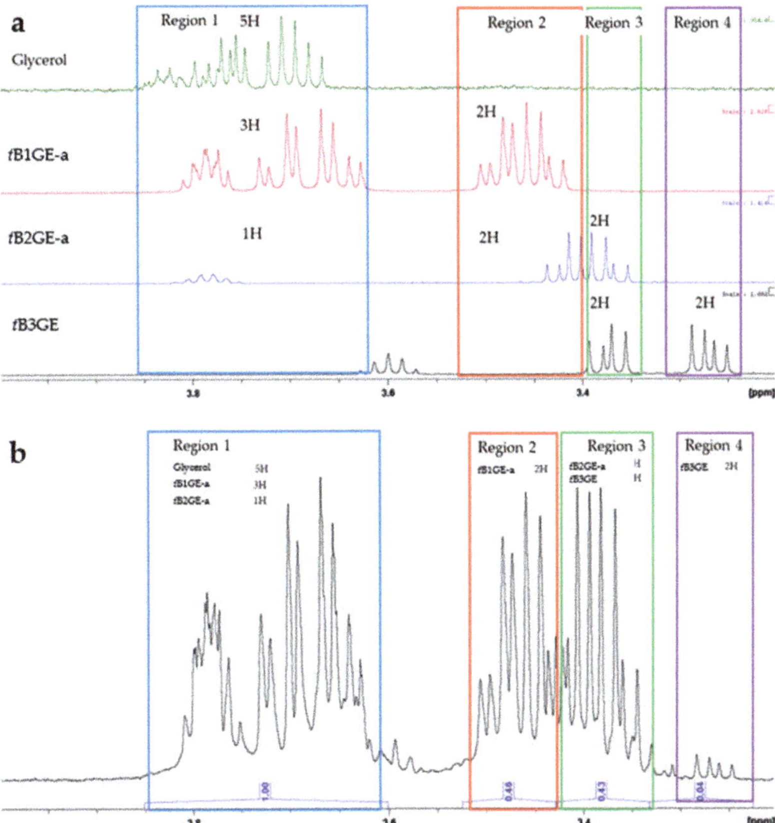

Figure 8. (**a**) Superimposed ^{1}H NMR spectra for glycerol, *t*B1GE-a, *t*B2GE-a, and *t*B3GE. (**b**) The ^{1}H NMR spectrum of a reaction sample.

Table 3. Contribution of the compounds indicated to the integration of the NMR spectra regions.

Region, R_j	δ (ppm)	Glycerol	*t*B1GE-a	*t*B2GE-a	*t*B3GE
1	3.850–3.604	5	3	1	-
2	3.524–3.429	-	2	-	-
3	3.429–3.333	-	-	4	2
4	3.333–3.225	-	-	-	2

Accordingly, the molar fractions of glycerol and the *t*BGEs were calculated from the values (n_i) given by Equations (1)–(6), which were proportional to the number of moles of each compound present in the sample. In these equations, IA_{Rj} (j = 1,2,3,4) are integration values corresponding to the region j according to Table 3. The conversion of glycerol (X_{Glyc}) and the *t*BGE selectivities (S_i) can be calculated according to Equations (5) and (6), respectively, considering that no products other than *t*BGEs and unreacted glycerol were present in the reaction mixture. Due to the abovementioned limitations of NMR analyses, in what follows, selectivities are reported for *t*B1GE and *t*B2GE that lump both monoethers and both diethers, respectively.

$$n_{t\text{B1GE}} = \frac{IA_{R2}}{2} \tag{1}$$

$$n_{tB2GE} = \frac{(IA_{R3} - IA_{R2})}{4} \tag{2}$$

$$n_{tB3GE} = IA_{R4} \tag{3}$$

$$n_{Glyc} = \frac{IA_{R1} - 3 \cdot n_{tB1GE} - n_{tB2GE}}{5} = \frac{(4 \cdot IA_{R1} - 6 \cdot IA_{R2} - IA_{R3} - IA_{R4})}{20} \tag{4}$$

$$X_{Glyc} = \frac{n_{tB_1GE} + n_{tB_2GE} + n_{tB_3GE}}{n_{Glyc} + n_{tB_1GE} + n_{tB_2GE} + n_{tB_3GE}} \tag{5}$$

$$S_i = \frac{n_i}{n_{tB_1GE} + n_{tB_2GE} + n_{tB_3GE}} \quad i = n_{tB1GE}, n_{tB2GE}, n_{tB3GE} \tag{6}$$

Figure 9 shows the relation between the selectivities to *tert*-butyl mono-, di-, and triethers of glycerol obtained, calculated from the results of the analyses performed by GC and ^{1}H NMR of the reaction samples. In general, a good agreement is observed; however, some samples led to larger discrepancies. The quantification of glycerol was identified as the main source of error, which reached ca. 5% and 7% for the NMR and GC analyses, respectively. In the case of the *t*BGEs, the errors were reduced to ca. 3% with both techniques. Higher errors could be associated to homogenization difficulties, particularly in samples with very low or very high glycerol conversions. In the first case, the high polarity and viscosity of glycerol complicated the sample manipulation. In the second case, the large difference in polarity between the reaction products, especially the di- and triethers, and that of the reactants led to the formation of micro-emulsions through phase segregation.

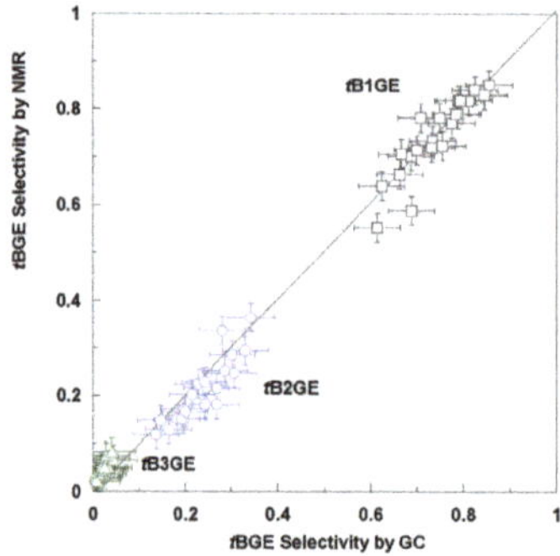

Figure 9. Selectivities to the *t*BGEs calculated from ^{1}H NMR and GC analyses.

Monitoring of the glycerol *tert*-butylation reactions through ^{1}H NMR has allowed illustrating the effects of some of the reaction conditions. In this regard, Figure 10 shows the influence on the glycerol conversion of the catalyst (PTSA) concentration after 24 h of reactions conducted at 70 °C and TBA/glycerol molar ratios of 4:1, 8:1, and 16:1. It can be seen that as expected, at a given TBA/glycerol ratio, the glycerol conversion increases at increasing PTSA concentration. For example, at the TBA/glycerol molar ratio of 4:1, the conversion increased from ca. 70% to 95% when the catalyst concentration passed from 8 wt.% to 32 wt.%. However, at a given catalyst content, the glycerol conversion decreased as the TBA/glycerol molar ratio increased. This was explained by the fact that the catalyst concentration was referred to the initial glycerol content of the reaction mixture. Therefore, the catalyst concentration over the total reaction volume decreased as the TBA/glycerol ratio increased due to the dilution caused by increasing amounts of TBA. For instance, when the catalyst concentration was fixed at 32 wt.% referred to the glycerol amount, the overall catalyst concentration decreased from 6.4 wt.% to 1.9 wt.% and finally 1.0% when the TBA/glycerol ratio increased from 4:1 to 8:1 and 16:1, respectively.

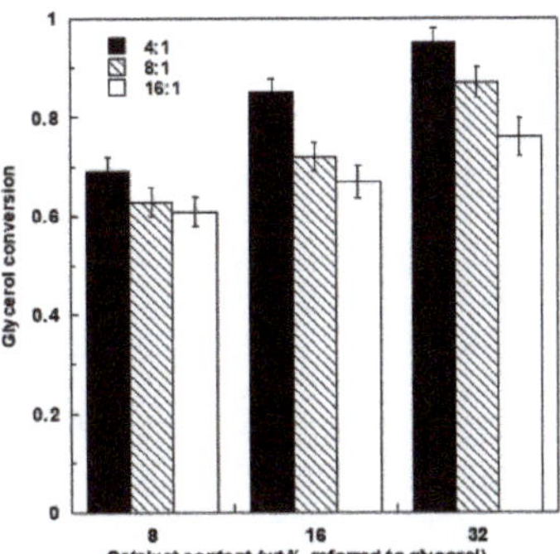

Figure 10. Glycerol conversion after 24 h of reaction as a function of the catalyst content.

Figure 11 shows the *t*BGE selectivities for the conversion points included in Figure 7. It can be seen that the selectivities were dictated by their own reaction progress, that is, the glycerol conversion. The catalyst content and TBA/glycerol ratio affected the conversion that could be achieved in a given reaction time, in this case, 24 h, but did not seem to influence the selectivity. In other words, the highest glycerol conversions attained corresponded to the reactions performed at the lowest TBA/glycerol ratio (4:1) and the highest PTSA concentration (32 wt.% referred to the initial glycerol amount) considered. High glycerol conversions were necessary to obtain the highest possible di- and triether selectivities. The first ones reached values of ca. 35% at their highest, whereas in the conditions of the present study, maximum *t*B3GE selectivities of ca. 8% were obtained. As concerns the monoethers, maximum conversions were obtained at the lowest glycerol conversion. In accordance with the in-series scheme that followed the *tert*-butylation reaction, the first products were proportionally more abundant at short reaction times (in batch processes), when they had little opportunity of being converted into higher ethers. As for the temperature, an effect similar to the rest of reaction variables was found, having a positive influence on the glycerol conversion but not affecting the *t*BGE selectivities.

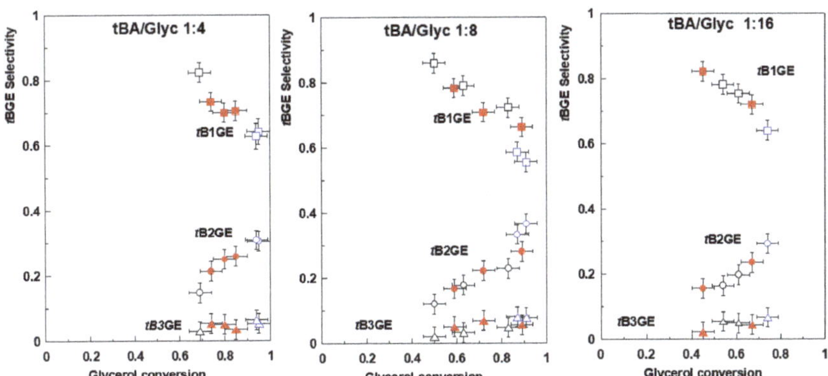

Figure 11. *t*BGE selectivities for glycerol etherification reaction conducted at TBA/glycerol molar ratios of 4:1 (**left**), 8:1 (**center**), and 16:1 (**right**). Catalyst (PSA) concentrations referred to the glycerol content were: 8 wt.% (open black symbols), 16 wt.% (filled red symbols), and 32 wt.% (open blue symbols).

2.4. Etherification of the Tert-Butyl Glycerol Monoether

With the purpose of increasing the yield of the higher *t*BGEs, the *tert*-butylation reaction was carried out starting from *t*B1GE instead of glycerol as indicated in Section 3.2. Figure 12 shows the evolution with reaction time of the *t*BGE molar fractions monitored through ^{1}H NMR (Figure 12a) and GC (Figure 12b). It was clear that the monoether converted into the diether without having a significant impact on the triether concentration. After the fifth day of reaction, the monoether conversion reached 38%; at that time, a new charge of TBA and catalyst was performed, aiming at further converting the monoether. The conversion increased up to 56% after two additional days of reaction. The GC analyses allowed distinguishing between both diethers, and the obtained results (Figure 12b) suggested that the triether was mainly formed from *t*B2GE-b. This seemed reasonable in view of the much stronger steric hindrance that would entail its formation from *t*B2GE-a. However, *t*B2GE-b was much less abundant than *t*B2GE-a, that is, the diether with both glycerol primary hydroxyl groups etherified. This showed that there were intrinsic difficulties in obtaining high selectivities to the *tert*-butyl glycerol triether through the homogeneously acid-catalyzed *tert*-butylation of glycerol with TBA.

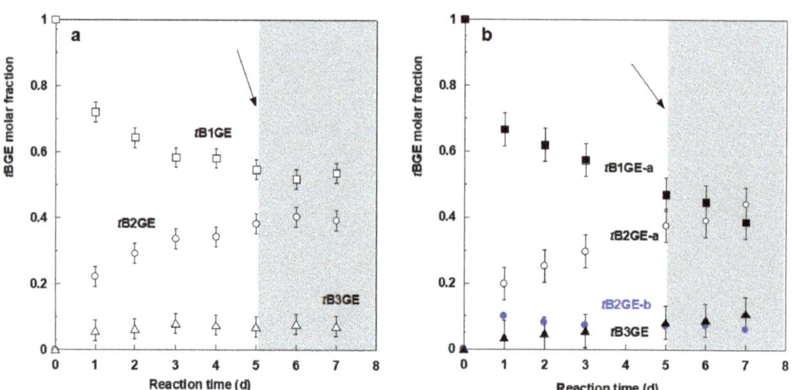

Figure 12. Evolution of *t*BGE concentration during the etherification with TBA of the *tert*-butyl glycerol monoether as monitored by (**a**) ^{1}H NMR and (**b**) GC. The arrows indicate the addition of a new charge of TBA and catalyst.

3. Materials and Methods

3.1. Materials and Analytical Techniques

tert-Butanol (TBA), anhydrous glycerol (99.5%), and 1,3,5-trimethoxybenzene (as the internal standard) were purchased from Acros Organics (Fairlawn, NJ, USA) *p*-Toluenesulfonic acid (PTSA) used as the homogeneous catalyst was purchased from Panreac S.L. (Darmstadt, Germany) CDCl$_3$ was purchased from Carlo-Erba (Val de Reuill, France) and used as received.

The attenuated total reflectance (ATR) infrared (IR) spectra were recorded on an Avatar 360 FT-IR spectrometer (ThermoFisher Scientific, Walthman, MA, USA). The gas chromatography (GC) analyses were performed on a Shimadzu gas chromatograph equipped (Kyoto, Tapan) with a flame ionization detector (FID) and a DB-23 (30 m, 0.32 mm ID, 0.25 μm) column. During the analyses, the oven temperature was kept for 10 min at 90 °C, then it was raised from 90 to 150 °C at a rate of 25 °C/min, and finally, it was maintained for 8 min at 150 °C. The samples for the GC analysis were prepared from 0.040 g of the reaction mixture that were diluted with 2.5 mL of a 2 g/L solution of 1,3,5-trimetoxybenzene in acetonitrile.

The nuclear magnetic resonance (NMR) analyses were performed on a Bruker Ascend 400 spectrometer (Rheinstetten, Germany) operated at 400 MHz and equipped with a PA

BBO 5 mm probe. All ^{1}H and ^{13}C chemical shifts were reported using the δ scale and were referenced to the residual signal of CHCl$_3$ at 7.26 ppm and that of CDCl$_3$ at 77.16. The pulse programs were the previously installed *zg30* for ^{1}H with 16 scans. CDCl$_3$ was the solvent of choice after discarding DMSO-d_6 and CD$_3$OD due to the overlapping of residual signals from CD$_2$HOD and H$_2$O, respectively, from those of the reaction products.

3.2. Tert-Butylation Reactions

Glycerol etherification reactions were performed in a 100 mL stainless steel stirred autoclave at 30 bar, TBA/glycerol molar ratios ranging between 4:1 and 16:1, catalyst concentrations of 8–32 wt.% PTSA referred to the glycerol loaded into the reactor, and temperatures within 70–90 °C. The samples were withdrawn from the reactors during the course of the reaction by means of the appropriate recirculating valves to maintain the pressure and agitation conditions.

An experiment was carried out using 1.5 g of the monoethers (*t*B1GEs), 10 g TBA and 0.24 g glycerol (TBA/glycerol molar ratio of 13.5:1), 70 °C, and 16 wt.% of PTSA catalyst referred to the *t*B1GEs. After five days of reactions, a new charge into the reactor of 10 g TBA and 0.24 g glycerol was carried out.

3.3. Isolation of the Tert-Butyl Ethers

In order to obtain the *t*BGEs for the NMR identification, a glycerol etherification reaction was conducted on the stainless steel autoclave at 90 °C and 30 bar with a TBA/glycerol molar ratio of 4:1 and 8 wt.% PTSA referred to the glycerol loaded into the reactor. After 24 h of reaction, the resulting mixture was concentrated in a rotary evaporator to remove the unreacted alcohol. Afterward, ca. 17 g of the *t*BGE/glycerol mixture were charged into a chromatography column using M60 silica as stationary phase. *t*B3GE, *t*B2GE-a, and *t*B2GE-b (see Figure 1) were separated using hexane/ethyl acetate (9:1) as the mobile phase, as reported by González et al. [42]. On the other hand, the *t*B1GE-a and *t*B1GE-b monoethers were eluted using a hexane/ethyl acetate (1:9) mixture.

4. Conclusions

Glycerol *tert*-butylation is a complex reaction leading to the formation of five glycerol *tert*-butyl ethers (*t*BGEs). All of them have practical interest: the monoethers as surfactants and components of cosmetics and pharmaceuticals and the di- and triethers as fuel additives. In order to suitably monitor the progress of the reaction between glycerol and *tert*-butanol (TBA), a method based on ^{1}H NMR analyses was developed in the present work that allowed for the quantification of unreacted glycerol and the *t*BGEs in only 90 s without the need for equipment calibration. These features are clear advantages compared with conventional GC analyses when fast, almost real-time, monitoring of the reaction is required. In contrast, the method was not able to distinguish between both monoether and both diether isomers, which were lumped into two groups of reaction products. For that reason, it was necessary to combine ^{1}H NMR and GC analyses to obtain a complete characterization of the reaction mixture.

The set of results available for the development of the analytical methods provided information of interest as concerns the formation of higher ethers. Glycerol *tert*-butylation is a consecutive reaction in which primarily formed monoethers lead to diethers that are finally converted into the triether. According to our results, the triether seemed to be formed from *t*B2GE-b instead of *t*B2GE-a due to easier access of the third *tert*-butyl moiety to a primary carbon atom than to a secondary one. However, the fact that *t*B2GE-a was much easier to form than *t*B2GE-b due to the higher reactivity of primary hydroxyls compared with the secondary ones, the double number of primary as compared with secondary hydroxyls present in glycerol, and steric effects explained the difficulties in forming the *tert*-butyl glycerol triether through this synthetic route.

Author Contributions: Conceptualization, L.M.G. and G.A.; methodology, A.C., I.R., and G.A.; formal analysis, A.C., I.R., and G.A.; investigation, A.C., I.R., I.C., G.A., and L.M.G.; resources, A.C., G.A., and L.M.G.; data curation, A.C., I.R., I.C., and G.A.; writing—original draft preparation, A.C., I.R., and G.A.; writing—review and editing, L.M.G. All authors have read and agreed to the published version of the manuscript.

Funding: This research received no external funding.

Data Availability Statement: Data will be made available upon request.

Conflicts of Interest: The authors declare no conflict of interest.

Appendix A

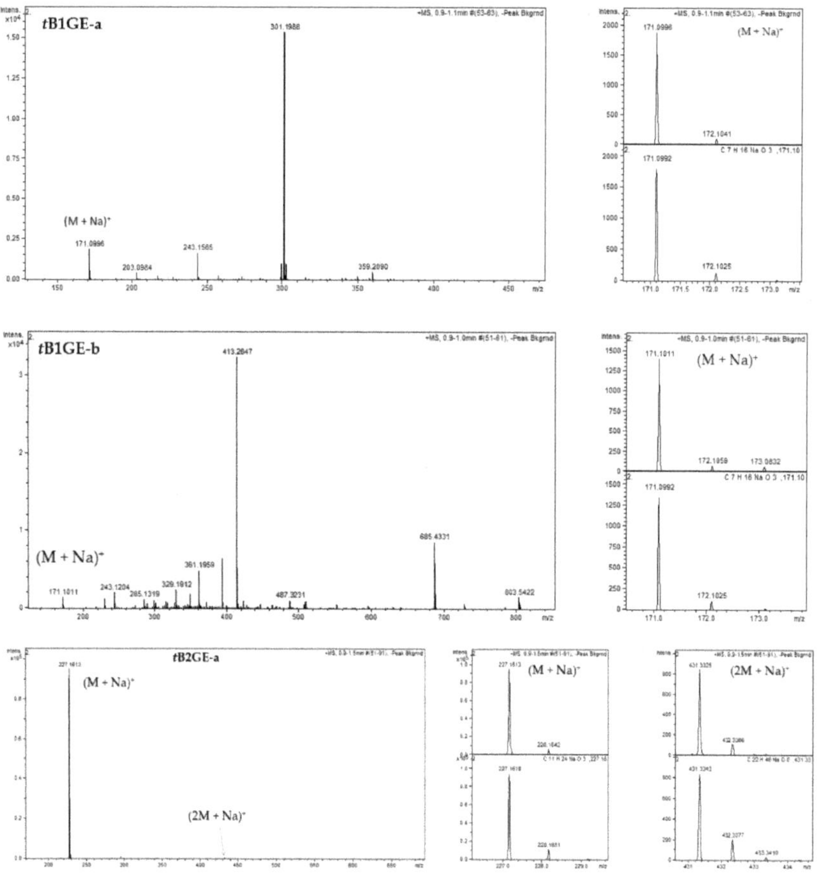

Figure A1. *Cont.*

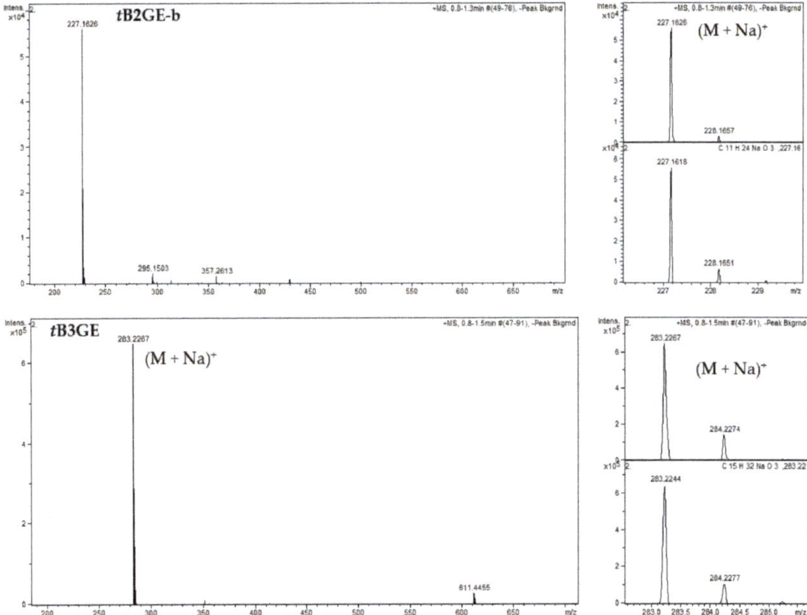

Figure A1. MS-ESI$^+$ spectra of *tert*-butyl ethers of glycerol.

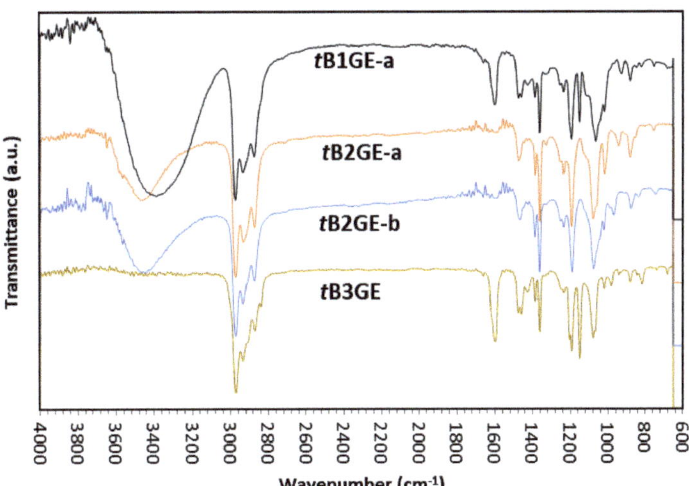

Figure A2. FTIR spectra of *tert*-butyl ethers of glycerol.

Appendix B

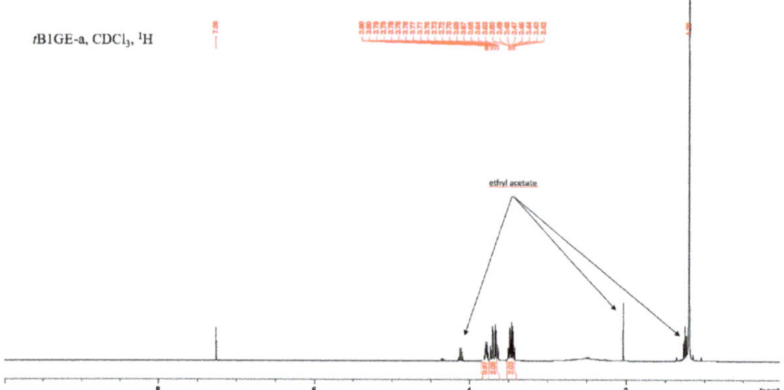

Figure A3. The ^{1}H NMR spectrum of *t*B1GE-a in CDCl$_3$.

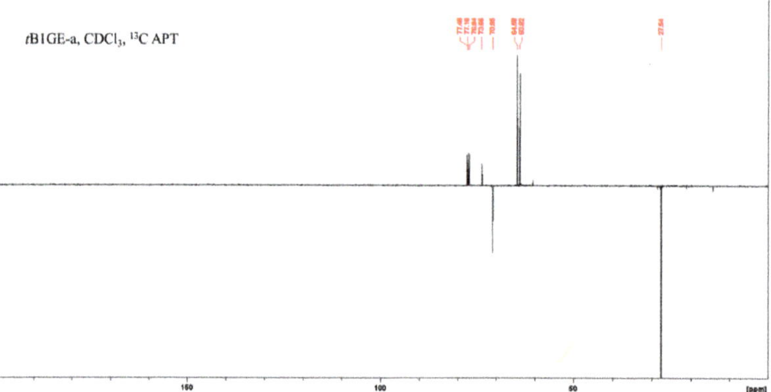

Figure A4. The ^{13}C APT NMR spectrum of *t*B1GE-a in CDCl$_3$.

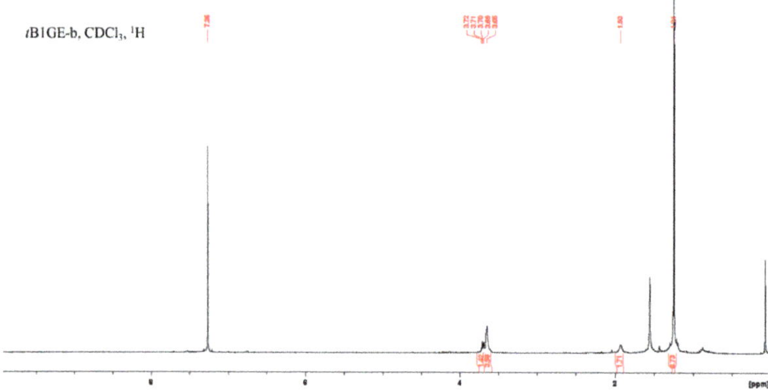

Figure A5. The ^{1}H NMR spectrum of *t*B1GE-b in CDCl$_3$.

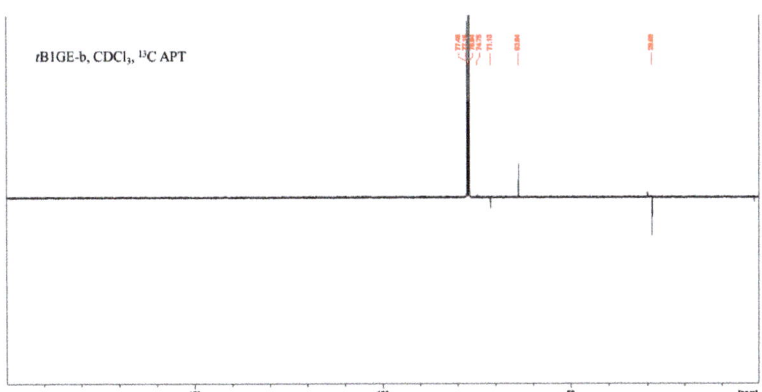

Figure A6. The ^{13}C APT NMR spectrum of *t*B1GE-b in CDCl$_3$.

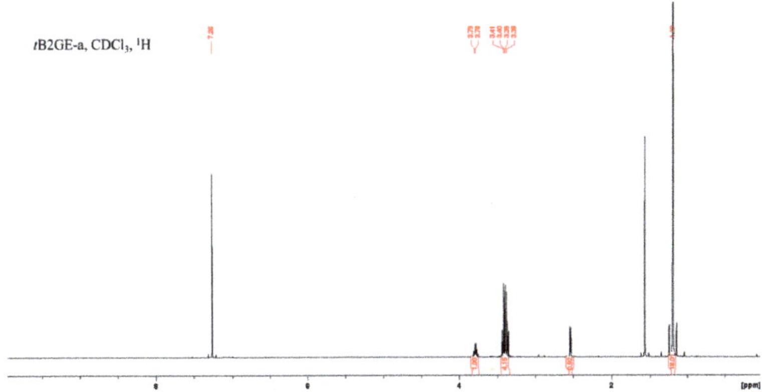

Figure A7. The ^{1}H NMR spectrum of *t*B2GE-a in CDCl$_3$.

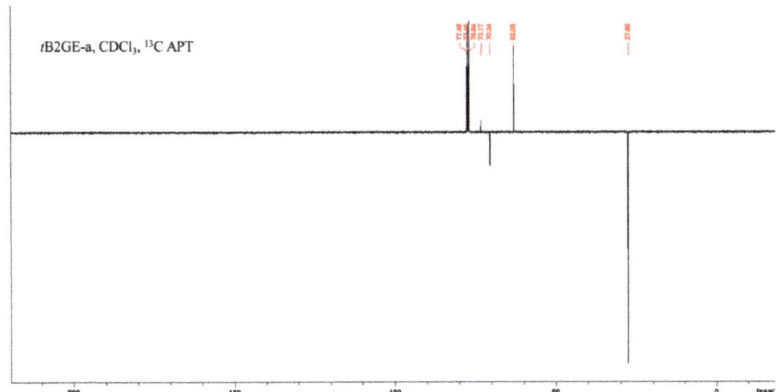

Figure A8. The ^{13}C APT NMR spectrum of *t*B2GE-a in CDCl$_3$.

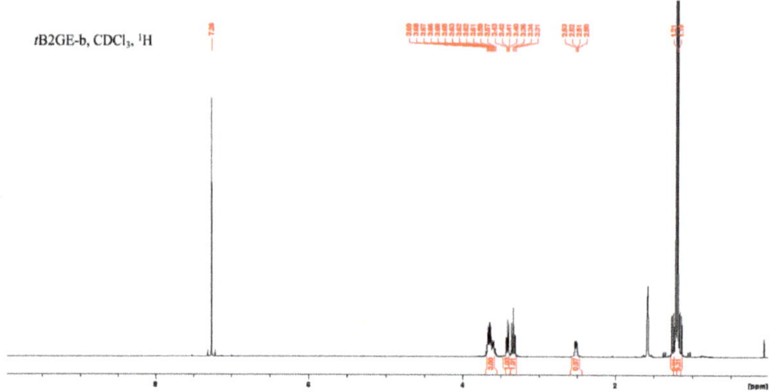

Figure A9. The ^{1}H NMR spectrum of *t*B2GE-b in CDCl$_3$.

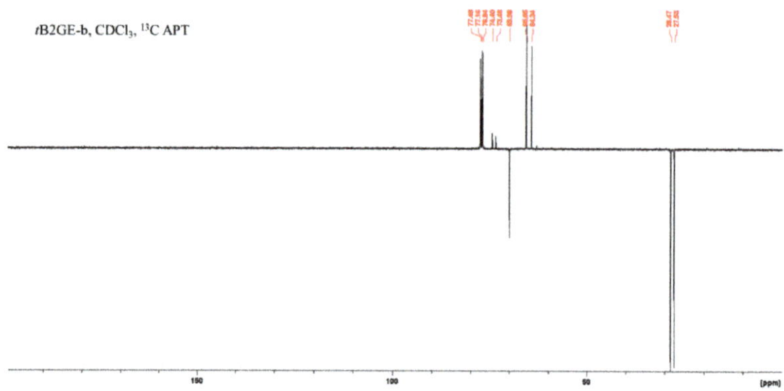

Figure A10. The ^{13}C APT NMR spectrum of *t*B2GE-b in CDCl$_3$.

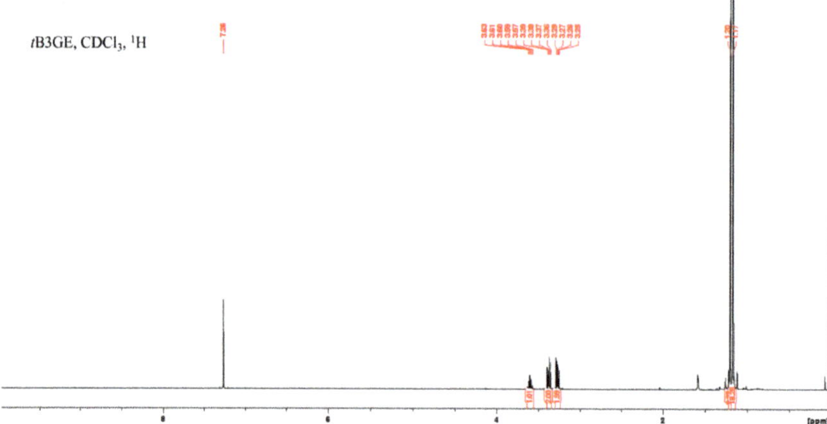

Figure A11. The ^{1}H NMR spectrum of *t*B3GE in CDCl$_3$.

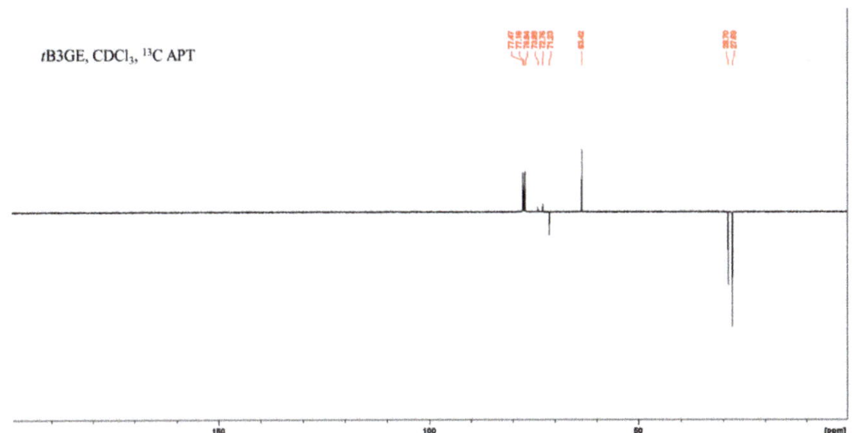

Figure A12. The ^{13}C APT NMR spectrum of *t*B3GE in CDCl$_3$.

References

1. ChemAnalyst. Glycerine Market Analysis. Available online: https://www.chemanalyst.com/industry-report/glycerine-market-635 (accessed on 7 August 2023).
2. International Energy Agency. *Renewables 2022. Analysis and Forecast to 2027*; International Energy Agency: Paris, France, 2022.
3. Morais Lima, P.J.; da Silva, R.M.; Chaves Girão Neto, C.A.; Câmara Gomes e Silva, N.; da Silva Souza, J.E.; Nunes, Y.L.; Sousa dos Santos, J.C. An overview on the conversion of glycerol to value-added products via chemical and biochemical routes. *Biotechnol. Appl. Biochem.* **2022**, *69*, 2794–2818. [CrossRef] [PubMed]
4. Checa, M.; Nogales-Delgado, S.; Montes, V.; Encinar, J.M. Recent Advances in Glycerol Catalytic Valorization: A Review. *Catalysts* **2020**, *10*, 1279. [CrossRef]
5. Pirzadi, Z.; Meshkani, F. From glycerol production to its value-added uses: A critical review. *Fuel* **2022**, *329*, 125044. [CrossRef]
6. Kaur, J.; Sarma, A.K.; Jha, M.K.; Gera, P. Valorisation of crude glycerol to value-added products: Perspectives of process technology, economics and environmental issues. *Biotechnol. Rep.* **2020**, *27*, e00487. [CrossRef] [PubMed]
7. Gujar, J.P.; Modhera, B. A review on catalytic conversion of biodiesel derivative glycerol to bio-olefins. *Mater. Today Proc.* **2023**, *72*, 2723–2730. [CrossRef]
8. Zhang, J.; Wang, Y.; Muldoon, V.L.; Deng, S. Crude Glycerol and glycerol as fuel and fuel additives in combustion applications. *Renew. Sust. Energ. Rev.* **2022**, *159*, 112206. [CrossRef]
9. Liu, Y.; Zhong, B.; Lawal, A. Recovery and utilization of crude glycerol, a biodiesel byproduct. *RSC Adv.* **2022**, *12*, 27997. [CrossRef]
10. Cornejo, A.; Barrio, I.; Campoy, M.; Lázaro, J.; Navarrete, B. Oxygenated fuel additives from glycerol valorization. Main production pathways and effects on fuel properties and engine performance: A critical review. *Renew. Sust. Energ. Rev.* **2017**, *79*, 1400–1413. [CrossRef]
11. Olson, A.L.; Tunér, M.; Verhelst, S. A concise review of glycerol derivatives for use as fuel additives. *Heliyon* **2023**, *9*, e13041. [CrossRef]
12. Queste, S.; Boudin, P.; Touraud, D.; Kunz, W.; Aubry, J.-M. Short chain glycerol 1-monoethers—A new class of solvo-surfactants. *Green Chem.* **2006**, *8*, 822–830. [CrossRef]
13. Gu, Y.; Azzouzi, A.; Pouilloux, Y.; Jérôme, F.; Barrault, J. Hetrogeneously catalyzed etherification of glycerol: New pathways for transformation of glycerol to more valuable chemicals. *Green Chem.* **2008**, *10*, 164–167. [CrossRef]
14. Sutter, M.; Da Silva, E.; Duguet, N.; Raoul, Y.; Métay, E.; Lemaire, M. Glycerol Ether Synthesis: A Bench Test for Green Chemistry Concepts and Technologies. *Chem. Rev.* **2015**, *115*, 8609–8651. [CrossRef] [PubMed]
15. Palanychamy, P.; Lim, S.; Yap, Y.H.; Leong, L.K. Critical Review of the Various Reaction Mechanisms for Glycerol Etherification. *Catalysts* **2022**, *12*, 1487. [CrossRef]
16. Behr, A.; Obendorf, L. Development of a Process for the Acid-Catalyzed Etherification of Glycerine and Isobutane Forming Glycerine Tertiary Butyl Ethers. *Eng. Life Sci.* **2002**, *2*, 179–208. [CrossRef]
17. Liu, J.; Yang, B. Liquid–Liquid–Solid Mass Transfer and Phase Behavior of Heterogeneous Etherification of Glycerol with Isobutene. *AIChE J.* **2018**, *64*, 2526–2535. [CrossRef]

18. Dominguez, C.M.; Romero, A.; Santos, A. Improved Etherification of Glycerol with *Tert*-Butyl Alcohol by the Addition of Dibutyl Ether as Solvent. *Catalysts* **2019**, *9*, 378. [CrossRef]
19. Klepáčová, K.; Mravec, D.; Bajus, M. *tert*-Butylation of glycerol catalysed by ion-exchange resins. *Appl. Catal. A Gen.* **2005**, *294*, 141–147. [CrossRef]
20. Melero, J.A.; Vicente, G.; Morales, G.; Paniagua, M.; Moreno, J.M.; Roldán, R.; Ezquerro, A.; Pérez, C. Acid-catalyzed etherification of bio-glycerol and isobutylene over sulfonic mesostructured silicas. *Appl. Catal. A Gen.* **2008**, *346*, 44–51. [CrossRef]
21. Zhao, W.; Yang, B.; Yi, C.; Lei, Z.; Xu, J. Etherification of Glycerol with Isobutylene to Produce Oxygenate Additive Using Sulfonated Peanut Shell Catalyst. *Ind. Eng. Chem. Res.* **2010**, *49*, 12399–12404. [CrossRef]
22. Zhao, W.; Yi, C.; Yang, B.; Hu, J.; Huang, X. Etherification of glycerol and isobutylene catalyzed over rare earth modified Hβ-zeolite. *Fuel Process. Technol.* **2013**, *112*, 70–75. [CrossRef]
23. González, M.D.; Salagre, P.; Taboada, E.; Llorca, J.; Cesteros, Y. Microwave-assisted synthesis of sulfonic acid-functionalized microporous materials for the catalytic etherification of glycerol with isobutene. *Green Chem.* **2013**, *15*, 2230–2239. [CrossRef]
24. González, M.D.; Salagre, P.; Mokaya, R.; Cesteros, Y. Tuning the acidic and textural properties of ordered mesoporous silicas for their application as catalysts in the etherification of glycerol with isobutene. *Catal. Today* **2014**, *227*, 171–178. [CrossRef]
25. Bozkurt, Ö.D.; Bağlar, N.; Çelebi, S.; Uzun, A. Screening of solid acid catalysts for etherification of glycerol with isobutene under identical conditions. *Catal. Today* **2020**, *357*, 483–494. [CrossRef]
26. Klepáčová, K.; Mravec, D.; Bajus, M. Etherification of Glycerol with *tert*-Butyl Alcohol Catalysed by Ion-Exchange Resins. *Chem. Pap.* **2006**, *60*, 224–230. [CrossRef]
27. Frusteri, F.; Arena, F.; Bonura, G.; Cannilla, C.; Spadaro, L.; Di Blasi, O. Catalytic etherification of glycerol by *tert*-butyl alcohol to produce oxygenated additives for diesel fuel. *Appl. Catal. A Gen.* **2009**, *367*, 77–83. [CrossRef]
28. Pico, M.P.; Romero, A.; Rodríguez, S.; Santos, A. Etherification of Glycerol by *tert*-Butyl Alcohol: Kinetic Model. *Ind. Eng. Chem. Res.* **2012**, *51*, 9500–9509. [CrossRef]
29. Celdeira, P.A.; Gonçalves, M.; Figueiredo, F.C.A.; Dal Bosco, S.M.; Mandelli, D.; Carvalho, W.A. Sulfonated niobia and pillared clay as catalysts in etherification reaction of glycerol. *Appl. Catal. A Gen.* **2014**, *478*, 98–106. [CrossRef]
30. Gonçalves, M.; Soler, F.C.; Isoda, N.; Carvalho, W.A.; Mandelli, D.; Sepúlveda, J. Glycerol conversion into value-added products in presence of a green recyclable catalyst: Acid black carbon obtained from coffee ground wastes. *J. Taiwan Inst. Chem. Eng.* **2016**, *60*, 294–301. [CrossRef]
31. Srinivas, M.; Raveendra, G.; Parameswaram, G.; Sai Prasad, P.S.; Lingaiah, N. Cesium exchanged tungstophosphoric acid supported on tin oxide: An efficient solid acid catalyst for etherification of glycerol with *tert*-butanol to synthesize biofuel additives. *J. Mol. Catal. A Chem.* **2016**, *413*, 7–14. [CrossRef]
32. Estevez, R.; López, M.I.; Jiménez-Sanchidrián, C.; Luna, D.; Romero-Salguero, F.J.; Bautista, F.M. Etherification of glycerol with *tert*-butyl alcohol over sulfonated hybrid silicas. *Appl. Catal. A Gen.* **2016**, *526*, 155–163. [CrossRef]
33. Melero, J.A.; Vicente, G.; Morales, G.; Paniagua, M.; Bustamante, J. Oxygenated Compounds Derived from Glycerol for Biodiesel Formulation: Influence on EN 14214 Quality Parameters. *Fuel* **2010**, *89*, 2011–2018. [CrossRef]
34. Lee, H.J.; Seung, D.; Jung, K.S.; Kim, H.; Filimonov, I.N. Etherification of glycerol by isobutylene: Tuning the product composition. *Appl. Catal. A Gen.* **2010**, *390*, 235–244. [CrossRef]
35. Altaner, C.M.; Saake, B. Quantification of the chemical composition of lignocellulosics by solution ^{1}H NMR spectroscopy of acid hydrolysates. *Cellulose* **2016**, *23*, 1003–1010. [CrossRef]
36. De Souza, A.C.; Rietkerk, T.; Selin, C.G.M.; Lankhorst, P.P. A robust and universal NMR method for the compositional analysis of polysaccharides. *Carbohydr. Polym.* **2013**, *95*, 657–663. [CrossRef] [PubMed]
37. Mittal, A.; Scott, G.M.; Amidon, T.E.; Kiemle, D.J.; Stipanovic, A.J. Quantitative analysis of sugars in wood hydrolyzates with ^{1}H NMR during the autohydrolysis of hardwoods. *Bioresour. Technol.* **2009**, *100*, 6398–6406. [CrossRef] [PubMed]
38. Cornejo, A.; Alegria-Dallo, I.; García-Yoldi, Í.; Sarobe, Í.; Sánchez, D.; Otazu, E.; Funcia, I.; Gil, M.J.; Martínez-Merino, V. Pretreatment and enzymatic hydrolysis for the efficient production of glucose and furfural from wheat straw, pine and poplar chips. *Bioresour. Technol.* **2019**, *288*, 121583. [CrossRef]
39. Cornejo, A.; Bimbela, F.; Moreira, R.; Hablich, K.; García-Yoldi, Í.; Maisterra, M.; Portugal, A.; Gandía, L.M.; Martínez-Merino, V. Production of Aromatic Compounds by Catalytic Depolymerization of Technical and Downstream Biorefinery Lignins. *Biomolecules* **2020**, *10*, 1338. [CrossRef]
40. Talavera-Prieto, N.M.C.; Ferreira, A.G.M.; Moreira, R.J.; Portugal, A.T.G. Monitoring of the Transesterification Reaction by Continuous Off-Line Density Measurements. *Fuel* **2020**, *264*, 116877. [CrossRef]
41. Kundu, R.; De, S. Characterization and analysis of the triglyceride transesterification process. *Biomass Convers. Biorefinery* **2023**, *13*, 4933–4948. [CrossRef]
42. González, M.D.; Cesteros, Y.; Llorca, J.; Salagre, P. Boosted selectivity toward high glycerol tertiary butyl ethers by microwave-assisted sulfonic acid-functionalization of SBA-15 and beta zeolite. *J. Catal.* **2012**, *290*, 202–209. [CrossRef]
43. Veiga, P.M.; Dias, A.G.; Henriques, C.A. Identification of Ethyl and t-Butyl Glyceryl Ethers Using Gas Chromatography Coupled with Mass Spectrometry. *J. Braz. Chem. Soc.* **2018**, *29*, 1328–1335. [CrossRef]

44. Jamróz, M.E.; Jarosz, M.; Witowska-Jarosz, J.; Bednarek, E.; Tecza, W.; Jamróz, M.H.; Dobrowolski, J.C.; Kijeński, J. Mono-, Di-, and Tri-Tert-Butyl Ethers of Glycerol: A Molecular Spectroscopic Study. *Spectroc. Acta Part A Mol. Biomol. Spectr.* **2007**, *67*, 980–988. [CrossRef] [PubMed]

45. Reich, H.J. WinDNMR:Dynamic NMR Spectra for Windows. *J. Chem. Educ.* **1995**, *72*, 1086. [CrossRef]

 *catalysts*

Article

Optimizing Al and Fe Load during HTC of Water Hyacinth: Improvement of Induced HC Physicochemical Properties

Mara Olivares-Marin [1], Silvia Román [2,*], Beatriz Ledesma [2] and Alfredo Álvarez [3]

[1] Department of Mechanical, Energetic and Materials Engineering, Merida Universitary Center, University of Extremadura, Santa Teresa de Jornet Avenue, 38, 06800 Mérida, Spain

[2] Department of Applied Physics, Industrial Engineering School, University of Extremadura, Elvas Avenue, 06006 Badajoz, Spain

[3] Department of Electrical Engineering, Industrial Engineering School, University of Extremadura, Elvas Avenue, 06006 Badajoz, Spain

* Correspondence: sroman@unex.es; Tel.: +34-924289600

Abstract: Nowadays, several alternatives have been proposed to increase the porosity and/or modify the surface groups of hydrochars from biomasses as well as to develop additional features on them. These alternatives can include specific modifications for the process, as previous steps or as postreatments, and the wide variety of forms in which they can be made can substantially affect the product distribution and properties. In this study, the hydrothermal carbonization process of an invasive floating plant (Water hyacinth) has been modified by introducing different amounts of iron ($FeCl_3$) and aluminium alloy (shaving scrap waste) during the hydrothermal reaction. The effects on process reactivity, phase distribution, and physicochemical properties of the samples obtained were studied by means of different characterization techniques such as thermogravimetry (TG-DTG), physical adsorption/desorption of N_2 at -196 °C, FT-IR spectroscopy, and scanning electron microscopy (SEM). In the case of iron-catalyzed reactions, the magnetite formation and magnetic behavior of the prepared hydrochars after a pyrolytic treatment was also estimated. The results obtained indicate that the porosity of the hydrochars was clearly improved to different extents by the addition of Al or Fe during direct synthesis. In addition, porous carbons with a moderate magnetic character were obtained.

Keywords: hydrothermal carbonization; magnetism; catalyst load; enhancement of porosity

Citation: Olivares-Marin, M.; Román, S.; Ledesma, B.; Álvarez, A. Optimizing Al and Fe Load during HTC of Water Hyacinth: Improvement of Induced HC Physicochemical Properties. *Catalysts* **2023**, *13*, 506. https://doi.org/10.3390/catal13030506

Academic Editor: Michael Renz

Received: 29 December 2022
Revised: 22 February 2023
Accepted: 27 February 2023
Published: 28 February 2023

1. Introduction

Hydrothermal carbonization (HTC) is an exciting valorization technique for converting organic solid waste into valuable products (a solid carbonaceous solid product called hydrochar (HC) and a liquid solution containing many valuable compounds) at relatively low temperatures (150–250 °C) and autogenous pressure [1]. HC can be compared to the char obtained by pyrolysis, because it is a coal-like material with incipient porosity. However, both techniques differ in several aspects, such as energy involved, need of water or activating agent, and properties of liquid phase. In relation to the carbon material, one important difference between a char and a HC is that in general, the former has cleaner pores, as the high temperatures usually associated with the flow of an inert gas help to remove the volatile content of the precursor, leaving behind an incipient porosity (which can be less or cleaner or have variable pore size depending on operating conditions but is also susceptible to being developed by subsequent activation). Chars obtained by pyrolysis usually have a greater C content than HC (i.e., greater heating value), mainly because more oxygen is removed compared to HTC. In addition, the thermal removal of oxygen functionalities and the aromatization developed in char can contribute to the development of a surface chemistry in which, in general, the electrons are more delocalized, and the number of acidic groups is reduced [2].

In contrast, HTC takes place not in an inert atmosphere but rather in a very reactive one in which water becomes a solvent and hydrolyzes the biomass, giving rise to an acidic system where acids are generated and enhance many different biomass degradation reactions (decarboxylation, demethanation, decarbonylation, etc.) so that HCs are generally highly acidic [1]. In addition, during HTC a significant part of the molecules that are formed at the liquid phase can recombine and yield macromolecules that eventually gather to form microspheres that are adsorbed on the HC [3]. This in turn can partially block porosity and also can contribute to increasing the solid yield, especially if long times are used [4]. To improve the low porosity volume of HCs, these materials are often subjected to postheating steps, sometimes under an inert agent [5] or by physical or chemical activation [6].

As the number of researchers working on HTC increases, the number of studies to better understand the nature of the degradation and recombination processes underlying the formation of primary or secondary HCs and the effect of variables apart from the typical (biomass load, time, and temperature) on the product distribution and properties also increases. One aspect that has recently attracted the attention of scientists is how the presence of mineral matter can affect the HTC process [7,8]. This mineral matter can either be part of the starting material or can be added to the process following diverse methodologies (preaddition to biomass by blending or adsorption and incorporation to process water). Mineral matter has been found to affect the process kinetics, reaction pathways, and heat of reactions as well as physicochemical properties of liquid and HC [8].

The addition of metals to the HTC reaction media as well as the introduction of postreatments such as pyrolysis and activation has been proven to affect the structural properties, elemental composition, and chemical surface functionalities; all these changes aimed to upgrade the solid carbon material towards subsequent applications. For example, HCs to be used as fuels should have low N or S and adequate ash composition [6]; if the material is designed to be used as an adsorbent, the development of specific chemical groups and porosity can be specifically interesting for favoring adsorption selectivity [9]; if the final use of the product is as soil amendment, the capture of certain nutrients on the HC is the aim [10].

Previous research has concluded that adding iron(III) chloride ($FeCl_3$) to a HTC system can be an effective way to incorporate Fe to a biomass HC, as compared to other ways of incorporating this metal such as ferromagnetic fluid and Fe particles. In a previous work, the authors found that almond shell $FeCl_3$-assisted HTC followed by pyrolysis could yield microporous magnetic carbon materials [5]. Other sources of Fe, such as Fe_3O_4 particles, have also been embedded to biomass HCs during hydrothermal treatment, and their presence was positive during further CO_2 activation, enhancing the development of microporosity [11]. Both $FeCl_3$ and Fe_3O_4 not only improved the carbon structure but also conferred the HCs with magnetism, after thermal treatment. Magnetism is a very desirable property in an adsorbent, especially when it is going to be used in aqueous solutions. Removing the activated carbon when it is saturated by just approaching a magnet facilitates management and has attracted scientific research [12].

In relation to the use of Al as a catalyst or porosity enhancement agent during HTC, very few scientific references are found. Al has been usually added in the form of salts, such as $Al_2(SO_4)_3$ [6] and $AlCl_3$ [13]. To the authors' knowledge, Al alloys, as a waste generated in the mining industry (particulate matter), have not been used as an additive in HTC processes before.

Eichhornia crassipes, also known as water hyacinth (WH), has been listed as one of the most dangerous invasive plants in the world. This weed is present in many fluvial areas all over warm places of Africa, Asia, and America and absorbs both nutrients and oxygen from water while also preventing sunlight from reaching the deep into bodies of water, resulting in the death of many living species [14]. In addition, the plant obstructs fishing operations in many areas, affecting their economy, which is very significant in the case of African countries in the vicinity of Victoria Lake, where it has also been associated with the propagation of plagues. Finally, WH's presence has damaged devices from hydric plants [13]. The challenge of making WH disappear without using chemicals that are toxic to other animals has moved the scientific community to find ways of managing this biomass involving thermal treatment to avoid spore dissemination. In Spain, fighting this weed has become a national priority, as is the case also in many other areas of the world, and the solution does not seem to be easy because it is extremely resistant (seeds can live for more than 20 years), has a very high growing rate, and can reproduce and spread very quickly [15]. In the basin of Guadiana River in southwest Spain, EUR 50 million was spent between 2004 and 2020 where WH had expanded over a length of 185 km along the river (total perimeter affected = 630 km) [16].

In this study, the use of this novel precursor (WH) has been investigated to produce, by Al- and Fe-catalyzed HTC, porous carbon materials. In both cases, for the first time how the catalyst load can affect process reactivity and structural and chemical changes of HCs and, in the case of Fe-assisted processes, additional potential magnetization due to further pyrolysis of the HCs was evaluated.

2. Results

2.1. Reactivity of Processes

Tables 1 and 2 display the solid yield values (SY, %) obtained for uncatalyzed Fe and Al-catalyzed HTC experiments. These results can be better analyzed in the precursor biomass degradation analyses that have been plotted in Figure 1.

Table 1. Solid yield (SY, %) and N_2 adsorption textural parameters of uncatalyzed HC and Al-catalyzed HCs.

	Solid Yield (%)	S_{BET}, m^2g^{-1}	V_{mi}, cm^3g^{-1}	V_{me}, cm^3g^{-1}	V_T, cm^3g^{-1}	S_{ext}, m^2g^{-1}
WH-HC	26.8	23	0.005	0.019	0.024	26
WH-HC-Al5	20.6	38	0.005	0.023	0.028	42
WH-HC-Al10	28.6	35	0.005	0.033	0.038	39
WH-HC-Al20	27.0	49	0.010	0.045	0.055	53

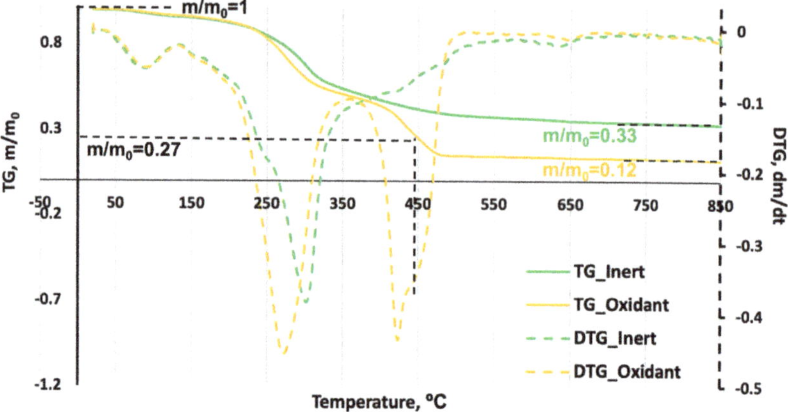

Figure 1. TGA and DTG curves of WH leaves under inert and oxidant atmosphere.

According to thermogravimetric analysis (TGA), under pyrolytic conditions, WH degrades mainly in the range 240–400 °C, with a residual degradation prevailing up to 550 °C and a very slight mass loss up to 850 °C (fixed carbon plus ash content can be estimated as residual mass at this temperature, that is, around 33%). When air is used as carrier gas, the degradation profiles change significantly; at the TGA curve, there are several slopes with two main significant degradation rates in the temperature ranges: 244–320 °C (this is close to the inert analysis) and 390–472 °C (here, the degradation behavior clearly differs from inert analyses). In fact, after this drop the WH residual mass reaches a value that is similar to the one found for HTC (26.8%). Final residual mass percentage upon combustion is 12%.

This suggests that HTC under the conditions here studied represents a middle situation between pyrolysis and combustion. This is consistent with the fact that volatile matter (most of it) was degraded during HTC, but additional reactions removed a part of the biomass that was resistant to pyrolysis because of the joint effect of water under subcritical conditions (H_3O^+) and many reactive molecules resulting from hemicellulose and cellulose degradation (mainly organic and inorganic acids). The presence of such molecules makes the system highly reactive and, as a whole, involve a complex combination of many types of degradation reactions (dehydration, decarboxylation, demethanation, decarbonylation, etc.) [1]. Other biomass materials (including water hyacinth stem and many lignocellulosic precursors) have given, under the same temperature, time, load, and reactor conditions, greater values of HTC solid yield [17].

2.2. Aluminium-Catalyzed HTC

Figure 2 shows the adsorption/desorption isotherms of the HCs obtained from uncatalyzed and Al-catalyzed WH leaves through HTC, while typical porosity parameters have been included in Table 1. According to these results, the addition of the Al alloy during HTC caused a slight change on the porous structure of the HCs. In the first place, the uncatalyzed process, as expected from most previously published studies on HTC, does not yield a carbon material with noticeable porosity development (S_{BET} of 23 m^2g^{-1}, with an almost negligible contribution of microporosity because of pore blockage by adsorbed, condensed, or polymerized biomass fragments) [1]. The addition of Al resulted in a slight improvement in porosity, which was more significant with the addition of a greater amount of metal (20 g) during HTC. The rise in N_2 adsorption at greater relative pressure values (Figure 2) suggest that Al was specifically effective in opening mesopores and external areas; total pore volume (V_T) was almost doubled by adding 20 g of Al, as compared to the pristine HC (0.055 vs. 0.024 cm^3g^{-1}). In fact, if one focuses on the lower relative pressure zone, it can be observed that the increase in N_2 adsorption is more abrupt for the noncatalyzed sample.

Other aluminum compounds, such as $AlCl_3$, added during the HTC of other biomass materials have also been effective in opening micropores, yielding mesoporous materials, with N_2 adsorption isotherms growing along the whole range of relative pressure, as is reported in the work of Liu et al. [13]. These authors, however, subjected the materials to subsequent aerobic calcination to favor the formation of Al-oxide active sites.

Al in the liquid phase under subcritical conditions has been reported to be converted into hydroxide ($2Al + 6H_2O \rightarrow 2Al(OH)_3 + 3H_2$), which in turn might have catalyzed the hydrogasification of the HC amorphous surface carbon ($C + 2H_2 \rightarrow CH_4$) [18].

Surface chemical functionalities, as determined by FT-IR spectrometry, suggested that the hydrothermal process involved the breaking up of cellulose and hemicellulose (compare signals of WH leaves, uncatalyzed HC, and HCs obtained using different amount of Al during HTC, plotted in Figure 3). Assignation of bands was based on previous studies [19].

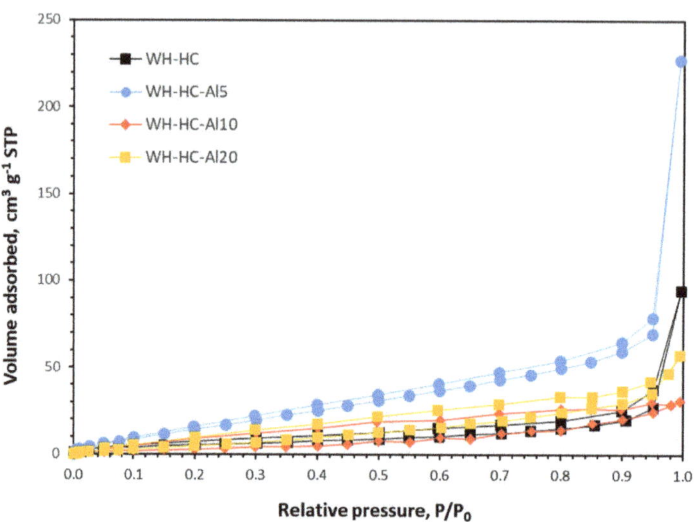

Figure 2. N_2 adsorption/desorption isotherms at $-196\ ^\circ$C of uncatalyzed HC and HCs prepared by Al-catalyzed HCs.

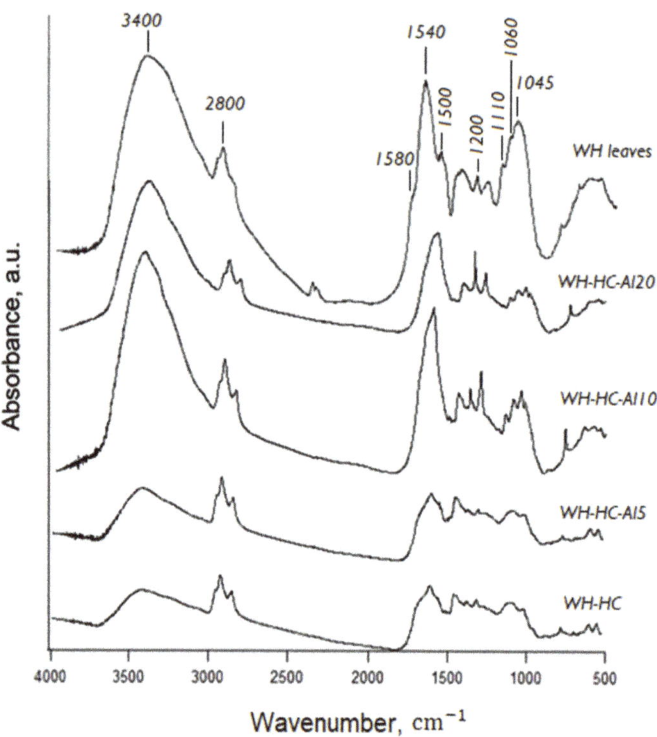

Figure 3. FT-IR spectra of WH, uncatalyzed HC and HCs prepared by Al-assisted HTC.

For instance, the loss of cellulose-type spectral bands at the pristine precursor are removed after HTC; that is the case of C-O bonds in alcohols of pyranose rings 1110, 1060, and 1045 cm^{-1} (see Figure 3). Moreover, the signals at wave numbers in the range

1600–1500 cm^{-1} found for hydrothermal-treated samples (and not for WH leaves) can be assigned to C=O vibrations on olefinic and/or aromatic structures, as also happens for C-O vibrations, typical for eter structures (1200 cm^{-1}), that are only visible for HCs.

The addition of Al alloy under these conditions to the reaction medium do not have a remarkable effect on the number, intensity, and position of the spectral bands of the HCs under any concentration, as compared to the uncatalyzed HTC run.

SEM analyses (Figure 4) revealed that the presence of Al alloy on the liquid–solid system altered the surface morphology of the HCs. In the case of the uncatalyzed reaction (WH-HC), the surface is full of pleats, with a low opening of pores and the presence of some microspheres. Other research [17] made under similar uncatalyzed HTC conditions have shown greater abundancy of these spheres; this might indicate that condensation reactions were less prominent in the present case or that their deposition on the primary HC was less facilitated because of diffusion of electrostatic resistance.

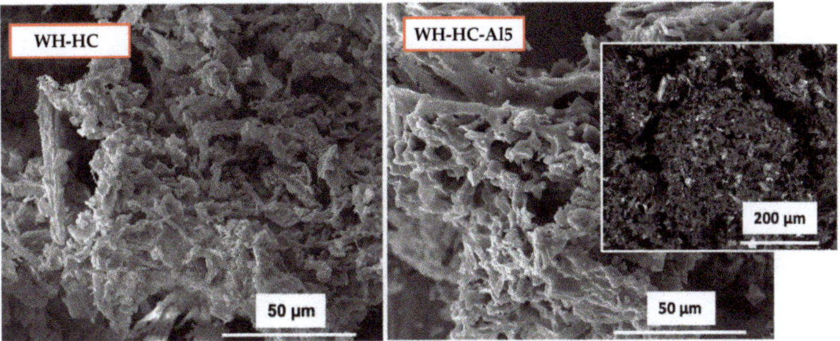

Figure 4. SEM micrographs of uncatalyzed HC and WH-HC-Al5. Embebed image on the right is a retro-dispersed electron image of WH-HC-Al5.

The images obtained with the retro-dispersed electron detector show small bright particles of different size and shape heterogeneously incorporated on the carbon matrix (see image inserted on the right in Figure 4). EDAX quantification analyses of the WH-HC-Al5 particles showed that apart from carbon (62.0%, wt./wt.) and oxygen (22.6%), metals such as calcium (1.5%), aluminum (1.7%), and phosphorous (0.9%) were present in the material.

2.3. Iron-Catalyzed HTC

Table 2 lists the SY values (both after HTC, SY, and after subsequent pyrolysis, WSY) as well as HC typical porosity features, as determined from N$_2$ adsorption isotherms at −196 °C (plotted in Figure 5). The addition of Fe and the further thermal treatment resulted in a greater reactivity and lower SY in relation to the uncatalyzed run except for the experiment made under the lowest Fe concentration (WH-HC-Fe0.25-600), which showed a rise in SY.

Table 2. Solid yield (SY, %) and N$_2$ adsorption textural parameters of Fe-catalyzed HTC.

	SY (%)	WSY (%)	S_{BET}, m^{2}g^{-1}	V_{mi}, cm^{3}g^{-1}	V_{me}, cm^{3}g^{-1}	V_T, cm^{3}g^{-1}	S_{ext}, m^{2}g^{-1}
WH-HC	26.8	-	23	0.005	0.019	0.024	33
WH-HC-Fe0.25-600	33.2	13.73	229	0.101	0.099	0.200	80
WH-HC-Fe0.5-600	22.3	11.15	76	0.023	0.075	0.098	78
WH-HC-Fe0.75-600	22.7	11.55	66	0.018	0.054	0.072	69

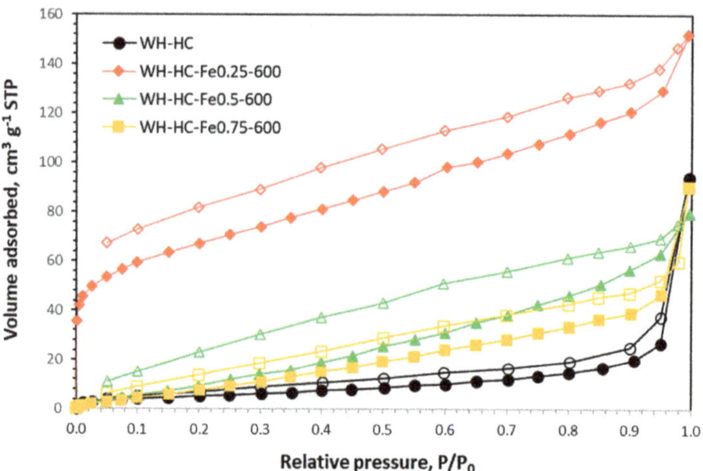

Figure 5. N_2 adsorption isotherms at $-196\ °C$ of WH-HC and HCs prepared by Fe-assisted HTC.

Increasing the $FeCl_3$ solution molarity might help bond breakage, as has been suggested in the specific case of cellulose [20], and could also inhibit adsorption of secondary HC that could be related to mass and energy transference resistance under such high concentrations.

A significantly better porosity improvement was found for Fe-catalyzed reactions in relation to Al-catalyzed samples. An apparent surface rise was found for the sample obtained at the lowest Fe concentration (0.25 M). In this way, the first point at the lowest relative pressure is much higher for sample WH-HC-Fe0.25-600, indicating a high contribution of primary micropores; this sample, however, also presents mesopores, as the N_2 gradually increases along the whole range of relative pressure.

Increasing metal concentration over 0.25 M is detrimental to pore development; not only do these samples have lower pore volumes, but their pore size distribution is also wider and larger. This result, joined to the lower SY found for samples made at 0.5 and 0.75 M, might indicate that pore destruction is taking place under such conditions (external burning could be postulated or maybe disorganized carbon removal could be inhibited under these conditions, taking into consideration that isotherms for these two runs are type III).

According to X. Zhu et al. [20], during pyrolysis, Fe^{3+} ions are hydrolyzed to amorphous Fe species ($Fe(OH)_3$ and $FeO(OH)$) at temperatures lower than $350\ °C$. Subsequently, these species are converted into Fe_2O_3 at temperatures lower than $400\ °C$. At higher temperatures ($500–700\ °C$), Fe_2O_3 should be reduced to Fe_3O_4 with the aim of reducing components such as amorphous carbon and gaseous CO. Finally, both Fe_2O_3 and Fe_3O_4 with the amorphous carbon yield the metal Fe.

Figure 6 shows a particle of sample WH-HC-Fe0.25-600. One can clearly see the homogeneously distributed particles on the carbon structure of the HC, whose EDX analyses confirm that they consist of Fe (this metal was with difference the most abundant, with a contribution of 9.3% wt./wt., being the proportion of C and O equal to 80.0 and 9.2, respectively). The image also allows identifying slit-shaped open pores that developped after the calcination process. Hao et al. [11] found both effects when hydrothermally treating several biomass materials to Fe_3O_4 (during HTC) and further CO_2 activation.

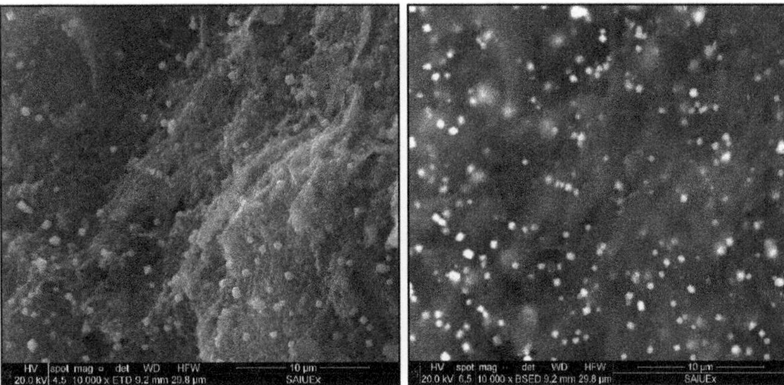

Figure 6. SEM micrographs of WH-HC-Fe0.25-600 (**left**) and its respective retro-dispersed electron image (**right**).

The FT-IR surface chemistry analysis made on the HCs have been included in Figure 7. As expected, the pyrolytic stage involved the removal of a significant quantity of the HC functional groups due to the high temperature applied. At the same time, these bonds suggest that the aromatization of the surface was intensified. In general, the three spectra are very similar, and the bands at 1600 y 1200 cm^{-1} typical of C=C bonds in aromatic rings and C-O in alcohols, ethers, or esters can be highlighted [19].

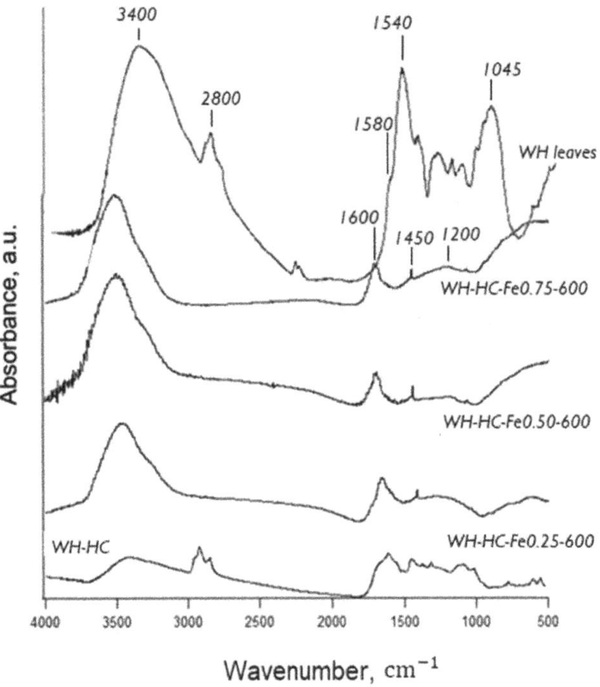

Figure 7. FT-IR spectra of WH, WH-HC, HCs, and HCs prepared by Fe-assisted HTC.

2.4. Magnetic Measurements

The magnetic behavior of the HCs prepared by Fe-assisted HTC was measured using a homemade installation (see details and equations employed in Section 3.3.3). This setup allowed the determination of magnetic susceptibility (X_m) of samples. Magnetization curves of Fe-catalyzed samples under a current of 8.5 A have been plotted in Figure 8.

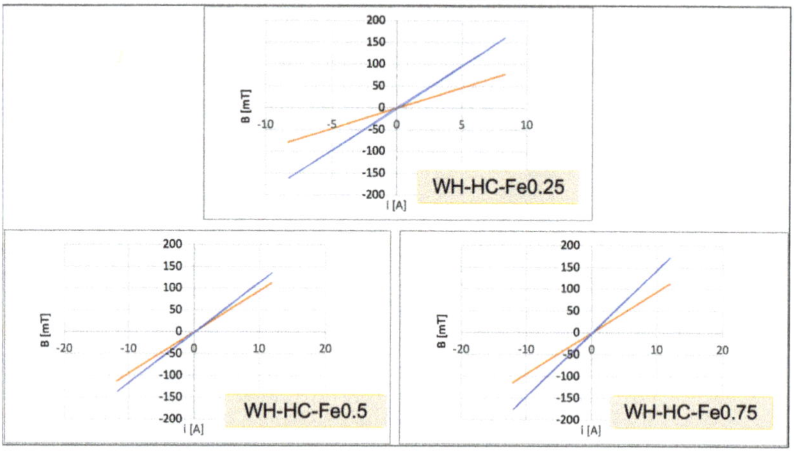

Figure 8. Magnetization curves applied to Fe-catalyzed HCs.

A correlation was found between the values of magnetic susceptibility found for the HCs and the porous development that was created upon the hydrothermal plus pyrolytic stage. In this way, the sample made using an $FeCl_3$ concentration of 0.25 M that showed the highest S_{BET} (229 m^2/g) also gave the greatest X_m (1.048). This value was much higher than that found for almond shell HCs obtained under the same reaction conditions and same installation (0.514), as reported in a previous work [9]. In that study, the authors found better results if pyrolysis was made at 800 °C than at 600 °C; however, for WH, making runs at 800 °C resulted not only in a porosity destruction but also in a significant drop of X_m (these results have not been shown here for the sake of brevity). These findings demonstrate that general assumptions cannot be made, and each biomass precursor must be specifically studied.

The other two samples made in this study, WH-HC-Fe0.5 and WH-HC-Fe0.75 that had a very poor and a similar pore volume, respectively, showed much lower values of X_m (0.27 and 0.46, respectively, under the same conditions).

3. Materials and Methods

3.1. Materials

Water hyacinth (WH) was gathered from the Guadiana River basin at Badajoz (southwest Spain, GPS coordinates 38.883889, −6.976824). The weed was pulled up and carried in hermetic plastic 15 L drums, and the three main parts (leaves, stem, and roots) were manually separated. Only leaves were used in this study because of their most favorable carbon density. After being oven-dried (110 °C, 12 h), WH leaves were crushed and sieved to a particle size of 1–2 mm (CISA 200/50 sieve, Norm ISO-3310.1); the particle size was chosen based on previous research with other lignocellulosic materials to guarantee that obtained HCs would have a granular character. The elemental analysis of the dried biomass showed proportions of C, O, H, and N equal to 41.14%, 4.93%, 49.63%, and 4.16%, respectively.

Aluminum alloy (AA2011 alloy) scraws were supplied from a local mining factory, as waste from their operation, in the form of particles with sizes of 2–4 mm. $FeCl_3$ was used in the form of iron (III) chloride 6-hydrate, as purchased from Panreac Applichem (Barcelona, Spain).

3.2. Methods

3.2.1. Standard HTC Processes

The standard HTC processes were performed in a 0.2 stainless steel autoclave (Berghof, Berlin, Germany) provided with a Teflon vessel. The dried biomass (10 g) was added to 150 mL of tap water, and the solution was kept under stirring conditions for 1 h. Thereafter, it was transferred to the Teflon vessel and introduced on the autoclave, which was then placed into a preheated oven (230 °C). The total time from the moment it was introduced in the oven and taken out was 20 h; this time period has been chosen based on previous experience of the research group [17], which demonstrated that this dwell time is enough to guarantee reaching a solid yield plateau that, in turn, indicates that biomass degradation for those temperature conditions is guaranteed. When the reaction time was reached, the autoclave was removed from the oven and subsequently placed in a cold-water bath and allowed to cool to room temperature.

After cooling, the solid phase was separated from the liquid by vacuum filtration and subsequently dried at 80 °C to remove residual moisture. The dried HC was stored in closed flasks that were placed into a desiccator until further analysis. The experiments were carried out under autogenous pressure in an autoclave without the possibility of measuring interior conditions, but according to previous studies, the pressure inside the vessel equaled that of the water at saturated conditions.

3.2.2. Modified Hydrothermal Carbonization Processes

For aluminum-catalyzed hydrothermal processes, a prefixed amount of Al particles (5, 10, or 20 g) was added to the water–biomass solution (100 mL of water and 3 g of CAM; that is, Al to biomass ratio of 5:3, 10:3, and 20:3) and subjected to HTC as in the case of the standard reaction. After filtrating the slurry, containing both the metal particles and the HC, the former was manually separated with the aim of pliers, and only HC weight accounted for SY.

These samples were named according to the nomenclature WH-HC-AlX, where X represents the amount of metal added to the system (for instance, WH-HC-Al5 stands for a sample in which 5 g of Al was added to HTC).

In the case of the processes catalyzed with iron, an Fe-containing mixture ($FeCl_3 \cdot 6H_2O$) of different concentration (0.25 M, 0.5 M and 0.75 M) was added (50 mL) to the dried weed (3 g) and directly subjected to HTC. These samples were named according to the nomenclature WH-HC-FeX-600, where X represents the amount of metal added to the system (for instance, WH-HC-Fe0.25-600 stands for a sample in which the concentration of $FeCl_3 \cdot 6H_2O$ mixture added to HTC was 0.25 M).

Then, Fe-catalyzed HCs were subjected to pyrolysis at 600 °C (N_2, 100 mL min^{-1}) in a vertical stainless steel piece of equipment described elsewhere [21].

3.3. Characterization Techniques

3.3.1. HC Reactivity and Thermal Behavior

Solid yield (SY, %) was calculated as the amount of solid product (i.e., HC) obtained after HTC in relation to the initial mass of precursor, expressed in percentage. For those runs catalyzed with Fe in which a second pyrolysis step was required, the whole solid yield (WSY) was calculated by considering both yield; that is, both fractions values were multiplied. In this way, the SY represents the final mass of char after HTC and pyrolysis in relation to the initial biomass mass.

Thermal analyses (TGA and DTG), were performed using a thermobalance (Setsys Evolution Setaram, Madrid, Spain). Argon or air (100 mL min^{-1} in both cases) were used as carrier agents, and a heating rate of 10 °C min^{-1} was applied.

3.3.2. HC Porosity and Surface Chemistry

The porosity of the HCs was explored by N_2 adsorption/desorption at −196 °C using a semiautomatic adsorption unit (AUTOSORB-1, Quantachrome, Tallahassee, FL,

USA). Prior to analyses, the samples were outgassed at 120 °C for 12 h. Experimental adsorption data were analyzed using suitable methods [22] to calculate (a) the value of the BET-specific surface (S_{BET}); (b) the external surface (S_{EXT}), calculated by the α_s method, using the reference nonporous solid proposed by Carrott et al. [23]; (c) the percentage of internal surface (S_{INT}), calculated as the difference between S_{BET} and S_{EXT}; (d) the volume of micropores through the Dubinin–Radushkevich equation (V_{miDR}); and (e) the volume of mesopores (V_{me}), calculated as the difference between the pore volume at p/p0 = 0.95 and p/p0 = 0.10.

In addition, the surface morphology of the samples was analyzed by scanning electron micrography (SEM, Hitachi, S-3600N, Krefeld, Germany) observation. The SEM samples were prepared by depositing about 50 mg of sample on an aluminum stud covered with conductive adhesive carbon tape and then coating with Rd–Pd for 1 min to prevent charging during observations. Imaging was done in the high vacuum mode at an accelerating voltage of 20 kV using secondary electrons.

Finally, the surface chemistry of the HCs was evaluated by means of FTIR spectroscopy. FTIR spectra were recorded with a Perkin Elmer model Paragon 1000PC spectrophotometer (Waltham, MA, USA) using the KBr disc method, with a resolution of 4 cm^{-1} and 100 scans (Perkin–Elmer 1720, Waltham, MA, USA). The composition of the cristaline part of the HCs was also analyzed by X-ray diffraction using Bruker equipment (Warwick, RI, USA).

3.3.3. Experimental Setup for the Study of Magnetic Behavior

The equipment used to measure the magnetic performance of materials has been described elsewhere [5]. Briefly, the HC was introduced on a cylindrical plastic tube (0.005 m diameter), a copper coil was wound 100 turns to be used as a pickup coil. The sample was then placed in the center of a long excitation coil (at the central part, far away from the ends) and subjected to external AC magnetic fields, H, created by different currents.

Under these conditions, H is quite uniform in the sample and can be written as

$$H = ni_{ex} \tag{1}$$

where n is the number of turns per meter of the excitation coil, that is,

$$n = \frac{N_{ex}}{L}. \tag{2}$$

in which L is the length of the coil.

As is known, the magnetic susceptibility is

$$\chi_m = \frac{1}{\mu_0 H}(B_{HC} - B_0) \tag{3}$$

where m_0 is the vacuum permeability (4p $\times$ 10^{-7} H/m) and B_{HC} and B_0 are the magnetic flux density in the sample and the air, respectively.

In order to determine the flux densities, two measurements are carried out as shown in Figure 9:

The *fem* measured in Figure 9 (leftside image) permits us to find B_{HC} by applicating the Faraday's law,

$$B_{HC} = -\frac{1}{N_{pu}S}\int emf\, dt \tag{4}$$

where S is the cross section of the sample, approximately equal to that of the pickup coil, and N_{HC} is the pickup coil number of turns.

The magnetic flux density measured by the Hall probe in Figure 9 (right-side image) is B_0. Substituting this value and the result of Equation (4) in Equation (3), magnetic susceptibility can be obtained.

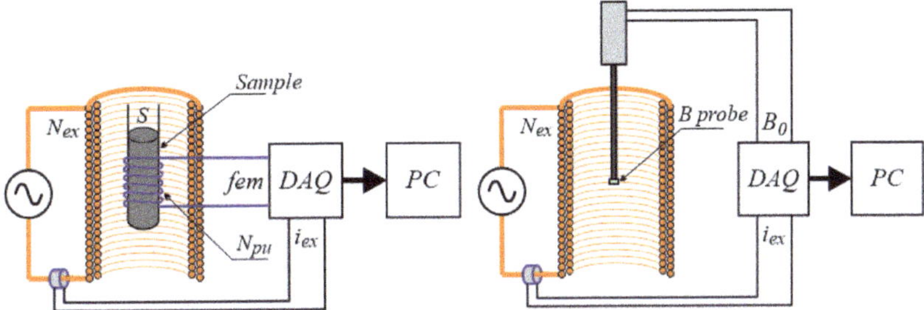

Figure 9. Setups to determine the magnetic susceptibility of carbons as B_{HC} (**left**) and B_0 (**right**).

4. Conclusions

Water hyacinth leaf HTC can yield a carbon material whose porosity is susceptible to being improved by means of the addition of Al or Fe to the reaction medium, although to a different extent. While Al addition induced a little porosity development and a slight widening of microporosity for smaller quantities of catalyst, the total pore volume was almost doubled in reference to the uncatalyzed HTC run (without further thermal treatment); the best result was found when the largest amount of Al alloy (20 g) was added to the system.

On the other hand, Fe, added as $FeCl_3$, clearly was more beneficial when a lower mixture concentration was used; for this metal, a concentration of 0.25 M yielded a carbon with a S_{BET} of 229 m^2/g, about 100 times higher than the noncatalyzed run. In this reaction, the use of a further pyrolytic step not only improved the HC activation but also induced a magnetic character to the carbon particles (magnetic susceptibility of 1.048) that could be easily isolated from their fluid medium by application of a magnetic field.

Author Contributions: Conceptualization, M.O.-M.; methodology, S.R., B.L. and A.Á.; validation, M.O.-M. and A.Á.; formal analysis, M.O.-M. and S.R.; investigation, B.L. and A.Á.; resources, M.O.-M. and S.R.; writing—original draft preparation, S.R. and A.Á.; writing—review and editing, M.O.-M.; funding acquisition, S.R. and M.O.-M. All authors have read and agreed to the published version of the manuscript.

Funding: This research was funded by the Agencia Estatal de Investigación (MINCIN), grant number PID2020-116144RB-I00/AEI/10.13039/501100011033XXX.

Acknowledgments: The authors thank the SAIUEX (Servicios de Apoyo a la Investigación de la Universidad de Extremadura) for support during the porosimetry characterization of carbon materials.

Conflicts of Interest: The authors declare no conflict of interest. The funders had no role in the design of the study; in the collection, analyses, or interpretation of data; in the writing of the manuscript; or in the decision to publish the results.

References

1. Román, S.; Libra, J.; Berge, N.; Sabio, E.; Ro, K.; Li, L.; Ledesma, B.; Álvarez, A.; Bae, S. Hydrothermal Carbonization: Modeling, Final Properties Design and Applications: A Review. *Energies* **2018**, *11*, 216. [CrossRef]
2. Moreno-Castilla, C.; López-Ramón, M.V.; Carrasco-Marín, F. Changes in surface chemistry ofactivated carbons by wet oxidation. *Carbon* **2000**, *38*, 1995–2001. [CrossRef]
3. Sevilla, M.; Fuertes, A.B. The production of carbon materials by hydrothermal carbonization of cellulose. *Carbon* **2009**, *47*, 2281–2289. [CrossRef]
4. Jung, D.; Zimmermann, M.; Kruse, A. Hydrothermal carbonization of fructose: Growth mechanisms and kinetic model. *ACS Sustain. Chem. Eng.* **2018**, *6*, 13877. [CrossRef]
5. Olivares, M.; Román, S.; Ledesma, B.; Álvarez, A. Magnetic Behavior of Carbon Materials Made from Biomass by Fe-Assisted Hydrothermal Carbonization. *Molecules* **2019**, *24*, 3996. [CrossRef]
6. Haj Yahia, S.; Lee, K.K.; Ayed, B.; Heddin, N.; Church, T.L. Activated Carbons from Hydrochars Prepared in Milk. *Sci. Rep.* **2019**, *9*, 16956. [CrossRef] [PubMed]
7. Mäkelä, M.; Fullana, A.; Yoshikawa, K. Ash behavior during hydrothermal treatment for solid fuel applications. Part 1: Overview of different feedstock. *Energy Convers. Manag.* **2016**, *121*, 402–408. [CrossRef]

8. Hammerton, J.M.; Ross, A.B. Inorganic Salt Catalysed Hydrothermal Carbonisation (HTC) of Cellulose. *Catalysts* **2022**, *12*, 492. [CrossRef]
9. Román, S.; Valente Nabais, J.M.; Ledesma, B.; Laginhas, C.; Titirici, M.-M. Surface Interactions during the Removal of Emerging Contaminants by Hydrochar-Based Adsorbents. *Molecules* **2020**, *25*, 2264. [CrossRef] [PubMed]
10. Chen, Y.; Sun, Z.; Su, Y.; Yang, J.; Li, M.; Hong, B.; Chen, G. Hydrochar Derived from Spent Mushroom Substrate Ameliorates Soil Properties and Nutrient Levels in Saline–Sodic Soil: An Incubation Study. *Sustainability* **2022**, *14*, 12958. [CrossRef]
11. Hao, W.; Björkman, E.; Yun, Y.; Lilliestråle, M.; Hedin, N. Iron Oxide Nanoparticles Embedded in Activated Carbons Prepared from Hydrothermally Treated Waste Biomass. *ChemSusChem* **2014**, *7*, 875. [CrossRef] [PubMed]
12. Phouthavong, V.; Yan, R.; Nijpanich, S.; Hagio, T.; Ichino, R.; Kong, L.; Li, L. Magnetic Adsorbents for Wastewater Treatment: Advancements in Their Synthesis Methods. *Materials* **2022**, *15*, 1053. [CrossRef] [PubMed]
13. Liu, J.; Yang, L.; Shuang, E.; Jin, C.; Gong, C.; Sheng, K.; Zhang, X. Facile one-pot synthesis of functional hydrochar catalyst for biomass valorization. *Fuel* **2022**, *315*, 123172–123183. [CrossRef]
14. NASA Reports. Earth Observatory: Water Hyacinth Re-Invades Lake Victoria. Available online: https://earthobservatory.nasa.gov/images/7426/water-hyacinth-re-invades-lake-victoria (accessed on 10 February 2023).
15. García-De-Lomas, J.; Dana, E.; Borrero, J.; Yuste, J.; Corpas, A.; Boniquito, J.; Castilleja, F.; Martínez, J.; Rodríguez, C.; Verloove, F. Rapid response to water hyacinth (Eichhornia crassipes) invasion in the Guadalquivir river branch in Seville (southern Spain). *Manag. Biol. Invasions* **2022**, *13*, 724–736. [CrossRef]
16. MITECO. Available online: https://www.miteco.gob.es/es/prensa/ultimas-noticias/el-camalote-est%C3%A1-controlado-en-todos-los-tramos-del-r%C3%ADo-guadiana/tcm:30-520159 (accessed on 10 February 2023).
17. Román, S.; Ledesma, B.; Álvarez, A.; Coronella, C.; Qaramaleki, S.V. Suitability of hydrothermal carbonization to convert water hyacinth to added-value products. *Renew. Energy* **2020**, *146*, 1649–1658. [CrossRef]
18. Shmelev, V.; Nikolaev, V.; Lee, J.H.; Yim, C. Hydrogen production by reaction of aluminum with water. *Int. J. Hydrogen Energy* **2016**, *41*, 16664. [CrossRef]
19. Fanning, P.E.; Vannice, M.A. A DRIFTS study of the formation of surface groups on carbon by oxidation. *Carbon* **1993**, *31*, 721–730. [CrossRef]
20. Zhu, X.; Qian, F.; Liu, Y.; Matera, D.; Wu, G.; Zhang, S.; Chen, J. Controllable synthesis of magnetic carbon composites with high porosity and strong acid resistance from hydrochar for efficient removal of organic pollutants: An overlooked influence. *Carbon* **2016**, *99*, 338–347. [CrossRef]
21. González, J.F.; Encinar, J.M.; Canito, J.L.; Sabio, E.; Chacón, M. Pyrolysis of cherry stones: Energy uses of the different fractions and kinetic study. *J. Anal. Appl. Pyrolysis* **2003**, *67*, 165–190. [CrossRef]
22. Sing, K.S.W.; Everett, D.H.; Haul, R.A.W.; Moscou, L.; Pierotti, R.A.; Siemieinewska, T. Reporting physisorption data for gas/solid systems with special reference to the determination of surface area and porosity. *Pure Appl. Chem.* **1985**, *57*, 613–619. [CrossRef]
23. Carrott, P.J.M.; Roberts, R.A.; Sing, K.S.W. Standard nitrogen adsorption data for nonporous carbons. *Carbon* **1987**, *25*, 769–770. [CrossRef]

 *catalysts*

Article

Conversion of Glucose to 5-Hydroxymethylfurfural Using Consortium Catalyst in a Biphasic System and Mechanistic Insights

Geraldo Ferreira David [1], Daniela Margarita Echeverri Delgadillo [2], Gabriel Abranches Dias Castro [2], Diana Catalina Cubides-Roman [1], Sergio Antonio Fernandes [2,*] and Valdemar Lacerda Júnior [1,*]

[1] Laboratório de Química Orgânica, Departamento de Química, Universidade Federal do Espírito Santo (UFES), Avenida Fernando Ferrari, 514, Goiabeiras, Vitória 29075-910, ES, Brazil; geraldo_david@yahoo.com.br (G.F.D.); dianacubides@yahoo.com (D.C.C.-R.)

[2] Grupo de Química Supramolecular e Biomimética (GQSB), Departamento de Química, Universidade Federal de Viçosa, Viçosa 36570-900, MG, Brazil; danielamargarita.e@gmail.com (D.M.E.D.); castrogabrielabranches@gmail.com (G.A.D.C.)

* Correspondence: santonio@ufv.br (S.A.F.); valdemar.lacerda@ufes.br (V.L.J.)

Abstract: We found an effective catalytic consortium capable of converting glucose to 5-hydroxy methylfurfural (HMF) in high yields (50%). The reaction consists of a consortium of a Lewis acid ($NbCl_5$) and a Brønsted acid (*p*-sulfonic acid calix[4]arene ($CX4SO_3H$)), in a microwave-assisted reactor and in a biphasic system. The best result for the conversion of glucose to HMF (yield of 50%) was obtained with $CX4SO_3H/NbCl_5$ (5 wt%/7.5 wt%), using water/NaCl and MIBK (1:3), at 150 °C, for 17.5 min. The consortium catalyst recycling was tested, allowing its reuse for up to seven times, while maintaining the HMF yield constant. Additionally, it proposed a catalytic cycle by converting glucose to HMF, highlighting the following two key points: the isomerization of glucose into fructose, in the presence of Lewis acid ($NbCl_5$), and the conversion of fructose into HMF, in the presence of $CX4SO_3H/NbCl_5$. A mechanism for the conversion of glucose to HMF was proposed and validated.

Keywords: biorefinery; niobium; calix[n]arenes

Citation: David, G.F.; Delgadillo, D.M.E.; Castro, G.A.D.; Cubides-Roman, D.C.; Fernandes, S.A.; Lacerda Júnior, V. Conversion of Glucose to 5-Hydroxymethylfurfural Using Consortium Catalyst in a Biphasic System and Mechanistic Insights. *Catalysts* **2023**, *13*, 574. https://doi.org/10.3390/catal13030574

Academic Editors: José María Encinar Martín and Sergio Nogales Delgado

Received: 14 February 2023
Revised: 7 March 2023
Accepted: 10 March 2023
Published: 12 March 2023

1. Introduction

Biomass is a renewable source of carbohydrates from which important chemicals can be obtained, such as 5-hydroxymethylfurfural (HMF) [1]. HMF is a versatile substance with a high market value, used in several industries as a fine chemical, medicine, energy, degradable plastic, and others [2]. Its price ranges from 500 to 1500 USD/kg and it has an expected growth of roughly 1.4% over the next years, possibly reaching USD 61 million by 2024 [3].

Glucose is present in large amounts in vegetable biomass, which can also be obtained from agricultural and/or forest residues, and is a promising substrate for HMF production [1,4]. The conversion of glucose to HMF occurs in a two-step process that involves the glucose to fructose isomerization and the subsequent dehydration of fructose to HMF. The direct conversion of glucose to HMF with high yields depends on combining the Lewis acid catalyst, for glucose isomerization in fructose, and the Brønsted acid catalyst, for fructose dehydration to HMF [1,5–7]. The development of a consortium catalyst with both properties (Lewis–Brønsted) is one of the major challenges in the conversion of biomass to HMF.

Several efforts have been made to develop a catalytic system with those features that is simple, efficient, not aggressive to equipment and environment, reproducible on a large scale, and that allows for its recovery and reuse [8–11]. The niobium catalyst and *p*-sulfonic acid calix[4]arene (CX4SO3H) organocatalyst are promising catalysts with the potential to convert carbohydrates, or biomass, into HMF.

Calix[n]arenes are used in several chemical transformations and have many advantages, such as high selectivity, easy manipulation, non-corrosiveness, low toxicity, and good thermal and chemical stability [4,12–20].

Recently, the great potential of calix[n]arenes has been adopted, and in some cases, for thin biorefinery processes, for the synthesis of Julolidines, HMF, levulinate esters, and biomass pretreatment, followed by fast pyrolysis to obtain levoglucosan [4,12,21].

Niobium catalysts are versatile, easy to handle, inexpensive, chemically stable, and commercially available [22,23]. To obtain HMF, the niobium catalyst has been used either on its own or in association with other catalytic systems. The use of niobium phosphate (NbP) as the only catalyst has shown a 15% yield of HMF from glucose at 145 °C and a reaction time of 180 min [24]. Carniti et al. used NbP for the direct conversion of cellobiose to HMF and reached yields between 5% and 10% in the temperature interval 110–130 °C after 3000 min of reaction [25]. The use of niobium acid (NbO) has also been reported. In the work of Catrinck et al., for example, 26% of the HMF yield was obtained from glucose at 152 °C and 120 min [26].

Moreover, when used in association with other systems, niobium has been reported as a promoter or active phase, support, solid acid catalyst, or redox material [5,6,27–32]. Some studies show a higher concentration of acid sites in niobium species, which is a key factor for obtaining HMF, along with high reaction temperatures and long reaction times [6,7,33,34].

In the work of Kreissl et al., a yield of 36% was reported when using mesoporous Nb_2O_5 as catalyst, and the reaction was performed in an autoclave at 500 rpm stirring speed, at 180 °C, for 180 min [35], whereas Liu et al. obtained a 26.8% yield in the conversion of glucose into HMF using Nb_2O_5, at 180 °C, for 180 min [36]. Using niobic acid, Huang et al. (2020) immobilized on regenerated cellulose, reaching 27.8% of the HMF and yield using glucose as substrate in a glass reactor at 150 °C for 250 min [37]. Meanwhile, Torres-Olea et al. evaluated the glucose dehydration to HMF, and found a yield of 44% using Nb_3Zr_7 as an acid catalyst after 90 min at 175 °C [6].

Herein, we report an efficient method for the conversion of glucose into HMF using the consortium catalyst, $CX4SO_3H/NbCl_5$, in a biphasic system (water/NaCl and methyl isobutyl ketone (MIBK)) in a MW reactor. To achieve the best results, different conditions such as temperature, reaction time, catalytic load, and different Lewis acids were evaluated.

2. Results and Discussion

2.1. Evaluation of Different Niobium Catalysts in HMF Synthesis

Different types of niobium catalysts ($NbCl_5$, Nb_2O_5, $NbOPO_4$, and HNb_3O_8) were evaluated under the initial reaction conditions: 0.25 mmol glucose (45 mg), $CX4SO_3H$ 5 wt%, niobium-based catalyst 7.5 wt%, MW, 150 °C, and 10 min of time reaction using water/NaCl and MIBK as the biphasic system.

In our results, when using $CX4SO_3H/Nb_2O_5$ or HNb_3O_8, a yield of 19% of HMF was obtained for the two consortium catalysts (Figure 1). When using $CX4SO_3H/NbOPO_4$ or $NbCl_5$ consortium catalysts for the conversion of glucose to HMF, yields of 23% and 42% were obtained, respectively (Figure 1).

Once it was determined that the $CX4SO_3H/NbCl_5$ consortium catalyst was the best system for the conversion of glucose into HMF, we further investigated the proportion between Brønsted and Lewis acids in the catalytic system.

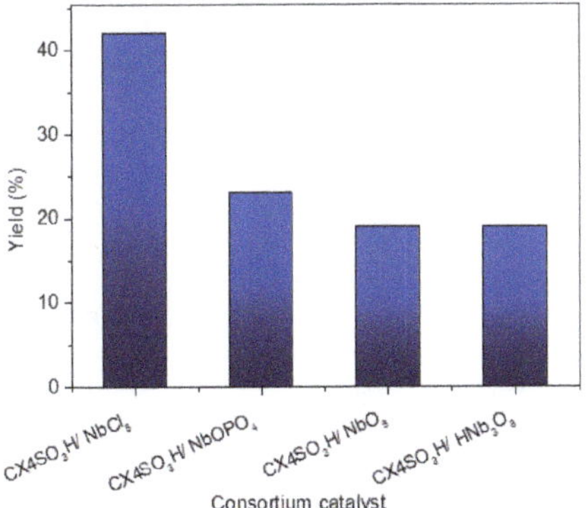

Figure 1. Comparison of different Lewis acids (niobium) fixed holding of CX4SO₃H Brønsted acid catalysts.

2.2. Evaluation of CX4SO₃H/NbCl₅ Consortium Catalyst Ratio

To investigate the need for a consortium catalyst (CX4SO₃H/NbCl₅) for the conversion of glucose into HMF, different ratios between Brønsted and Lewis acids were evaluated (Table 1). Initially, we evaluated CX4SO₃H (5 wt%) and different proportions of NbCl₅ (0–10 wt%) (Table 1, Entries 1–5). In the absence of NbCl₅, a marginal yield of only trace was obtained (Table 1, Entry 1). For amounts of 2.5, 5.0, and 7.5 wt% NbCl₅, the yield of HMF increased to 42%, and for 10 wt%, the yield decreased to 29% (Table 1, Entries 2–5). After establishing that 7.5 wt% NbCl₅ is the best ratio for the conversion of glucose into HMF, we evaluated different proportions of CX4SO₃H (Table 1, Entries 6–9). In the absence of CX4SO₃H, a 20% yield of HMF was obtained (Table 1, Entry 6). With the addition of increasing amounts of CX4SO₃H, the conversion of glucose into HMF reached a maximum value of 42% with the CX4SO₃H/NbCl₅ consortium (Table 1, Entry 4). For CX4SO₃H amounts greater than 5 wt%, the HMF yield decreased to 39% and 32% (Table 1, Entries 8 and 9) and the formation of humins was observed.

Table 1. Evaluation of ratio CX4SO₃H/NbCl₅ consortium catalyst.

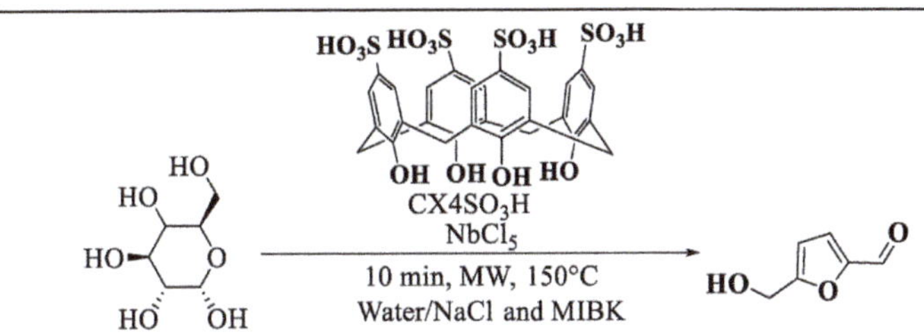

Entry	CX4SO₃H (% wt)	NbCl₅ (% wt)	Temperature (°C)	Yield (%)
1	5.0	0.0	150	trace
2	5.0	2.5	150	27
3	5.0	5.0	150	31
4	5.0	7.5	150	42
5	5.0	10.0	150	29
6	0.0	7.5	150	20
7	2.5	7.5	150	30
8	7.5	7.5	150	39
9	10.0	7.5	150	32

Reagents and conditions: 0.25 mmol of glucose (45 mg), water/NaCl and MIBK/(1/3 *v/v*).

Several works [38–41] have been published seeking to understand the association between Brønsted and Lewis acid and its good performance in the conversion of glucose into HMF. The first step toward glucose to HMF conversion involves the isomerization of glucose to fructose, followed by the dehydration of fructose, resulting in HMF. The isomerization of glucose has been described as the most difficult step of the process and a Lewis acid is usually used to effectively promote the isomerization; a Brønsted or Lewis acid is then used to promote the dehydration of fructose [38,39,42–45]. Consequently, we can infer that the $NbCl_5$ acts as Lewis acid, effectively promoting the isomerization of glucose, and partially helping in the dehydration of fructose; it is the $CX4SO_3H$, however, that improves the production of HMF due to its Brønsted acidic nature.

2.3. Evaluation of Temperature and Time

Following the evaluation of the catalyst charge, we tested different reaction temperatures keeping the other reaction parameters unchanged (0.25 mmol of glucose, 10 min, water/NaCl, and MIBK (1/3 *v*/*v*)). For temperatures below 150 °C, the HMF yield decreased to 21% and 30%, respectively (Table 2, Entries 1 and 2). However, when the temperature was greater than 150 °C, the HMF yield decreased from 42% to 33% (Table 2, Entries 3 and 4).

Table 2. Evaluation of reaction temperature.

Entry	CX4SO$_3$H (% wt)	NbCl$_5$ (% wt)	Temperature (°C)	Yield (%)
1	5.0	7.5	130	21
2	5.0	7.5	140	30
3	5.0	7.5	150	42
4 [a]	5.0	7.5	160	33

Reagents and conditions: 0.25 mmol of glucose (45 mg), water/NaCl and MIBK (1/3 *v*/*v*). [a] Formation of humin was observed.

This behavior has been reported in other studies on the dehydration of sugars to HMF. One study [46], for example, obtained an HMF yield of 51.5% in a MW glucose to HMF conversion, performed at 180 °C; however, when the temperature was increased to 190 °C, the HMF yield decreased to 40%. It is known that, although increasing the temperature promotes the dehydration of fructose to HMF, a very high temperature can lead to the formation of secondary compounds [47]. In our work, we noticed the same behavior in the formation of humin when the reaction was conducted at 160 °C.

The effect of the reaction time on the formation of HMF from glucose was evaluated by performing experiments between 5.0 and 22.5 min (Figure 2). We observed that the yield of HMF improved with the increase in the reaction time, until obtaining a maximum yield of 50% at 17.5 min of reaction (Figure 2). Further increases in the reaction time, however, resulted in a yield decrease, from 50% to 42% (Figure 2). It has been reported in the literature that long periods of reactions decrease the selectivity, leading to secondary reactions, such as humin formation and other undesirable products [48,49].

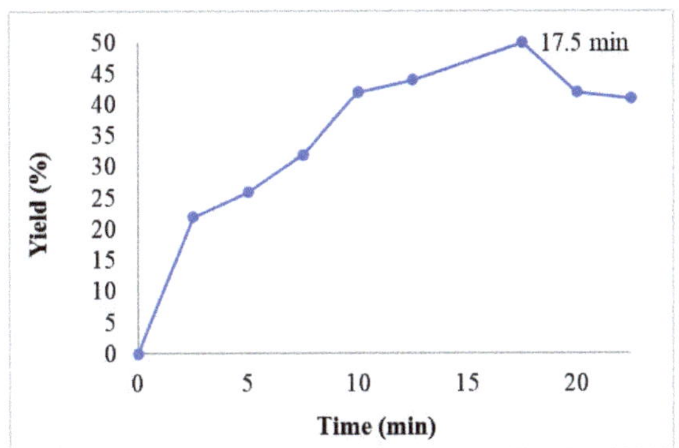

Figure 2. Evaluation of HMF yield over time. Reagents and conditions: 0.25 mmol glucose (45 mg), CX4SO$_3$H/NbCl$_5$ (5/7.5 wt%), MW, 150 °C, water/NaCl, and MIBK (1/3 v/v).

2.4. Evaluation of the Addition of Different Salt in Biphasic System

In order to determine how the composition of the biphasic system could affect the conversion of glucose to HMF, experiments were carried out with different salts using an organic phase MIBK (Table 3).

Table 3. Evaluation of the addition of different salt in biphasic system.

| Entry | Phase | | Yield (%) |
	Aqueous	Organic	
1	NaCl	MIBK	50
2	KCl	MIBK	29
3	CaCl$_2$	MIBK	24
4	MgCl$_2$	MIBK	11
5	NaCl	-	3
6	-	MIBK	-

Reagents and conditions: 0.25 mmol glucose (45 mg), CX4SO$_3$H/NbCl5 (5/7.5 wt%), MW, 17.5 min, 150 °C, water/NaCl, and MIBK (1/3 v/v).

The effect of adding different salts to an aqueous phase was verified, and the best performance was obtained with NaCl (Table 3, Entry 1). When using KCl, CaCl$_2$, and MgCl$_2$, the yield was 29, 24, and 11%, respectively (Table 3, Entries 2–4). To verify the effect of the extracting solvent and the aqueous phase, experiments were carried out in the absence of MIBK, obtaining only 3% of HMF; whereas, without an aqueous phase, no product was detected.

The use of a biphasic system allows the reaction to be carried out in an aqueous phase and the organic solvent to act as an extractor for the product [50]. HMF is easily extracted in organic solvents that are immiscible with water. The biphasic reaction mixture is used for the continuous removal of HMF from the aqueous phase by an organic phase to prevent its side reactions (humin) [46–49,51,52].

It is known from the literature that the addition of salt, such as NaCl, KCl, CaCl$_2$, and MgCl$_2$, helps to modify the aqueous phase by modifying the partition coefficient of the system, allowing a larger fraction of a product to migrate toward the organic phase. This is called the salting-out effect, which is generated when the ions of the salts modify the intermolecular forces between the liquids in equilibrium, allowing greater immiscibility between them [53].

The lower yield observed in the presence of bivalent cations (Table 3, Entries 3 and 4) can be explained by the hydration radius size of these species. The smaller the ion's hydration radius, the greater the salting-out effect [54]. Additionally, since Ca^{2+} and Mg^{2+} have the largest hydration radium among the studied cations, their salting-out effect is smaller, and the extraction efficiency is lower [55,56].

2.5. Evaluation of Catalytic Effect of Other Lewis Acids

To demonstrate the efficiency of Lewis ($NbCl_5$) and Brønsted ($CX4SO_3H$) acids acting as a consortium, other metallic chlorides were also evaluated: $CrCl_3$, $CoCl_2$, $MnCl_2$, $AlCl_3$, $NiCl_2$, $FeCl_3$, $FeCl_2$, $ZnCl_2$, $SnCl_2$, and $CuCl_2$ (Figure 3). The highest HMF yield continued to occur when using $NbCl_5$ as the Lewis acid. However, the $AlCl_3$ and $CrCl_3$ salts presented moderate yields of 41% and 39%, respectively (Figure 3). The other Lewis acids that were evaluated showed yields between 31 and 14% (Figure 3).

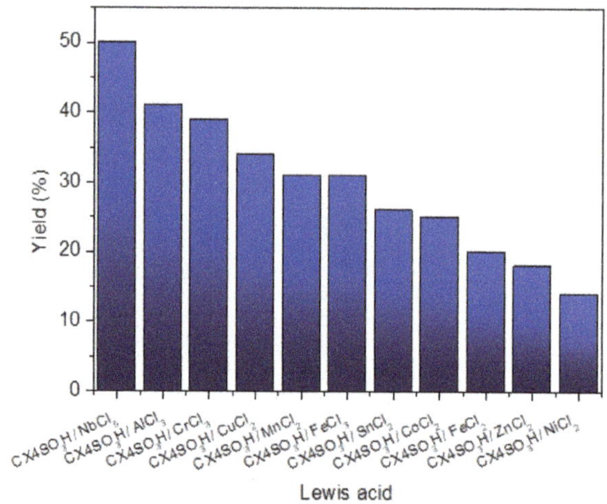

Figure 3. Evaluation of different Lewis acids. Reagents and conditions: 0.25 mmol glucose (45 mg), $CX4SO_3H/NbCl_5$ (5/7.5 wt%), MW, 150 °C, 17.5 min, water/NaCl, and MIBK (1/3 *v/v*).

2.6. Literature Comparison of Methods for the Conversion Glucose to HMF

In Table 4, some parameters of the methodology developed in this study, for the conversion of glucose into HMF, are compared with other studies found in the literature. References shown in Table 4 used, as a catalyst, a mixture of a Lewis and Brønsted acid (Table 4, Entries 2–4) or a bifunctional catalyst with Lewis–Brønsted acids. Viera et al. (Table 4, Entry 2) reported that the niobium catalyst acts in the formation of mannose and fructose, allowing the formation of HMF when combined with the Brønsted catalyst (HCl) in a biphasic THF/water system, using a rate of the catalyst of 1:2 wt% [7]. The biphasic THF/water system was also reported by Choudhary et al., (Table 4, Entry 3) who used chromium chloride ($CrCl_3$) as the Lewis acid and HCl as the Brønsted catalyst [57].

Table 4. Comparison of methodology developed for the conversion of glucose to HMF with data from the literature.

Entry	Organic Phase/Reaction Phase (Ratio)	Catalyst	Experimental Conditions	Yield (%)	Ref
1	MIBK/Water [a] (3:1)	$CX_4SO_3H/$ $NbCl_5$	T = 150 °C cat = 1/1.5 wt% time = 17.5 min	50	This work
2	THF [b]/Water (4:1)	Nb_2O_5/HCl	T = 130 °C cat = $\frac{1}{2}$ wt% time = 120 min	47	[58]
3	THF [b]/Water [a] (2:1)	$CrCl_3/HCl$	T = 140 °C cat = 3/10 wt% time = 180 min	59	[59]
4	SCB [c]/Water [a] (2:1)	$AlCl_3/HCl$	T = 170 °C cat = not reported * time = 40 min	62	[60]
5	MIBK/water [a] (6:1)	PTSA-Ca/AC	T = 180 °C cat = 1/1 time = 1440 min	57	[61]

[a] NaCl saturated solution; [b] THF: Tetrahydrofuran; and [c] SCB: 2-sec-butylphenol. * 5 mM $AlCl_3$ was added in the reaction and HCl was added until the pH of the solution was 2.5.

On the other hand, Pagán-Torres also used HCl as a Brønsted catalyst, but with $AlCl_3$ as a Lewis catalyst (3:10 catalyst ratio) and a 2-sec-butylphenol/water biphasic system. Bounoukta et al. reported a yield of 57% of HMF (Table 4, Entry 5), using the bifunctional catalyst of *p*-toluenesulfonic acid on an activated carbon surface and functionalized with activated charcoal (1:1) [60]. The catalyst ratio that they used is close to the one used in our study, 1:1.5 of $CX4SO_3H/NbCl_5$; this parameter varies greatly between published studies (Table 4). Although the reported yields are generally good, the use of THF (Table 4, Entries 2–3) as the organic phase is not recommended, since it is a problematic solvent according to the principles of green chemistry [62]. Moreover, the use of HCl as the Brønsted acid (Table 4, Entries 2–4) presents difficulties for an industrial scale production due to its corrosive qualities as a highly toxic reagent. Finally, while it is possible to use a green solvent, such as MIBK, as the organic phase for the bifunctional catalyst, its reaction time is too long (1440 min or 24 h) to be viable.

2.7. Evaluation of Other Carbohydrates to Produce HMF

Other carbohydrates, in addition to glucose, were also evaluated: sucrose, mannose, maltose, raffinose melibiose, galactose, and cellulose, the yields of which were, respectively, 50, 42, 32, 31, 19, 19, 17, and 15% (Figure 4). In general, to obtain 5-HMF directly from sugars, it is necessary that a sequence of the reaction involves catalysts for hydrolysis, isomerization of glucose to fructose, and finally, the acid-catalyzed dehydration of fructose. For example, the direct conversion of cellulose to 5-HMF using heterogeneous catalysts is difficult due to the low reactivity of cellulose and the high instability of 5-HMF [48,57,60,63,64].

Sucrose showed to have the best HMF performance; since it is a dimer formed by the union of *α-D*-glucose and *β-D*-fructose through a glycosidic bond, sucrose can undergo a hydrolysis in which the carbohydrate units that make it up are separated. Additionally, it is more prone to undergo dehydration directly to HMF, since the glucose that must undergo an initial isomerization to fructose is converted into HMF [47]. On the other hand, the evaluation of cellulose is also interesting; its result was comparable with that of a more complex biomass system [65]. The 15% yield obtained from cellulose is not far from the 20% HMF obtained by one study [66] that achieved this performance using formic acid and betaine as catalysts, with 60 min of reaction at 190 °C. Another study [67] obtained a 35% yield of HMF from microcrystalline cellulose using $CrCl_3$, in addition to using ionic liquid

([EMIM]Cl) as the reaction medium. Despite being efficient in converting cellulose into HMF, ionic liquids are expensive and toxic reagents.

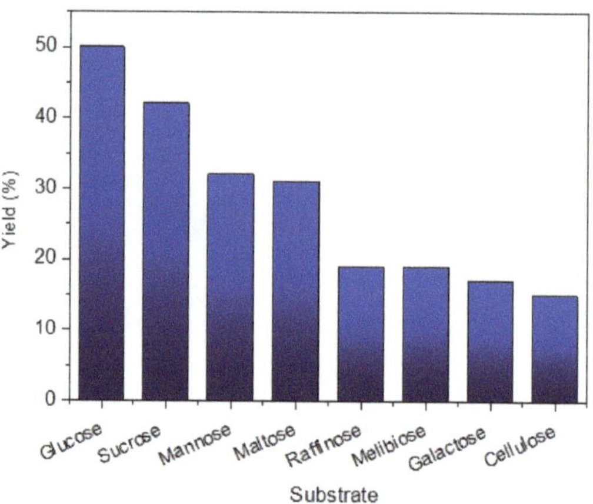

Figure 4. Evaluation of different carbohydrates.

2.8. Catalyst Recycling

The evaluation of catalyst recycling is a key-factor from a commercial and environmental point of view. Thus, if the catalyst can be used more times without decreasing its effectiveness while maintaining a constant yield of the product, less investment is needed in the purchase of reagents, and, from an environmental perspective, less chemical waste occurs [68,69].

At the end of the reaction, the system was cooled, and the organic phase was separated. The aqueous phase containing the $CX4SO_3H/NbCl_5$ consortium catalyst remained in the tube, and a new load of glucose and MIBK were added and used in a new reaction. In our work, it was possible to reuse the $CX4SO_3H/NbCl_5$ catalyst consortium up to six times while keeping the HMF yield constant. The yield decreased from 49% in the sixth cycle to 39% and 38% in the respective seventh and eighth cycles of the reuse of the system catalyst, as shown in Figure 5.

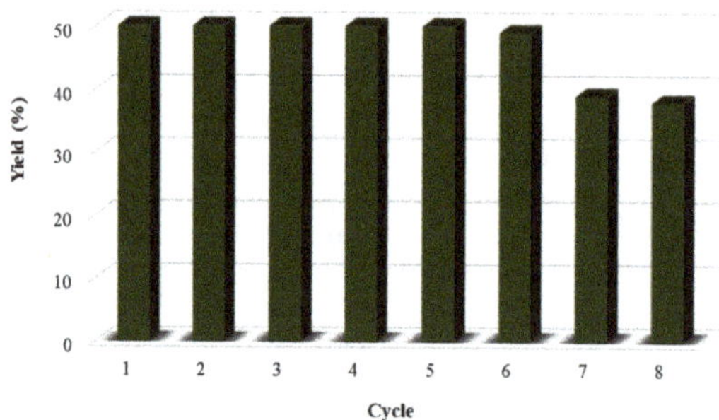

Figure 5. Evaluation of system catalyst and aqueous phase recycle. Reagents and conditions: 0.25 mmol glucose (45 mg), $CX4SO_3H/NbCl_5$ (5/7.5 wt%), MW, 150 °C, 17.5 min, water/NaCl, and MIBK (1/3 *v/v*).

2.9. Reaction Mechanism

The mechanism of glucose for fructose isomerization is still a matter of debate. There are two mechanistic proposals for the key step of the isomerization of glucose to fructose: one proceeding via a 1,2-enediol intermediate [59,70–73] and an intramolecular hydride shift from C-2 to C-1 [11,57,74–76] (Scheme 1). A study with isotopic labeling using D_2O was performed under the same optimized conditions (0.25 mmol glucose (45 mg), 150 °C, 17.5 min, water/NaCl, and MIBK (1/3 v/v) for $NbCl_5$ (7.5 wt%), $CX4SO_3H$ (5 wt%), and $CX4SO_3H/NbCl_5$ (5/7.5 wt%), except for the solvent (H_2O was replaced by D_2O).

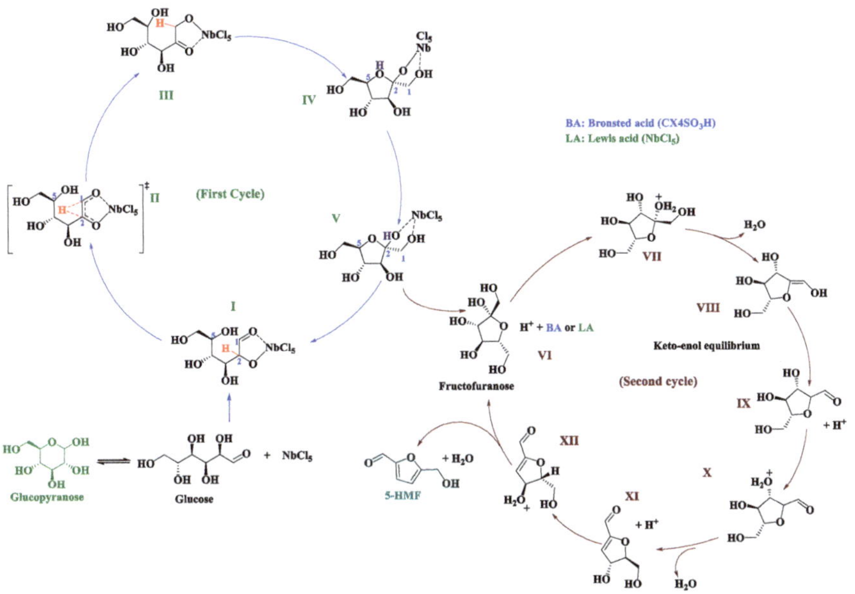

Scheme 1. Mechanistic proposal for the conversion of glucose to HMF. The first catalytic cycle is based on the reference [75].

Analysis of the mass spectrum (GC-MS) showed that there was no incorporation of deuterium atoms into the HMF structure for none of the catalytic systems evaluated: $NbCl_5$, $CX4SO_3H$, and $CX4SO_3H/NbCl_5$. Since D_2O was used as the solvent, we propose a mechanistic step of the isomerization of glucose to fructose through simultaneously activating the carbonyl group at the C-1 position and the hydroxyl group at the C-2 position on glucose (intermediate **I**, Scheme 1) proceeding an intramolecular hydride shift from C-2 to C -1 catalyzed by $NbCl_5$ (state transition **II** and intermediate **III**, Scheme 1). Then, cyclization occurs through the attack of the hydroxyl on the carbonyl (intermediates **IV** and **V**, Scheme 1). The subsequent dehydration of fructose occurs with the $CX4SO_3H/NbCl_5$ consortium with protonation of the hydroxyl bound to the anomeric carbon of fructofuranose, followed by the loss of a water molecule and the formation of the enol (intermediate **VIII**, Scheme 1). The enol that is in tautomeric equilibrium with the keto (aldehyde) form (intermediate **IX**, Scheme 1). From the aldehyde, there is the protonation of a second hydroxyl, followed by the loss of a water molecule, leading to the formation of an α,β-unsaturated aldehyde (intermediate **X**, Scheme 1). Finally, the protonation of the secondary hydroxyl (intermediate **XI**, Scheme 1) followed by the loss of a water molecule leads to the formation of the aromatic ring of the HMF.

3. Materials and Methods

3.1. Materials

$Nb_2O_5 \cdot nH_2O$ (NbO), $NbOPO_4 \cdot nH_2O$ (NbP), and HNb_3O_8 were supplied by Companhia Brasileira de Metalurgia e Mineração (CBMM, Araxá, Minas GeraisBrazil). The niobium pentachloride ($NbCl_5$) and the standard HMF were purchased from Sigma-Aldrich (Saint

Louis, United States), as well as all other materials and reagents that were necessary for the development of this work.

3.2. Synthesis of CX4SO₃H

The CX4SO$_3$H was synthesized according to the procedures reported in the literature [77–79].

3.3. General Procedure for Conversion of Glucose into HMF

In a Pyrex® glass tube suitable for microwave, the following were added: 0.25 mmol glucose (45 mg), 5.0 wt% of CX4SO$_3$H, 7.5 wt% of niobium chloride (NbCl$_5$), 1.0 mL of aqueous solution saturated with NaCl, and 3.0 mL of MIBK. After sealing the tube, this mixture was taken to a MW reactor (CEM Discovery), where it was heated to 150 °C under magnetic stirring for 17.5 min. At the end of the experiment, the mixture was cooled at room temperature and the organic phase was separated and dried over anhydrous sodium sulphate to remove residual water; the mixture was subsequently filtered and transferred to a 5.0 mL volumetric flask, which had its volume checked with methanol. From this solution, an aliquot of 100 µL was removed and transferred to another 5.0 mL volumetric flask. Finally, from this solution, an aliquot of 397 µL was removed into a vial, along with a 603 µL aliquot of methanol. The sample was analyzed by ultra-high-performance liquid chromatography (UHPLC).

3.4. Quantification of HMF by UHPLC

The chromatograms were obtained by UHPLC employing a Thermo Scientific Accela LC liquid chromatograph (diode array detector (DAD), autoinjector, and Accela pump) (Thermo Fischer Scientific, Austin, TX, USA). The column used for separation was a Hypersil GOLD reverse phase (50 × 2.1 mm, 1.9 µm particle size, and 175 Å pore) (Thermo Fischer Scientific, Austin, TX, USA). The mobile phase consisted of water and methanol (1:1), and elution was carried out in isocratic mode for two minutes. The applied amount was 200 µL min^{-1} and the injection volume was 1 µL (partial loop), with a temperature of 25 °C for the injector and column. The peak of HMF was detected at a wavelength of 280 nm.

HMF was quantified based on the external calibration technique. Standard solutions were prepared in methanol, with HMF at concentrations of 2–50 mg L^{-1} and injected into the UHPLC system. The calibration curve (R^2 = 0.9960) was obtained in relation to the area of each HMF standard. The HMF yield (%) was calculated based on a calibration curve.

3.5. Catalyst Recycling

The recycling of system catalysts was conducted using a model reaction employing 0.25 mmol glucose (45 mg), 5.0 wt% CX4SO$_3$H, 7.5 wt% NbCl$_5$, 1.0 mL saturated aqueous solution for NaCl, and 3.0 mL of MIBK. This mixture was heated to 150 °C for 17.5 min. At the end of this period, the system was cooled and the organic phase was separated. To the aqueous phase that stood in the tube, a new 0.25 mmol glucose load (45 mg) and MIBK (3.0 mL) were added, which were subjected to the reaction conditions. The recycling procedure was repeated another seven times.

4. Conclusions

In this study, we presented an efficient route to convert glucose into HMF with a high 50% yield using the consortium catalysts NbCl$_5$ and CX4SO$_3$H in an MW reactor. The reaction conditions optimized result was CX4SO$_3$H/NbCl$_5$ (5 wt%/7.5 wt%) as a consortium catalyst, using water/NaCl and MIBK (1:3 *v/v*) at 150 °C for a 17.5-minute reaction time. This catalyst system showed excellent recyclability, with its catalytic activity maintained for six cycles. Thus, the glucose isomerization is due to the Lewis acid (NbCl$_5$), followed by the fructose dehydration to HMF, which is due to CX4SO$_3$H/NbCl$_5$. The application of this consortium catalytic system is attractive, environmentally friendly, and cost-effective for the conversion of biomass into higher value-added products. Finally, isotopic labeling experiments suggested a mechanism for the glucose isomerization for fructose reactions involving an intramolecular hydride shift from C-2 to C-1 since the incorporation of a deuterium atom was not observed; the proposed mechanism was validated using mass spectrometry.

Author Contributions: Conceptualization: G.F.D., S.A.F. and V.L.J.; methodology: G.F.D., S.A.F., D.M.E.D., G.A.D.C., V.L.J. and D.C.C.-R.; investigation, G.F.D., D.M.E.D. and G.A.D.C.; resources: G.F.D., S.A.F., D.M.E.D., G.A.D.C., V.L.J. and D.C.C.-R.; writing—original draft preparation: G.A.D.C., S.A.F., D.M.E.D., G.A.D.C., V.L.J. and D.C.C.-R.; supervision: S.A.F. and V.L.J. All authors have read and agreed to the published version of the manuscript.

Funding: The authors would like to thank FAPES (Fundação de Amparo Pesquisa e Inovação do Espírito Santo; Profix process number 355/2018), CAPES (Coordenação de Aperfeiçoamento Pessoal de Nível Superior; Finance Code 001), CNPq (Conselho Nacional de Desenvolvimento Científico e Tecnológico), FAPEMIG (Fundação de Amparo à Pesquisa do Estado de Minas Gerais), National Council for Scientific and Technological Development (CNPq-Process no.: CNPq 315389/2020-6), and NCQP (Núcleo de Competências em Química do Petróleo) for their financial and technical support. S.A.F. and V.L.Jr. are supported by the Research Fellowships from CNPq.

Data Availability Statement: Data will be made available on request.

Conflicts of Interest: The authors declare no conflict of interest.

References

1. Yin, Y.; Ma, C.; Li, W.; Luo, S.; Liu, Y.; Wu, X.; Wu, Z.; Liu, S. Rapid Conversion of Glucose to 5-Hydroxymethylfurfural Using a MoCl$_3$ Catalyst in an Ionic Liquid with Microwave Irradiation. *Ind. Crops Prod.* **2021**, *160*, 113091. [CrossRef]
2. Zhang, L.; Tian, Y.; Wang, Y.; Dai, L. Enhanced Conversion of α-Cellulose to 5-HMF in Aqueous Biphasic System Catalyzed by FeCl$_3$-CuCl$_2$. *Chinese Chem. Lett.* **2021**, *32*, 2233–2238. [CrossRef]
3. Kong, Q.S.; Li, X.L.; Xu, H.J.; Fu, Y. Conversion of 5-Hydroxymethylfurfural to Chemicals: A Review of Catalytic Routes and Product Applications. *Fuel Process. Technol.* **2020**, *209*, 106528. [CrossRef]
4. Pereira, S.; Oliveira Santana Varejão, J.; de Fátima, Â.; Fernandes, S.A. P-Sulfonic Acid Calix[4]Arene: A Highly Efficient Organocatalyst for Dehydration of Fructose to 5-Hydroxymethylfurfural. *Ind. Crops Prod.* **2019**, *138*, 4–10. [CrossRef]
5. El Fergani, M.; Candu, N.; Tudorache, M.; Bucur, C.; Djelal, N.; Granger, P.; Coman, S.M. From Useless Humins By-Product to Nb@graphite-like Carbon Catalysts Highly Efficient in HMF Synthesis. *Appl. Catal. A Gen.* **2021**, *618*, 118130. [CrossRef]
6. Torres-Olea, B.; García-Sancho, C.; Cecilia, J.A.; Oregui-Bengoechea, M.; Arias, P.L.; Moreno-Tost, R.; Maireles-Torres, P. Influence of Lewis Acidity and CaCl2 on the Direct Transformation of Glucose to 5-Hydroxymethylfurfural. *Mol. Catal.* **2021**, *510*, 111685. [CrossRef]
7. Vieira, J.L.; Almeida-Trapp, M.; Mithöfer, A.; Plass, W.; Gallo, J.M.R. Rationalizing the Conversion of Glucose and Xylose Catalyzed by a Combination of Lewis and Brønsted Acids. *Catal. Today* **2020**, *344*, 92–101. [CrossRef]
8. Zhou, C.; Zhao, J.; Yagoub, A.E.G.A.; Ma, H.; Yu, X.; Hu, J.; Bao, X.; Liu, S. Conversion of Glucose into 5-Hydroxymethylfurfural in Different Solvents and Catalysts: Reaction Kinetics and Mechanism. *Egypt. J. Pet.* **2017**, *26*, 477–487. [CrossRef]
9. Li, L.; Ding, J.; Jiang, J.G.; Zhu, Z.; Wu, P. One-Pot Synthesis of 5-Hydroxymethylfurfural from Glucose Using Bifunctional [Sn,Al]-Beta Catalysts. *Cuihua Xuebao/Chinese J. Catal.* **2015**, *36*, 820–828. [CrossRef]
10. Teimouri, A.; Mazaheri, M.; Chermahini, A.N.; Salavati, H.; Momenbeik, F.; Fazel-Najafabadi, M. Catalytic Conversion of Glucose to 5-Hydroxymethylfurfural (HMF) Using Nano-POM/Nano-ZrO2/Nano-γ-Al2O3. *J. Taiwan Inst. Chem. Eng.* **2015**, *49*, 40–50. [CrossRef]
11. Wang, T.; Glasper, J.A.; Shanks, B.H. Kinetics of Glucose Dehydration Catalyzed by Homogeneous Lewis Acidic Metal Salts in Water. *Appl. Catal. A Gen.* **2015**, *498*, 214–221. [CrossRef]
12. Abranches, P.A.D.S.; De Paiva, W.F.; De Fátima, Â.; Martins, F.T.; Fernandes, S.A. Calix[n]Arene-Catalyzed Three-Component Povarov Reaction: Microwave-Assisted Synthesis of Julolidines and Mechanistic Insights. *J. Org. Chem.* **2018**, *83*, 1761–1771. [CrossRef]
13. Braga, I.B.; Castañeda, S.M.B.; Vitor De Assis, J.; Barros, A.O.; Amarante, G.W.; Valdo, A.K.S.M.; Martins, F.T.; Rosolen, A.F.D.P.; Pilau, E.; Fernandes, S.A. Anise Essential Oil as a Sustainable Substrate in the Multicomponent Double Povarov Reaction for Julolidine Synthesis. *J. Org. Chem.* **2020**, *85*, 15622–15630. [CrossRef] [PubMed]
14. Liberto, N.A.; Simões, J.B.; de Paiva Silva, S.; da Silva, C.J.; Modolo, L.V.; de Fátima, Â.; Silva, L.M.; Derita, M.; Zacchino, S.; Zuñiga, O.M.P.; et al. Quinolines: Microwave-Assisted Synthesis and Their Antifungal, Anticancer and Radical Scavenger Properties. *Bioorganic Med. Chem.* **2017**, *25*, 1153–1162. [CrossRef]
15. Natalino, R.; Varejão, E.V.V.; da Silva, M.J.; Cardoso, A.L.; Fernandes, S.A. P-Sulfonic Acid Calix[n]Arenes: The Most Active and Water Tolerant Organocatalysts in Esterification Reactions. *Catal. Sci. Technol.* **2014**, *4*, 1369–1375. [CrossRef]

16. Palermo, V.; Sathicq, A.; Liberto, N.; Fernandes, S.; Langer, P.; Jios, J.; Romanelli, G. Calix[n]Arenes: Active Organocatalysts for the Synthesis of Densely Functionalized Piperidines by One-Pot Multicomponent Procedure. *Tetrahedron Lett.* **2016**, *57*, 2049–2054. [CrossRef]

17. Rezende, T.R.M.; Varejão, J.O.S.; Sousa, A.L.L.D.A.; Castañeda, S.M.B.; Fernandes, S.A. Tetrahydroquinolines by the Multicomponent Povarov Reaction in Water: Calix[: N] Arene-Catalysed Cascade Process and Mechanistic Insights. *Org. Biomol. Chem.* **2019**, *17*, 2913–2922. [CrossRef]

18. Santos, L.S.; Fernandes, S.A.; Pilli, R.A.; Marsaioli, A.J. A Novel Asymmetric Reduction of Dihydro-β-Carboline Derivatives Using Calix[6]Arene/Chiral Amine as a Host Complex. *Tetrahedron Asymmetry* **2003**, *14*, 2515–2519. [CrossRef]

19. Simões, J.B.; De Fátima, Â.; Sabino, A.A.; De Aquino, F.J.T.; Da Silva, D.L.; Barbosa, L.C.A.; Fernandes, S.A. Organocatalysis in the Three-Component Povarov Reaction and Investigation by Mass Spectrometry. *Org. Biomol. Chem.* **2013**, *11*, 5069–5073. [CrossRef]

20. Simões, J.B.; De Fátima, Â.; Sabino, A.A.; Almeida Barbosa, L.C.; Fernandes, S.A. Efficient Synthesis of 2,4-Disubstituted Quinolines: Calix[n]Arene- Catalyzed Povarov-Hydrogen-Transfer Reaction Cascade. *RSC Adv.* **2014**, *4*, 18612–18615. [CrossRef]

21. David, G.F.; Ríos-Ríos, A.M.; de Fátima, Â.; Perez, V.H.; Fernandes, S.A. The Use of P-Sulfonic Acid Calix[4]Arene as Organocatalyst for Pretreatment of Sugarcane Bagasse Increased the Production of Levoglucosan. *Ind. Crops Prod.* **2019**, *134*, 382–387. [CrossRef]

22. Ziolek, M.; Sobczak, I. The Role of Niobium Component in Heterogeneous Catalysts. *Catal. Today* **2017**, *285*, 211–225. [CrossRef]

23. David, G.F.; Pereira, S.d.P.S.; Fernandes, S.A.; Cubides-Roman, D.C.; Siqueira-Roman, R.K.; Perez, V.H.; Lacerda, V. Fast Pyrolysis as a Tool for Obtaining Levoglucosan after Pretreatment of Biomass with Niobium Catalysts. *Waste Manag.* **2021**, *126*, 274–282. [CrossRef]

24. Junior, M.M.d.J.; Fernandes, S.A.; Borges, E.; Baêta, B.E.L.; Rodrigues, F.d.Á. Kinetic Study of the Conversion of Glucose to 5-Hydroxymethylfurfural Using Niobium Phosphate. *Mol. Catal.* **2022**, *518*, 112079. [CrossRef]

25. Carniti, P.; Gervasini, A.; Bossola, F.; Dal Santo, V. Cooperative Action of Brønsted and Lewis Acid Sites of Niobium Phosphate Catalysts for Cellobiose Conversion in Water. *Appl. Catal. B Environ.* **2016**, *193*, 93–102. [CrossRef]

26. Catrinck, M.N.; Ribeiro, E.S.; Monteiro, R.S.; Ribas, R.M.; Barbosa, M.H.P.; Teófilo, R.F. Direct Conversion of Glucose to 5-Hydroxymethylfurfural Using a Mixture of Niobic Acid and Niobium Phosphate as a Solid Acid Catalyst. *Fuel* **2017**, *210*, 67–74. [CrossRef]

27. Morales, A.; Santana, A.; Althoff, G.; Melendez, E. Host-Guest Interactions between Calixarenes and Cp2NbCl 2. *J. Organomet. Chem.* **2011**, *696*, 2519–2527. [CrossRef]

28. Redshaw, C.; Rowan, M.; Homden, D.M.; Elsegood, M.R.J.; Yamato, T.; Pérez-Casas, C. Niobium- and Tantalum-Based Ethylene Polymerisation Catalysts Bearing Methylene- or Dimethyleneoxa-Bridged Calixarene Ligands. *Chem.-A Eur. J.* **2007**, *13*, 10129–10139. [CrossRef] [PubMed]

29. Acho, J.A.; Doerrer, L.H.; Lippard, S.J. Pentamethylcyclopentadienyl and Cyclopentadienyl Tantalum and Niobium Calixarene Compounds and Their Water and Acetonitrile Inclusion Complexes. *Inorg. Chem.* **1995**, *34*, 2542–2556. [CrossRef]

30. Dürr, S.; Bechlars, B.; Radius, U. Calix[4]Arene Supported Group 5 Imido Complexes. *Inorganica Chim. Acta* **2006**, *359*, 4215–4226. [CrossRef]

31. Caselli, A.; Solari, E.; Scopelliti, R.; Floriani, C. A Synthetic Methodology to Niobium Alkylidenes: Reactivity of a Nb=Nb Double Bond Anchored to a Calix[4]Arene Oxo Surface with Ketones, Aldehydes, Imines, and Isocyanides. *J. Am. Chem. Soc.* **1999**, *121*, 8296–8305. [CrossRef]

32. Caselli, A.; Solari, E.; Scopelliti, R.; Floriani, C.; Re, N.; Rizzoli, C.; Chiesi-Villa, A. Dinitrogen Rearranging over a Metal-Oxo Surface and Cleaving to Nitride: From the End-on to the Side-on Bonding Mode, to the Stepwise Cleavage of the N≡N Bonds Assisted by Nb(III)-Calix[4]Arene. *J. Am. Chem. Soc.* **2000**, *122*, 3652–3670. [CrossRef]

33. Carniti, P.; Gervasini, A.; Biella, S.; Auroux, A. Niobic Acid and Niobium Phosphate as Highly Acidic Viable Catalysts in Aqueous Medium: Fructose Dehydration Reaction. *Catal. Today* **2006**, *118*, 373–378. [CrossRef]

34. Zhang, L.; Li, K.; Zhu, X. Study on Two-Step Pyrolysis of Soybean Stalk by TG-FTIR and Py-GC/MS. *J. Anal. Appl. Pyrolysis* **2017**, *127*, 91–98. [CrossRef]

35. Kreissl, H.T.; Nakagawa, K.; Peng, Y.K.; Koito, Y.; Zheng, J.; Tsang, S.C.E. Niobium Oxides: Correlation of Acidity with Structure and Catalytic Performance in Sucrose Conversion to 5-Hydroxymethylfurfural. *J. Catal.* **2016**, *338*, 329–339. [CrossRef]

36. Liu, Y.; Li, H.; He, J.; Zhao, W.; Yang, T.; Yang, S. Catalytic Conversion of Carbohydrates to Levulinic Acid with Mesoporous Niobium-Containing Oxides. *Catal. Commun.* **2017**, *93*, 20–24. [CrossRef]

37. Huang, F.; Jiang, T.; Dai, H.; Xu, X.; Jiang, S.; Chen, L.; Fei, Z.; Dyson, P.J. Transformation of Glucose to 5-Hydroxymethylfurfural Over Regenerated Cellulose Supported Nb2O5·nH2O in Aqueous Solution. *Catal. Letters* **2020**, *150*, 2599–2606. [CrossRef]

38. Hu, D.; Zhang, M.; Xu, H.; Wang, Y.; Yan, K. Recent Advance on the Catalytic System for Efficient Production of Biomass-Derived 5-Hydroxymethylfurfural. *Renew. Sustain. Energy Rev.* **2021**, *147*. [CrossRef]

39. Kim, Y.; Mittal, A.; Robichaud, D.J.; Pilath, H.M.; Etz, B.D.; St. John, P.C.; Johnson, D.K.; Kim, S. Prediction of Hydroxymethylfurfural Yield in Glucose Conversion through Investigation of Lewis Acid and Organic Solvent Effects. *ACS Catal.* **2020**, *10*, 14707–14721. [CrossRef]

40. Swift, T.D.; Nguyen, H.; Erdman, Z.; Kruger, J.S.; Nikolakis, V.; Vlachos, D.G. Tandem Lewis Acid/Brønsted Acid-Catalyzed Conversion of Carbohydrates to 5-Hydroxymethylfurfural Using Zeolite Beta. *J. Catal.* **2016**, *333*, 149–161. [CrossRef]

41. Wrigstedt, P.; Keskiväli, J.; Leskelä, M.; Repo, T. The Role of Salts and Brønsted Acids in Lewis Acid-Catalyzed Aqueous-Phase Glucose Dehydration to 5-Hydroxymethylfurfural. *ChemCatChem* **2015**, *7*, 501–507. [CrossRef]
42. Megías-Sayago, C.; Navarro-Jaén, S.; Drault, F.; Ivanova, S. Recent Advances in the Brønsted/Lewis Acid Catalyzed Conversion of Glucose to Hmf and Lactic Acid: Pathways toward Bio-Based Plastics. *Catalysts* **2021**, *11*, 1395. [CrossRef]
43. Yu, I.K.M.; Tsang, D.C.W.; Yip, A.C.K.; Su, Z.; De Oliveira Vigier, K.; Jérôme, F.; Poon, C.S.; Ok, Y.S. Organic Acid-Regulated Lewis Acidity for Selective Catalytic Hydroxymethylfurfural Production from Rice Waste: An Experimental-Computational Study. *ACS Sustain. Chem. Eng.* **2019**, *7*, 1437–1446. [CrossRef]
44. Zhang, H.; Zhao, H.; Zhai, S.; Zhao, R.; Wang, J.; Cheng, X.; Shiran, H.S.; Larter, S.; Kibria, M.G.; Hu, J. Electron-Enriched Lewis Acid-Base Sites on Red Carbon Nitride for Simultaneous Hydrogen Production and Glucose Isomerization. *Appl. Catal. B Environ.* **2022**, *316*, 121647. [CrossRef]
45. Lu, S.; Lyu, J.; Han, X.; Bai, P.; Guo, X. Effective Isomerization of Glucose to Fructose by Sn-MFI/MCM-41 Composites as Lewis Acid Catalysts. *J. Taiwan Inst. Chem. Eng.* **2020**, *116*, 272–278. [CrossRef]
46. Das, B.; Mohanty, K. Sulfonic Acid-Functionalized Carbon Coated Red Mud as an Efficient Catalyst for the Direct Production of 5-HMF from D-Glucose under Microwave Irradiation. *Appl. Catal. A Gen.* **2021**, *622*, 118237. [CrossRef]
47. Tempelman, C.; Jacobs, U.; Hut, T.; Pereira de Pina, E.; van Munster, M.; Cherkasov, N.; Degirmenci, V. Sn Exchanged Acidic Ion Exchange Resin for the Stable and Continuous Production of 5-HMF from Glucose at Low Temperature. *Appl. Catal. A Gen.* **2019**, *588*, 117267. [CrossRef]
48. Huang, F.; Su, Y.; Tao, Y.; Sun, W.; Wang, W. Preparation of 5-Hydroxymethylfurfural from Glucose Catalyzed by Silica-Supported Phosphotungstic Acid Heterogeneous Catalyst. *Fuel* **2018**, *226*, 417–422. [CrossRef]
49. Fachri, B.A.; Abdilla, R.M.; Bovenkamp, H.H.V.D.; Rasrendra, C.B.; Heeres, H.J. Experimental and Kinetic Modeling Studies on the Sulfuric Acid Catalyzed Conversion of d -Fructose to 5-Hydroxymethylfurfural and Levulinic Acid in Water. *ACS Sustain. Chem. Eng.* **2015**, *3*, 3024–3034. [CrossRef]
50. Gomes, F.N.D.C.; Pereira, L.R.; Ribeiro, N.F.P.; Souza, M.M.V.M. Production of 5-Hydroxymethylfurfural (HMF) via Fructose Dehydration: Effect of Solvent and Salting-Out. *Brazilian J. Chem. Eng.* **2015**, *32*, 119–126. [CrossRef]
51. Abranches Dias Castro, G.; Lopes, N.; Fernandes, S.; da Silva, M. Copper Phosphotungstate-Catalyzed Microwave-Assisted Synthesis of 5-Hydroxymethylfurfural in a Biphasic System. *Cellulose* **2022**, *29*. [CrossRef]
52. Castro, G.A.D.; Fernandes, S.A. Green Synthesis of 5-Hydroxymethylfurfural in a Biphasic System Assisted by Microwaves. *Catal. Letters* **2022**. [CrossRef]
53. Román-Leshkov, Y.; Dumesic, J.A. Solvent Effects on Fructose Dehydration to 5-Hydroxymethylfurfural in Biphasic Systems Saturated with Inorganic Salts. *Top. Catal.* **2009**, *52*, 297–303. [CrossRef]
54. Görgényi, M.; Dewulf, J.; Van Langenhove, H.; Héberger, K. Aqueous Salting-out Effect of Inorganic Cations and Anions on Non-Electrolytes. *Chemosphere* **2006**, *65*, 802–810. [CrossRef]
55. Saha, B.; Abu-Omar, M.M. Advances in 5-Hydroxymethylfurfural Production from Biomass in Biphasic Solvents. *Green Chem.* **2014**, *16*, 24–38. [CrossRef]
56. Esteban, J.; Vorholt, A.J.; Leitner, W. An Overview of the Biphasic Dehydration of Sugars to 5-Hydroxymethylfurfural and Furfural: A Rational Selection of Solvents Using COSMO-RS and Selection Guides. *Green Chem.* **2020**, *22*, 2097–2128. [CrossRef]
57. Choudhary, V.; Mushrif, S.H.; Ho, C.; Anderko, A.; Nikolakis, V.; Marinkovic, N.S.; Frenkel, A.I.; Sandler, S.I.; Vlachos, D.G. Insights into the Interplay of Lewis and Brønsted Acid Catalysts in Glucose and Fructose Conversion to 5-(Hydroxymethyl)Furfural and Levulinic Acid in Aqueous Media. *J. Am. Chem. Soc.* **2013**, *135*, 3997–4006. [CrossRef]
58. Vieira, J.L.; Paul, G.; Iga, G.D.; Cabral, N.M.; Bueno, J.M.C.; Bisio, C.; Gallo, J.M.R. Niobium Phosphates as Bifunctional Catalysts for the Conversion of Biomass-Derived Monosaccharides. *Appl. Catal. A Gen.* **2021**, *617*. [CrossRef]
59. Ståhlberg, T.; Rodriguez-Rodriguez, S.; Fristrup, P.; Riisager, A. Metal-Free Dehydration of Glucose to 5-(Hydroxymethyl)Furfural in Ionic Liquids with Boric Acid as a Promoter. *Chem.-A Eur. J.* **2011**, *17*, 1456–1464. [CrossRef] [PubMed]
60. Pagán-Torres, Y.J.; Wang, T.; Gallo, J.M.R.; Shanks, B.H.; Dumesic, J.A. Production of 5-Hydroxymethylfurfural from Glucose Using a Combination of Lewis and Brønsted Acid Catalysts in Water in a Biphasic Reactor with an Alkylphenol Solvent. *ACS Catal.* **2012**, *2*, 930–934. [CrossRef]
61. Bounoukta, C.E.; Megías-Sayago, C.; Ammari, F.; Ivanova, S.; Monzon, A.; Centeno, M.A.; Odriozola, J.A. Dehydration of Glucose to 5-Hydroxymethlyfurfural on Bifunctional Carbon Catalysts. *Appl. Catal. B Environ.* **2021**, *286*, 119938. [CrossRef]
62. Prat, D.; Wells, A.; Hayler, J.; Sneddon, H.; McElroy, C.R.; Abou-Shehada, S.; Dunn, P.J. CHEM21 Selection Guide of Classical- and Less Classical-Solvents. *Green Chem.* **2015**, *18*, 288–296. [CrossRef]
63. Binder, J.B.; Cefali, A.V.; Blank, J.J.; Raines, R.T. Mechanistic Insights on the Conversion of Sugars into 5-Hydroxymethylfurfural. *Energy Environ. Sci.* **2010**, *3*, 765–771. [CrossRef]
64. Scarim, C.B.; Lira de Farias, R.; Vieira de Godoy Netto, A.; Chin, C.M.; Leandro dos Santos, J.; Pavan, F.R. Recent Advances in Drug Discovery against Mycobacterium Tuberculosis: Metal-Based Complexes. *Eur. J. Med. Chem.* **2021**, *214*. [CrossRef]
65. Zhao, H.; Holladay, J.E.; Brown, H.; Zhang, Z.C. Metal Chlorides in Ionic Liquid Solvents Convert Sugars to 5-Hydroxymethylfurfural. *Science* **2007**, *316*, 1597–1600. [CrossRef] [PubMed]
66. Delbecq, F.; Wang, Y.T.; Len, C. Various Carbohydrate Precursors Dehydration to 5-HMF in an Acidic Biphasic System under Microwave Heating Using Betaine as a Co-Catalyst. *Mol. Catal.* **2017**, *434*, 80–85. [CrossRef]

67. Abou-Yousef, H.; Hassan, E.B.; Steele, P. Rapid Conversion of Cellulose to 5-Hydroxymethylfurfural Using Single and Combined Metal Chloride Catalysts in Ionic Liquid. *Ranliao Huaxue Xuebao/J. Fuel Chem. Technol.* **2013**, *41*, 214–222. [CrossRef]
68. Silvestri, C.; Silvestri, L.; Forcina, A.; Di Bona, G.; Falcone, D. Green Chemistry Contribution towards More Equitable Global Sustainability and Greater Circular Economy: A Systematic Literature Review. *J. Clean. Prod.* **2021**, *294*, 126137. [CrossRef]
69. Anastas, P.; Eghbali, N. Green Chemistry: Principles and Practice. *Chem. Soc. Rev.* **2010**, *39*, 301–312. [CrossRef]
70. Gupta, D.; Saha, B. Dual Acidic Titania Carbocatalyst for Cascade Reaction of Sugar to Etherified Fuel Additives. *Catal. Commun.* **2018**, *110*, 46–50. [CrossRef]
71. Noma, R.; Nakajima, K.; Kamata, K.; Kitano, M.; Hayashi, S.; Hara, M. Formation of 5-(Hydroxymethyl)Furfural by Stepwise Dehydration over TiO2 with Water-Tolerant Lewis Acid Sites. *J. Phys. Chem. C* **2015**, *119*, 17117–17125. [CrossRef]
72. Parveen, F.; Upadhyayula, S. Efficient Conversion of Glucose to HMF Using Organocatalysts with Dual Acidic and Basic Functionalities-A Mechanistic and Experimental Study. *Fuel Process. Technol.* **2017**, *162*, 30–36. [CrossRef]
73. Li, G.; Pidko, E.A.; Hensen, E.J.M.; Nakajima, K. A Density Functional Theory Study of the Mechanism of Direct Glucose Dehydration to 5-Hydroxymethylfurfural on Anatase Titania. *ChemCatChem* **2018**, *10*, 4084–4089. [CrossRef]
74. Pidko, E.A.; Degirmenci, V.; Hensen, E.J.M. On the Mechanism of Lewis Acid Catalyzed Glucose Transformations in Ionic Liquids. *ChemCatChem* **2012**, *4*, 1263–1271. [CrossRef]
75. Loerbroks, C.; Van Rijn, J.; Ruby, M.P.; Tong, Q.; Schüth, F.; Thiel, W. Reactivity of Metal Catalysts in Glucose-Fructose Conversion. *Chem.-A Eur. J.* **2014**, *20*, 12298–12309. [CrossRef] [PubMed]
76. Román-Leshkov, Y.; Moliner, M.; Labinger, J.A.; Davis, M.E. Mechanism of Glucose Isomerization Using a Solid Lewis Acid Catalyst in Water. *Angew. Chemie-Int. Ed.* **2010**, *49*, 8954–8957. [CrossRef] [PubMed]
77. Gutsche, C.; Iqbal, M. P-Tert-Buthylcalix[4]Arene. *Org Synth* **1990**, *68*, 234.
78. Casnati, A.; Ca', N.D.; Sansone, F.; Ugozzoli, F.; Ungaro, R. Enlarging the Size of Calix[4]Arene-Crowns-6 to Improve Cs+/K+selectivity: A Theoretical and Experimental Study. *Tetrahedron* **2004**, *60*, 7869–7876. [CrossRef]
79. Shinkai, S.; Tsubaki, T.; Sone, T.; Manabe, O. A New Synthesis of P-Nitrocalix[6]Arene. *Tetrahedron Lett.* **1985**, *26*, 3343–3344. [CrossRef]

Article

Organically Functionalized Porous Aluminum Phosphonate for Efficient Synthesis of 5-Hydroxymethylfurfural from Carbohydrates

Riddhi Mitra, Bhabani Malakar and Asim Bhaumik *

School of Materials Sciences, Indian Association for the Cultivation of Science, 2A & 2B Raja S. C. Mullick Road, Jadavpur, Kolkata 700032, India; smsrm2633@iacs.res.in (R.M.); scsbm2857@iacs.res.in (B.M.)
* Correspondence: msab@iacs.res.in

Abstract: Naturally occurring fossil fuels are the major resource of energy in our everyday life, but with the huge technological development over the years and subsequent energy demand, the reserve of this energy resource is depleting at an alarming rate, which will challenge our net energy resources in the near future. Thus, an alternative sustainable energy resource involving biomass and bio-refinery has become the most emerging and demanding approach, where biofuels can be derived effectively from abundant biomass via valuable chemical intermediates like 5-hydroxymethylfurfural (5-HMF). 5-HMF is a valuable platform chemical for the synthesis of fuel and fine chemicals. Herein, we report the synthesis of the organically functionalized porous aluminum phosphonate materials: Ph-ALPO-1 in the absence of any template and Ph-ALPO-2 by using 1,3-diaminopropane-N,N,N′,N′-tetraacetic acid as a small organic molecule template and phenylphosphonic acid as a phosphate source. These hybrid phosphonates are used as acid catalysts for the synthesis of 5-HMF from carbohydrates derived from biomass resources. These Ph-ALPO-1 and Ph-ALPO-2 materials catalyzed the dehydration of fructose to 5-HMF with total yields of 74.6% and 90.7%, respectively, in the presence of microwave-assisted optimized reaction conditions.

Keywords: aluminum phosphate ($AlPO_4$); microporous and mesoporous materials; biomass conversion; 5-hydroxymethylfurfural (5-HMF); organic–inorganic hybrids

Citation: Mitra, R.; Malakar, B.; Bhaumik, A. Organically Functionalized Porous Aluminum Phosphonate for Efficient Synthesis of 5-Hydroxymethylfurfural from Carbohydrates. *Catalysts* **2023**, *13*, 1449. https://doi.org/10.3390/catal13111449

Academic Editors: José María Encinar Martín and Sergio Nogales Delgado

Received: 16 October 2023
Revised: 13 November 2023
Accepted: 16 November 2023
Published: 19 November 2023

1. Introduction

With the continuous depletion of fossil fuels as result of industrialization the global attention seeks to identify alternative renewable energy sources to overcome a future worldwide energy crisis [1–3]. Thus, the concept of "biomass valorization" has emerged as a pivotal solution at the intersection of science, technology, and sustainability [4–6]. Biomass is described as the renewable organics that can come from flora and fauna. Hence, biomass is the largest feedstock for alternative renewable energy resources available in nature. The majority of biomass resources consist of carbohydrates, bearing hexose (C6) and pentose (C_5)-based sugars [7]. Thus, carbohydrates derived from plants, agricultural waste, and forestry residues have gained considerable attention as a renewable and sustainable energy feedstock. These biomass-derived organic molecules can be converted into biofuels through various chemical and biochemical processes, offering an alternative to fossil fuels. The synthesis of biofuels from biomass involves transformation of the complex organic molecules present in biomass into simpler hydrocarbon compounds that can be used as 'fine platform chemicals' [8]. This process utilizes different technologies and pathways, each one having its own advantages and challenges.

Hence, carbohydrates have widespread potential to produce essential chemicals, which can otherwise be derived directly from petroleum resources. 5-HMF, consisting of a furan ring along with aldehyde and primary alcohol functional groups, is considered as one of the most essential biomass-derived platform chemicals by the US Department of

Energy [9]. This makes 5-HMF an effective intermediate, which can further be converted to many valuable products like 2,5-dimethylfuran, an essential biofuel [10,11]. That is why 5-HMF is also known as 'sleeping giant' in the field of intermediate chemistry [12]. Monosaccharides like fructose, glucose, etc., in the presence of acid catalysts, easily dehydrate to yield 5-HMF [13]. According to the literature to date, several mineral acids already showed high catalytic activity towards the synthesis of 5-HMF from different carbohydrates [14]. However, there are many obstacles of using these homogeneous catalysts, like separation from reaction mixture, recyclability, etc. Thus, heterogeneous catalysts have become an emerging choice for researchers in catalyzing these reactions [15–17]. Although many heterogeneous solid acid catalysts are reported in the literature for the 5-HMF synthesis from carbohydrate resources, organically functionalized porous materials can be the superior contenders over others due to their high specific surface area and inherent surface hydrophobicity for the guest molecules to react at the catalytic site [18–20]. Further, monodispersity in the particle size and Lewis acidity of the framework metals may improve catalytic performance of solid acid catalysis.

Crystalline microporous and mesoporous materials are intensively studied over the past half-century due to their high specific surface area, high thermal and chemical stability, and wide diversity in the framework architectures [21–23]. Among them phosphonate-based porous frameworks are of particular interest due to the chemistry involved between the trivalent metal cation (MO_4) tetrahedra and phosphate tetrahedral sites [24,25]. Open frameworks and high surface area (internal surface) could strengthen their catalytic and adsorption abilities. The construction of organic–inorganic hybrid porous materials is a rapidly expanding field of materials chemistry for the synthesis of the materials having specific structural entity and functionality [26]. Metal phosphonates, having organic functionality in the framework, is one such group of hybrid materials, which attracted extensive interest due to their emerging applications in many areas like ion-exchange, catalysis, charge storage, sensing, etc. [27]. Metal phenyl phosphonates represent a particularly important class of organic–inorganic hybrid materials having covalent bonds between the inorganic cluster and organic phenyl moieties [28]. To date, a large number of microporous $AlPO_4$ frameworks have been synthesized via a soft-templating route, containing 8–18 ring pore apertures (Figure 1), where small organic molecules like primary/secondary/tertiary amines and quaternary ammonium salts act as a template in directing the crystal structure and inbuilt nanoscale porosities [29].

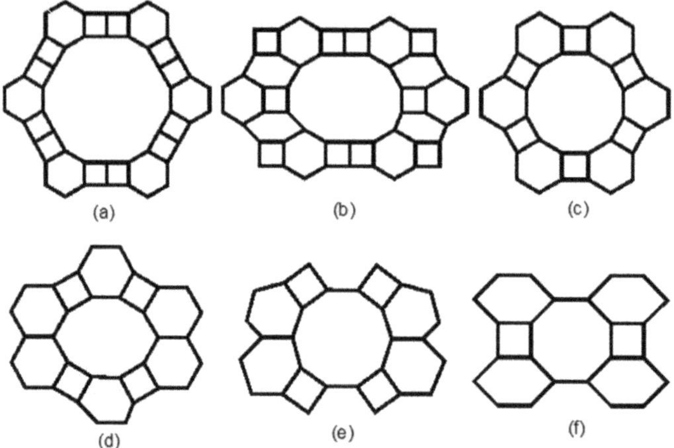

Figure 1. Different reported $AlPO_4$-n frameworks with various nanoscale pore openings (**a**) VPI-5 (18-ring), (**b**) $ALPO_4$-8 (14-ring), (**c**) $ALPO_4$-5 (12-ring), (**d**) $ALPO_4$-11 (10-ring), (**e**) $ALPO_4$-41 (10-ring), (**f**) $ALPO_4$-25 (8-ring).

In this context, it is pertinent to mention that the catalysis under microwave irradiation conditions is considered as a green chemical approach for the synthesis of small organic molecules. Microwave irradiation heats up the entire reaction mixture from its core through the polarization of the medium. So, generally polar solvents like DMSO, DMF, acetonitrile, etc., are used as medium of the microwave-assisted reactions. Further, aging and decomposition of 5-HMF in prolonged storing is a major issue [30], which could be considerably decreased if we could develop strategies for synthesizing 5-HMF from carbohydrates having no impurity of humins in the product.

In this work, we have synthesized phenyl functionalized organic–inorganic hybrid porous aluminum phosphonate materials Ph-ALPO-1 and Ph-ALPO-2, respectively, in the absence and presence of an organic additive 1,3-diaminopropane-N,N,N′,N′-tetraacetic acid in the synthesis mixture. These hybrid materials have then been used as heterogeneous catalyst for 5-HMF synthesis from carbohydrates under microwave-assisted heating conditions.

2. Results

The wide-angle powder X-ray diffraction (PXRD) of both Ph-ALPO-1 and Ph-ALPO-2 materials was studied for understanding the crystalline phase of these materials. Both samples exhibited sharp peak at $2\theta = 5.08°$. PXRD data collected in the range of $4°$ to $40°$ 2θ has been analyzed using the Expo2014 software package [31]. Ph-ALPO-1/2 materials are identified as lamellar structure with triclinic phase, P1 symmetry and the unit cell parameters are a =17.61 Å, b = 6.79 Å, and c = 4.88 Å, α = 106.690°, β = 94.204°, and γ = 80.948°. The diffraction peaks are indexed in Table S1 (Supporting Information). The resemblance of PXRD patterns for both Ph-ALPO-1 and Ph-ALPO-2 indicates the structure remains unchanged by using the template molecule in the synthesis (Figure 2A). However, a considerable improvement in the crystallinity was observed for Ph-ALPO-2; the sample was synthesized in the presence of the template. In order to measure the permanent porosity and the surface area of both the materials, we have analyzed N_2 adsorption–desorption analysis at 77 K. A high nitrogen uptake at low P/P_0 followed by slow and steady capillary condensation at higher P/P_0 indicates a mixture of type-I and type-IV isotherms has been observed (Figure 2B) [32]. Consequently, the pore size distribution (PSD) plots were calculated from these isotherms using NLDFT model (Figure 2C). These PSD plots suggested a hierarchical porosity with a peak micropore at 1.3 nm. Whereas Ph-ALPO-2 displayed a distinct mesopore at 5.8 nm, the peak for mesopores of Ph-ALPO-1 remains much distributed suggesting template plays a crucial role in organizing the overall framework rigid, organized and porous. Again, the pore volume of Ph-ALPO-1 and Ph-ALPO-2 comes out as 0.38 cc g^{-1} and 0.63 cc g^{-1}, respectively. Ph-ALPO-1 and Ph-ALPO-2 showed the Brunauer–Emmett–Teller (BET) specific surface areas of 73 and 106 m^2g^{-1}, respectively. Larger BET surface area of Ph-ALPO-2 over Ph-ALPO-1 again suggests the advantage of using organic template in the synthesis of this porous aluminum phosphonate framework. A greater surface area is always beneficial for the application in catalysis or sorption application. Fourier-transformed infrared spectra of both the materials were investigated for understanding the bonding connectivity of both the materials. Both of the samples displayed almost similar spectral features (Figure 2D). In the spectra, wavenumber of 4000 to 400 cm^{-1} was selected for detailed study of the bonding connectivity of the materials. An overlaying between the bands of the vibrational bending of H-O-H of the hydration water at ~1620 cm^{-1} and the sharp stretching band of the phenyl group (ν_{C-C}) at ~1595 cm^{-1} has been observed. The band at 3000 cm^{-1} could be attributed to the C-H vibrational stretching and the band at ~1490 and ~1440 cm^{-1} indicate the skeletal vibrations of the rings. The intense and broad band in between ~3600 and ~3200 cm^{-1} is attributed to the O-H vibrational stretching due to the presence of hydration water. Another two bands in this region have been observed, i.e., one very sharp at ~3640 cm^{-1} is probably indicating the presence of adsorbed water having no interaction by H-bonding and the other broader near ~3510 cm^{-1} is clearly indicating this water molecule interacting weakly through H-bonds [33]. The other stretching and bending modes of the Al-O-P bonds are shown in Figure 2D.

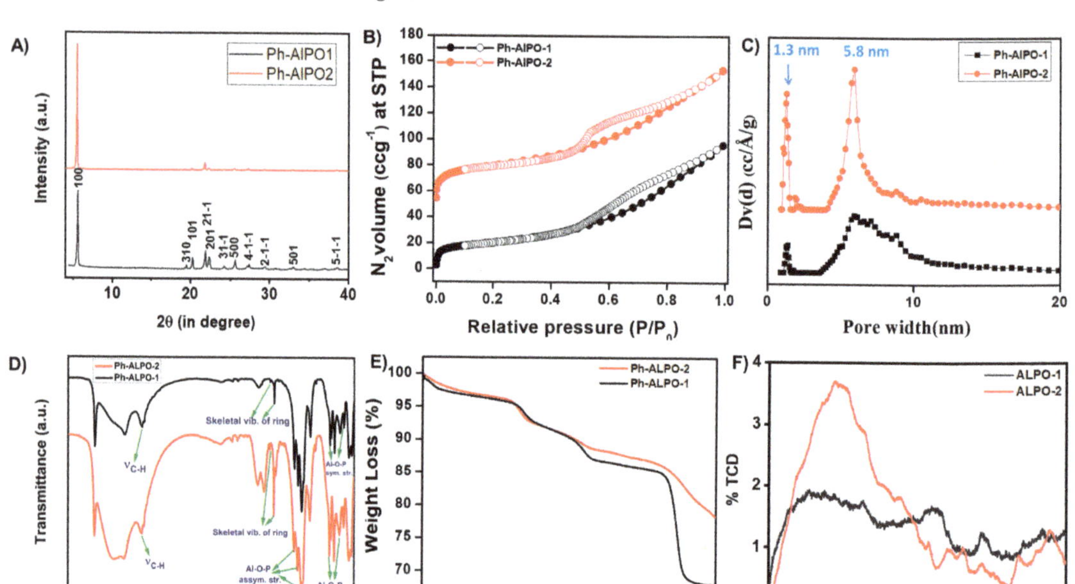

Figure 2. (**A**) Powder XRD patterns of Ph-ALPO-1 (black) and Ph-ALPO-2 (red), (**B**) N_2 adsorption/desorption isotherm of Ph-ALPO-1 (black) and Ph-ALPO-2 (red), (**C**) Non-local density functional theory (NLDFT) pore size distribution (PSD) of Ph-ALPO-1 (black) and Ph-ALPO-2 (red), (**D**) FT-IR spectra of Ph-ALPO-1 (black) and Ph-ALPO-2 (red), (**E**) Thermogravimetric curve of Ph-ALPO-1 (black) and Ph-ALPO-2 (red), (**F**) NH_3-TPD curve of Ph-ALPO-1 (black) and Ph-ALPO-2 (red).

Thermal stability of Ph-ALPO-1 and Ph-ALPO-2 materials was investigated through thermogravimetric analysis (TGA). The TGA analysis data was plotted in Figure 2E. Both the samples showed triplicate degradation between 25 and 800 °C. The first two mass losses between 6 and 15% occurred below 220 °C that are associated with an endothermic peak, which is because of desorption of physiosorbed and chemisorbed water molecules in the pores of the materials. The next mass loss occurs at temperatures higher than 250 °C, which is mainly due to the different degradation processes of organic matter. The mass loss recorded in this temperature range is about 8–10%. In the comparative TGA curves of both the materials, it clearly indicates that the final mass loss of Ph-ALPO-1 occurs at a faster rate than Ph-ALPO-2, which in turn clearly suggested the stability of the AlPO framework synthesized in the presence of the template molecule. Temperature-programmed desorption of ammonia (NH_3-TPD) analysis of both the materials was carried out under an inert He gas flow, and the corresponding 10% NH_3/He gas mixture desorption profile of ALPO-1 and ALPO-2 is shown in Figure 2F. From the NH_3-TPD profiles, it is seen that a broad NH_3 desorption peak centered around 80 °C was observed for both Ph-ALPO-1 and Ph-ALPO-2. This result suggested that NH_3 molecules are weekly bound in ALPO-1 and 2 surfaces. The observed total acidity of Ph-ALPO-1 and Ph-ALPO-2 were 92.59 and 285.82 μmolg^{-1}, respectively. Considerably higher acidity of Ph-ALPO-2 synthesized in the presence of the template has motivated us to explore its catalytic activity.

The morphology of both the hybrid AlPO materials has been investigated by transmission electron microscopic analysis (FEG-TEM). All the electron microscopic images are shown in Figure 3A–D. From the images, it was clearly shown that Ph-ALPO-1 shows irregular morphology (Figure 3A,B). Several 20–30 nm rod like particles are clearly visible for Ph-ALPO-1. However, these particles are irregularly nucleating to form several other morphologies as well (Figure 3A,B). But, one of our major objectives was to control the morphology of these nanoparticles, which could help in the catalytic reaction. We found that the small molecular template modification for the synthesis of Ph-ALPO-2 not only organize the particle morphology but also it hinders the agglomeration to generate monodisperse rod like nanoparticles of average size 20–30 nm (Figure 3C,D). The elemental mapping images shown in Figure 3E–H also confirms the template removal in Ph-ALPO-2 framework as no traces of "N" (comes from the template molecule) was found in the overall specimen. It is pertinent to mention that uniform distribution of C, O, P, and Al percentages in Ph-ALPO-2 (Figure 3E–H) also suggested the advantage of the phenyl functionalization of the AlPO framework via our above-mentioned process.

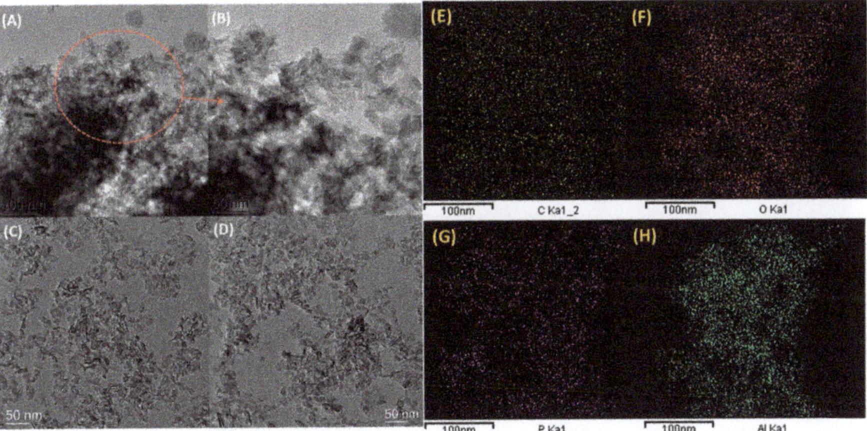

Figure 3. UHR-TEM images of Ph-ALPO-1 (**A,B**) and Ph-ALPO-2 (**C,D**). (**E–H**) are the C, O, P and Al elemental mapping data in Ph-ALPO-2.

3. Discussion

3.1. Catalytic Properties

3.1.1. Catalysis with Ph-ALPO-2

The 5-HMF synthesis from carbohydrates involve an acid catalyzed dehydration reaction and this has been extensively studied over the years. From the above characterizations and discussion of results, Ph-ALPO-2 could be a more promising material for acid catalysis compared to Ph-ALPO-1 as although both the materials have similar bonding connectivity and framework, Ph-ALPO-2 has more surface area than Ph-ALPO-1. Different carbohydrate precursors can be employed for the synthesis of 5-HMF by varying the monosaccharides. For optimizing the reaction conditions with maximum 5-HMF yield we have varied four major factors like the amount of catalyst loading, reaction temperature, reaction time, and choice of medium.

At first, taking fructose as the substrate and keeping all other parameters constant, we have performed the reaction by varying the catalyst amount from 7.08 wt % to 31.4 wt % to optimize the amount of catalyst needed for maximum 5-HMF yield. Figure 4A shows the different 5-HMF yield as a function of amount of given catalyst for fructose as the substrate. We have observed that upon increasing the catalyst amount from 2 mg up to 6 mg, the yield of the product has increased but further increases in the catalyst amount from 6 mg to 8 mg results in decreasing the product yield. From Figure 4A the relatively highest yield becomes 68.45% which was found for 6 mg of catalyst.

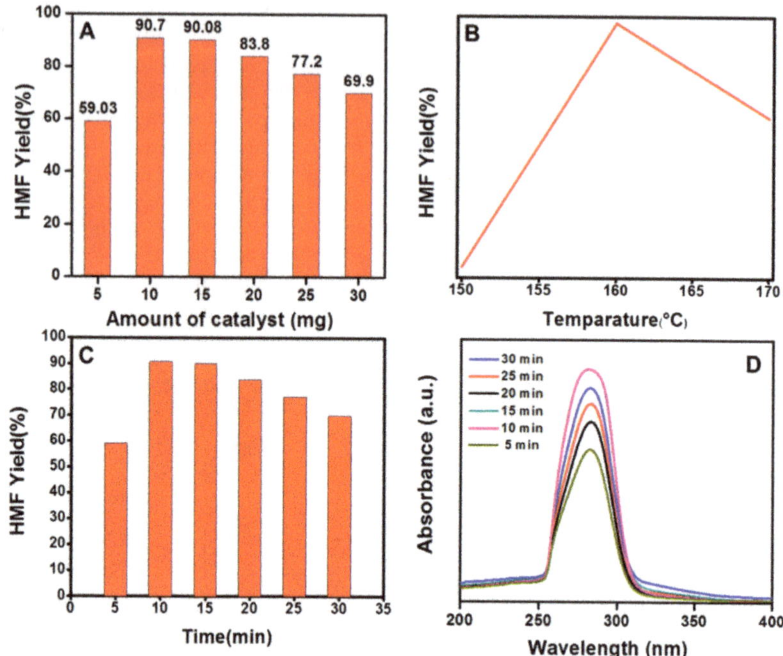

Figure 4. (**A**) Dependence of 5-HMF yield (%) on amount of catalyst; (**B**) Temperature dependence of 5-HMF yield (%); (**C**) Time dependence of 5-HMF yield (%) at the reaction temperature of 160 °C; (**D**) UV-Visible spectra of the product 5-HMF with reaction time (5–30 min). Typical standard deviation in the 5-HMF yield is ±1%.

Now we have carried out the reaction by varying the temperature from 140 °C to 170 °C with the similar carbohydrate substrate to observe the temperature dependence of the reaction on the product yield and the yield vs. temperature profile has been drawn from the 5-HMF yield (%) estimated through the UV-Visible spectrophotometric (shown in Figure 4B) studies. From this profile, the optimum temperature was 160 °C. At this temperature, fructose showed the highest yield for the 5-HMF in the presence of 6 mg of the catalyst for 15 min reaction time.

Then, we have plotted the 5-HMF yield as a function of reaction time from 5 min to 30 min at optimized temperature and catalyst amount by taking fructose as the substrate and we have drawn a suitable kinetic profile by calculating 5-HMF yield (%) through UV-Visible spectra using Beer–Lambert's law (Figure 4C). At 160 °C, 25.5 mg fructose shows the highest yield for the 5-HMF conversion reaction after 15 min reaction time with 6 mg aluminum phosphonate catalyst.

All these catalytic results are further tabularized in Table 1. Some observations seen from this table are, for example, that upon increasing the catalyst amount from 2 mg to 4 mg to 6 mg, the yield of the product has increased accordingly, but a further increase in the amount from 6 mg to 8 mg results in decreases in the product yield. The reason behind this observation may be that a certain increase in the catalyst gives a higher 5-HMF yield, as expected, according to an increasing number of acidic sites with the increasing amount of the catalyst, but a further increase in this catalyst amount after 6 mg reduces the product yield, which may be due to catalyst poisoning or further degradation of the main product by other byproducts. Again, a decreasing product yield was observed at further higher temperatures and longer reaction times. This may be attributed to the fact that a prolonged exposure time and higher reaction temperature may promote self- and cross-polymerization of 5-HMF with fructose, to yield the byproduct humins in the dehydration step as the appearance of a dark-colored product in these cases has been observed in the reaction vial.

Table 1. Catalytic results of Ph-ALPO-2 for the 5-HMF formation under different conditions.

Sl. No.	Substrate Amount (mg)	Catalyst Amount (mg)	Temperature (°C)	Time (min.)	Yield (%)
1.	25.5	2	150	20	42.8
2.	25.5	4	150	20	56.04
3.	25.5	6	150	20	68.45
4.	25.5	8	150	20	63.92
5.	25.5	6	160	20	83.3
6.	25.5	6	170	20	76.09
7.	25.5	6	160	5	59.03
8.	25.5	6	160	10	90.08
9.	25.5	6	160	15	90.7
10.	25.5	6	160	25	77.2
11.	25.5	6	160	30	69.9

The formation of 5-HMF was confirmed from the 600 Hz ^{1}H NMR spectra of crude product mixture and the solution after solvent extraction using 'DCM-H$_2$O' mixture, and these are shown in Figure S1. The 150 Hz ^{13}C-NMR spectrum has also provided further confirmation of 5-HMF formation (Figure S2). The λ_{max} for 5-HMF in UV-Visible spectrum arises at ~284 cm^{-1} (Figure S3). From that characteristic peak using the Beer–Lambert's law the yield of 5-HMF (shown in Figure 4D) has been calculated. The maximum amount of 5-HMF, 90.7%, was found for 6 mg of Ph-ALPO-2 catalyst at 160 °C in 15 min, from 25.5 mg of fructose as the substrate.

Then, we have estimated the catalytic activity by varying the substrates using some common monosaccharides like glucose and galactose in the above optimized condition as shown in Figure 5B. We observed that glucose and galactose give an almost similar yield for 5-HMF, which is quite low compared to fructose itself. We have tried this reaction using different solvents like 'DMSO + H$_2$O', DMF, 'acetonitrile + H$_2$O', and NMP. The yields are calculated from the respective UV-Visible spectra and presented as a bar diagram in Figure 5A. Compared with DMSO, other solvents showed low yields of 5-HMF. This result suggested that in the case of DMSO, the maximum amount of energy is converted from microwave radiation, which transforms maximum fructose to 5-HMF. The decreasing trend of the 5-HMF yields was DMSO > DMSO + H$_2$O > NMP > DMF> acetonitrile + H$_2$O.

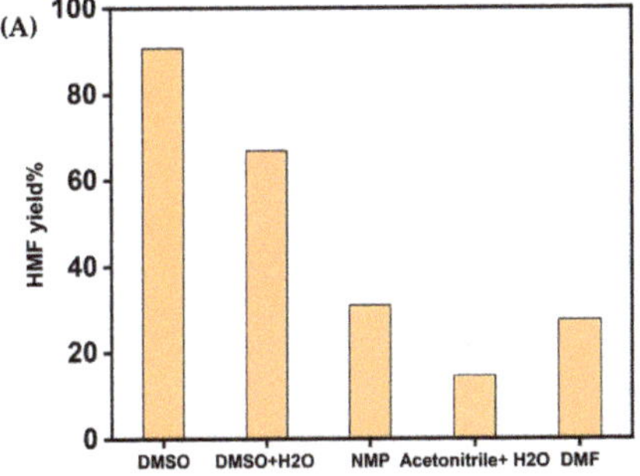

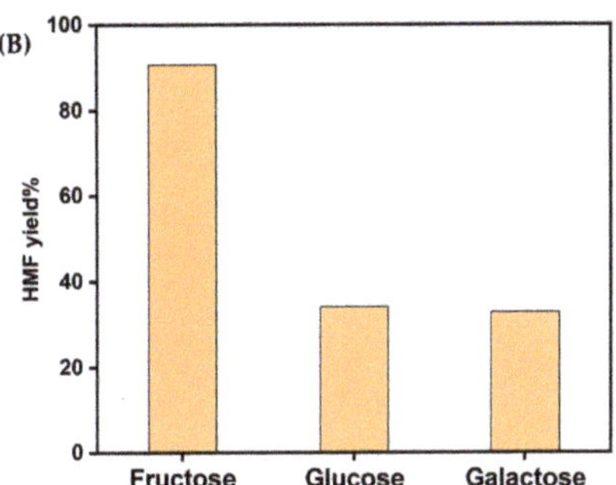

Figure 5. (**A**) Solvent dependence of substrate and (**B**) Selectivity of substrates for 5-HMF yield.

Further, we have performed the experiment with the lignocellulosic solid biomass (sugarcane pulp) also under optimized catalytic conditions. However, even after 30 min of microwave irradiation, we found a trace of 5-HMF only. This result suggested that Ph-ALPO-2 can catalyze the sugar content of the sugarcane pulp. Actually, when we use a solid biomass like lignin, after completion of the reaction, the lignin will form an insoluble physical mixture with the catalyst. As lignin has some solubility in DMSO, we tried to separate the catalyst through solvent extraction. But the wholesome biomass cannot be separated from the catalyst.

3.1.2. Catalysis with Ph-ALPO-1

With the above optimized condition, the 5-HMF yield from fructose in the case of Ph-ALPO-1 as the catalyst was 74.6% (as shown in Figure S4).

3.2. Role of Template in Synthesizing Phenyl Aluminophosphate Catalyst

From the nitrogen sorption isotherm, it is clear that Ph-ALPO-2, which has been synthesized using 1,3-diaminopropane-N,N,N′,N′-tetraacetic acid as a small-molecule template, has more surface area than Ph-ALPO-1, which has been synthesized in non-templated method. Again, from the TGA analysis, the framework of Ph-ALPO-2 shows higher stability than Ph-ALPO-1; i.e., upon using the template, the robustness of the framework of the AlPO material becomes much higher. Now, from the catalysis point of view, according to NH_3-TPD analysis, the template-based synthesized material shows higher active (acidic) sites than Ph-ALPO-1 (synthesized in non-templated method). As a result of all these above-mentioned facts, the yield of 5-HMF is also higher in the case of Ph-ALPO-2 in comparison with Ph-ALPO-1. These comparative results could be attributed to the fact that for generating porosity in the case of Ph-ALPO-2, the phenyl groups in phenylphosphonic acid is used as a spacer along with 1,3-diaminopropane-N,N,N′,N′-tetraacetic acid, which is used as a small-molecule template. But for Ph-ALPO-1, the pores are generated only due to the presence of phenyl spacer present in the phenylphosphonic acid. Thus, the higher BET surface area, crystallinity, and uniform particle morphology of the Ph-ALPO-2 catalyst synthesized in the presence of organic template are responsible for the higher yield of 5-HMF in this dehydration reaction.

3.3. Comparision of Ph-ALPO-2 with Other Catalysts in the Literature

The reaction parameters and corresponding yields of 5-HMF of this Ph-ALPO-2 material compared with those of related porous catalysts [34–38] available in the literature are tabularized in Table 2. From the results shown here, we can conclude that Ph-ALPO-2 displayed a considerably good catalytic efficiency for the synthesis of 5-HMF from biomass-derived fructose.

Table 2. Comparative study of catalytic efficiency in HMF conversion reaction between Ph-ALPO-1/2 and other zeo-type catalysts in the literature.

Catalyst	Temp. (°C)	Solvent	Time (min.)	Substrate	Yield (%)	Ref.
Al-KCC-1	162	DMSO	60	Fructose	92.9	[33]
NBPW-01	80	DMSO	180	Fructose	89	[34]
NBPW-06	80	DMSO	90	Fructose	96.7	[35]
Al-MCM-41	170	H_2O/MIBK	75	Fructose	42.3	[36]
LPSnP-1	120	H_2O/MIBK: 2-butanol	20	Fructose	77	[36]
TESAS-SBA-15	130	H_2O/MIBK: 2-butanol	141	Fructose	71	[37]
PHMS-2	160	DMSO	25	Fructose	83.7	[38]
Ph-ALPO-1	160	DMSO	15	Fructose	74.6	This work
Ph-ALPO-2	160	DMSO	15	Fructose	90.7	This work

4. Materials and Methods

4.1. Chemicals

Aluminum iso-propoxide, phenylphosphonic acid, liquid ammonia (25% aqueous), 1,3-diaminopropane-N,N,N′,N′-tetraacetic acid and fructose were purchased from Sigma Aldrich, Bangalore, India. All solvents were used after purification. Deionized water was used for synthesis and cleaning purposes.

4.2. Instrumentation

Powder X-ray diffraction data of aluminum phosphonates Ph-ALPO-1 and Ph-ALPO-2 were collected from a Bruker D-8 Advanced SWAX, Germany diffractometer using Ni-filtered Cu Kα (λ = 0.15406 nm) radiation. Specific surface area and porosity were analyzed from the N_2 sorption analysis at 77 K by using an Anton Paar Quanta Tec (USA) Isorb-HP-1 100 gas sorption analyzer. The pore-size distribution plots for both the materials were obtained from the non-local density functional theory (NLDFT) model. Both samples were dried and degassed at 150 °C for 5 h under high vacuum before the sorption analysis. Ultrahigh-resolution transmission electron microscopic (UHR-FEG TEM) system of JEOL JEM 2100F (Japan) working with at a 200 kV electron source, the particle size and nanostructure of the Ph-ALPO-1 and Ph-ALPO-2 frameworks were analyzed. Samples for these TEM analysis was prepared by dropping isopropanol solutions of both the materials into two separate carbon-coated copper grids and then drying them under a high vacuum. Using KBr pellets and by using a Shimadzu FT-IR 8400S (Japan) instrument, Fourier transform infrared (FT-IR) spectra of the samples were obtained. The Anton Paar Microwave Synthesis Reactor (Monowave 300, USA) was used for all of the catalysis experiments. For measuring the surface acidity of these aluminum phosphonate materials temperature-programmed desorption of ammonia (TPD-NH_3) analysis was carried out in an AMI-300 Lite Chemisorption Analyzer of Altramira Instruments LLC, USA.

4.3. Synthesis of Ph-ALPO-1

In a typical synthesis of Ph-ALPO-1, 204 mg (0.001 M) phenylphosphonic acid was dissolved in 18 mL deionized water and stirred for 30 min on a magnetic stirrer. Then, 158 mg (0.001 M) aluminum iso-propoxide was added in it and again stirred for another 15 min. After that liquid ammonia was added in dropwise manner until the pH of the solution becomes ~8.0. Then, it was again stirred for ~3 h. Then, the reaction mixture was transferred into a 30 mL Teflon-lined stainless-steel autoclave and placed in an oven at 170 °C for 3 days. After 3 days, aluminum phenylphosphonate material was formed. It was then separated and washed with water and methanol by centrifugation. The yield of the product becomes ~90 mg. Then, the product was collected in the Eppendorf and characterized thoroughly.

4.4. Synthesis of Ph-ALPO-2

In this stage, 306 mg (0.001M) 1,3-diaminopropane-N,N,N′,N′-tetraacetic acid was dissolved in 21 mL deionized water and stirred for 45 min. Then, 158 mg (0.001 M) phenylphosphonic acid was mixed in it and again stirred for another 30 min. Then, 204 mg (0.001 M) aluminum iso-propoxide was added in it and stirred for another 3 h. After that, liquid ammonia was added in a dropwise manner until the pH of the solution became ~8.0. Aluminum phenylphosphonate gel had been prepared. Now, this gel was transferred into a 30 mL Teflon-lined stainless-steel autoclave and placed in an oven at 170 °C for 3 days. After 3 days, an aluminum phenylphosphonate framework containing the template molecules was formed. It was then separated and washed with water and methanol by centrifugation. The yield of the product was ~120 mg.

4.5. Extraction of Template from Ph-ALPO-2

The as-synthesized product was taken in a beaker and stirred with 50% EtOH and 50% water for 6 h continuous stirring at room temperature. The extracted product was then separated and washed with water and EtOH via centrifugation, and the product was collected in the Eppendorf and used for further application.

4.6. Activation of Catalyst

In order to remove adsorbed solvent molecules, which may partially impede catalytic activity by blocking some active sites, the catalysts are activated prior to the catalytic reaction. We activated all the catalysts (Ph-ALPO-1 and Ph-ALPO-2) at 120 °C for 6 h in a hot air oven before putting them into the catalysis chamber. For all subsequent catalytic experiments, a similar activation route for the catalysts was followed.

4.7. 5-HMF Synthesis

Here, fructose, glucose, and galactose were used as carbohydrate sources for the synthesis of 5-HMF. In a typical synthesis, the carbohydrate was charged with activated catalyst and DMSO (2 mL) in a microwave-safe G-10 vial. The vial was then correctly sealed and kept at the desired temperature for the above-mentioned amount of time in a microwave reactor. The internal pressure of the vial was initially nil. However, after the steady-state condition, the internal pressure was elevated to 5 bar. The temperature of the microwave was controlled by a previously set program; briefly, the final temperature was set as 160 °C for a holding time of 15 min with a continuous stirring speed of 600 rpm, and the reaction mixture was cooled to 25 °C to obtain the product. At 160 °C, the temperature at which the catalysis was optimized, 24% of the catalyst, Ph-ALPO-2, produced 90.7% HMF yield in DMSO medium in 15 min reaction time. Water and DCM were used to extract 5-HMF from the reaction mixture. In DMSO-d6 medium, the extracted product was examined via ^{1}H-NMR and ^{13}C-NMR spectroscopic analysis (distinctive signals of 5-HMF shown in Figures S1 and S2). Then, 5-HMF yield was calculated from the solution using UV-Visible spectrophotometry.

5. Conclusions

In this work, we have demonstrated the synthesis of a new porous organic–inorganic hybrid aluminum phosphonate framework material having a hydrophobic phenyl moiety at the pore surface in the presence and absence of a small organic molecule 1,3-diaminopropane-N,N,N',N'-tetraacetic acid as a template via a hydrothermal synthetic route. We have successfully shown the crucial role of porosity and surface area, and the use of organic template in the hydrothermal synthesis for the catalytic application. The synthetic protocol and the catalytic results described here are cost-effective and environmentally friendly. Furthermore, these porous phenylphosphonates show the potential to synthesize 5-HMF from carbohydrates in a very high yield in a short reaction time using a green chemical route. Thus, the easy and convenient synthesis method reported herein could offer a new opportunity for researchers and industries for the sustainable synthesis of biofuel intermediates over organically functionalized aluminum phosphonates.

Supplementary Materials: The following supporting information can be downloaded at: https://www.mdpi.com/article/10.3390/catal13111449/s1, Indexing of powder XRD of Ph-AlPO-1/2 (Table S1), ^{1}H NMR (Figure S1) and ^{13}C-NMR (Figure S2) spectra of 5-HMF, UV-Visible characteristic peak of 5-HMF (Figure S3), and comparative catalytic activity of Ph-ALPO-1 and Ph-ALPO-2 through UV-Visible spectrophotometry (Figure S4).

Author Contributions: Experiments, investigation, and formal analysis were performed by R.M. B.M. involved in the formal analysis of the catalysts. A.B. provided the resources and conducted the investigation and overall supervision of this project. R.M. wrote the draft manuscript with the help of A.B. All authors have read and agreed to the published version of the manuscript.

Funding: BM wants to thank UGC, New Delhi (NTA Ref. No.: 221610080374), for a Junior Research Fellowship.

Data Availability Statement: Data is contained within the article.

Acknowledgments: RM wants to thank Anirban Ghosh for helping with the catalytic experiments and draft manuscript preparation and for contributing to useful scientific discussions. AB would like to acknowledge DST-SERB, New Delhi, for a Core Research Grant (Project no. CRG/2022/002812).

Conflicts of Interest: The authors declare no conflict of interest.

References

1. Höök, M.; Tang, X. Depletion of fossil fuels and anthropogenic climate change—A review. *Energy Policy* **2013**, *52*, 797–809. [CrossRef]
2. Ragauskas, A.J.; Williams, C.K.; Davison, B.H.; Britovsek, G.; Cairney, J.; Eckert, C.A.; Frederick, W.J.J.; Hallett, J.P.; Leak, D.J.; Liotta, C.L.; et al. The path forward for biofuels and biomaterials. *Science* **2006**, *311*, 484–489. [CrossRef] [PubMed]
3. Cunha, J.T.; Romani, A.; Domingues, L. Whole Cell Biocatalysis of 5-Hydroxymethylfurfural for Sustainable Biorefineries. *Catalysts* **2022**, *12*, 202. [CrossRef]
4. Barta, K.; Ford, P.C. Catalytic Conversion of Nonfood Woody Biomass Solids to Organic Liquids. *Acc. Chem. Res.* **2014**, *47*, 1503–1512. [CrossRef]
5. Corma, A.; Iborra, S.; Velty, A. Chemical Routes for the Transformation of Biomass into Chemicals. *Chem. Rev.* **2007**, *107*, 2411–2502.
6. Liu, Y.Z.; Chen, W.S.; Xia, Q.Q.; Guo, B.T.; Wang, Q.W.; Liu, S.X.; Liu, Y.X.; Li, J.; Yu, H.P. Efficient Cleavage of Lignin-Carbohydrate Complexes and Ultrafast Extraction of Lignin Oligomers from Wood Biomass by Microwave-Assisted Treatment with Deep Eutectic Solvent. *ChemSusChem* **2017**, *10*, 1692–1700. [CrossRef]
7. Chheda, J.N.; Roman-Leshkov, Y.; Dumesic, J.A. Production of 5-Hydroxymethylfurfural and Furfural by Dehydration of Biomass-Derived Mono- and Poly-Saccharides. *Green Chem.* **2007**, *9*, 342–350. [CrossRef]
8. Osatiashtiani, A.; Lee, A.F.; Brown, D.R.; Melero, J.A.; Morales, G.; Wilson, K. Bifunctional SO$_4$/ZrO$_2$ Catalysts for 5-Hydroxymethylfufural (5-HMF) Production from Glucose. *Catal. Sci. Technol.* **2014**, *4*, 333–342. [CrossRef]
9. Bozell, J.J.; Petersen, G.R. Technology Development for the Production of Biobased Products from Biorefinery Carbohydrates—The US Department of Energy's "Top 10" Revisited. *Green Chem.* **2010**, *12*, 539–554. [CrossRef]
10. Solanki, B.S.; Rode, C.V. Selective Hydrogenation of 5-HMF to 2,5-DMF Over a Magnetically Recoverable Non-Noble Metal Catalyst. *Green Chem.* **2019**, *21*, 6390–6406.
11. Cortez-Elizalde, J.; Córdova-Pérez, G.E.; Silahua-Pavón, A.A.; Pérez-Vidal, H.; Cervantes-Uribe, A.; Cordero-García, A.; Arévalo-Pérez, J.C.; Becerril-Altamirano, N.L.; Castillo-Gallegos, N.C.; Lunagómez-Rocha, M.A.; et al. 2,5-Dimethylfuran Production by Catalytic Hydrogenation of 5-Hydroxymethylfurfural Using Ni Supported on Al$_2$O$_3$-TiO$_2$-ZrO$_2$ Prepared by Sol-Gel Method: The Effect of Hydrogen Donors. *Molecules* **2022**, *27*, 4187. [CrossRef] [PubMed]
12. Tong, X.L.; Ma, Y.; Li, Y.D. Biomass into Chemicals: Conversion of Sugars to Furan Derivatives by Catalytic Processes. *Appl. Catal. A Gen.* **2010**, *385*, 1–13. [CrossRef]
13. Tempelman, C.H.L.; Oozeerally, R.; Degirmenci, V. Heterogeneous Catalysts for the Conversion of Glucose into 5-Hydroxymethyl Furfural. *Catalysts* **2021**, *11*, 861. [CrossRef]
14. Bains, R.; Kumar, A.; Chauhan, A.S.; Das, P. Dimethyl carbonate solvent assisted efficient conversion of lignocel lulosic biomass to 5-hydroxymethylfurfural and furfural. *Renew. Energy* **2022**, *197*, 237–243. [CrossRef]
15. Mondal, S.; Mondal, J.; Bhaumik, A. Sulfonated Porous Polymeric Nanofibers as an Efficient Solid Acid Catalyst for the Production of 5-Hydroxymethylfurfural from Biomass. *ChemCatChem* **2015**, *7*, 3570–3578. [CrossRef]
16. Yin, Y.; Ma, C.H.; Li, W.; Luo, S.; Zhang, Z.S.; Liu, S.X. Insights into Shape Selectivity and Acidity Control in NiO-Loaded Mesoporous SBA-15 Nanoreactors for Catalytic Conversion of Cellulose to 5-Hydroxymethylfurfural. *ACS Sustain. Chem. Eng.* **2022**, *10*, 17081–17093. [CrossRef]
17. Zhang, Y.Z.; Zhao, B.W.; Das, S.; Degirmenci, V.; Walton, R.L. Tuning the Hydrophobicity and Lewis Acidity of UiO-66-NO$_2$ with Decanoic Acid as Modulator to Optimise Conversion of Glucose to 5-Hydroxymethylfurfural. *Catalysts* **2022**, *12*, 1502. [CrossRef]
18. Elhamifar, D.; Nasr-Esfahani, M.; Karimi, B.; Moshkelgosha, R.; Shábani, A. Ionic Liquid and Sulfonic Acid Based Bifunctional Periodic Mesoporous Organosilica (BPMO-IL-SO$_3$H) as a Highly Efficient and Reusable Nanocatalyst for the Biginelli Reaction. *ChemCatChem* **2014**, *6*, 2593–2599.
19. Herbst, A.; Janiak, C. MOF Catalysts in Biomass Upgrading Towards Value-Added Fine Chemicals. *CrystEngComm* **2017**, *19*, 4092–4117.
20. Chongdar, S.; Bhattacharjee, S.; Bhanja, P.; Bhaumik, A. Porous organic–inorganic hybrid materials for catalysis, energy and environmental applications. *Chem. Commun.* **2022**, *58*, 3429–3460.
21. Lewis, D.W.; Willock, D.J.; Catlow, C.R.A.; Thomas, J.M.; Hutchings, G.J. De Novo Design of Structure-Directing Agents for the Synthesis of Microporous Solids. *Nature* **1996**, *382*, 604–606. [CrossRef]
22. Cundy, C.S.; Cox, P.A. The Hydrothermal Synthesis of Zeolites: Precursors, Intermediates and Reaction Mechanism. *Microporous Mesoporous Mater.* **2005**, *82*, 1–78.
23. Jiang, J.X.; Yu, J.H.; Corma, A. Extra-Large-Pore Zeolites: Bridging the Gap between Micro and Mesoporous Structures. *Angew. Chem. Int. Ed.* **2010**, *49*, 3120–3145. [CrossRef] [PubMed]

24. Wilson, S.T.; Lok, B.M.; Messina, C.A.; Cannan, T.R.; Flanigen, E.M. Aluminophosphate Molecular Sieves: A New Class of Microporous Crystalline Inorganic Solids. *J. Am. Chem. Soc.* **1982**, *104*, 1146–1147. [CrossRef]

25. Davis, M.E.; Saldarriaga, C.; Montes, C.; Garces, J.; Crowder, C. A Molecular-Sieve With 18-Membered Rings. *Nature* **1988**, *331*, 698–699. [CrossRef]

26. Mizoshita, N.; Tani, T.; Inagaki, S. Syntheses, Properties and Applications of Periodic Mesoporous Organosilicas Prepared from Bridged Organosilane Precursors. *Chem. Soc. Rev.* **2011**, *40*, 789–800. [CrossRef]

27. Bhanja, P.; Na, J.; Jing, T.; Lin, J.J.; Wakihara, T.; Bhaumik, A.; Yamauchi, Y. Nanoarchitectured Metal Phosphates and Phosphonates: A New Material Horizon toward Emerging Applications. *Chem. Mater.* **2019**, *31*, 5343–5362. [CrossRef]

28. Gagnon, K.J.; Perry, H.P.; Clearfield, A. Conventional and Unconventional Metal-Organic Frameworks Based on Phosphonate Ligands: MOFs and UMOFs. *Chem. Rev.* **2012**, *112*, 1034–1054. [CrossRef]

29. Li, J.Y.; Qi, M.; Kong, J.; Wang, J.Z.; Yan, Y.; Huo, W.F.; Yu, J.H.; Xu, R.R.; Xu, Y. Computational Prediction of the Formation of Microporous Aluminophosphates with Desired Structural Features. *Microporous Mesoporous Mater.* **2010**, *129*, 251–255. [CrossRef]

30. Konstantin, I.; Galkin, K.I.; Krivodaeva, E.A.; Romashov, L.V.; Zalesskiy, S.S.; Kachala, V.V.; Burykina, J.V.; Ananikov, V.P. Critical Influence of 5-Hydroxymethylfurfural Aging and Decomposition on the Utility of Biomass Conversion in Organic Synthesis. *Angew. Chem. Int. Ed.* **2016**, *55*, 8338–8342.

31. Altomare, A.; Cuocci, C.; Giacovazzo, C.; Moliterni, A.; Rizzi, R.; Corriero, N.; Falcicchio, A. EXPO2013: A Kit of Tools for Phasing Crystal Structures from Powder Data. *J. Appl. Cryst.* **2013**, *46*, 1231–1235. [CrossRef]

32. Chowdhury, A.; Bhattacharjee, S.; Chatterjee, R.; Bhaumik, A. A New Nitrogen Rich Porous Organic Polymer for Ultra-High CO_2 Uptake and as An Excellent Organocatalyst for CO_2 Fixation Reactions. *J. CO2 Util.* **2022**, *65*, 102236. [CrossRef]

33. Shahangi, F.; Chermahini, A.N.; Saraji, M. Dehydration of fructose and glucose to 5-hydroxymethylfurfural over Al-KCC-1 silica. *J. Energy Chem.* **2018**, *27*, 769–780. [CrossRef]

34. Qiu, G.; Wang, X.; Huang, C.; Li, Y.; Chen, B. Niobium phosphotungstates: Excellent solid acid catalysts for the dehydration of fructose to 5-hydroxymethylfurfural under mild conditions. *RSC Adv.* **2018**, *8*, 32423–32433. [CrossRef] [PubMed]

35. Jiang, C.W.; Su, A.X.; Li, X.M. Preparation of Aluminosilicate Mesoporous Catalyst and its Application for Production 5-Hydroxymethyl Furfural Dehydration from Fructose. *Adv. Res. Mater.* **2011**, *396–398*, 1190–1193. [CrossRef]

36. Dutta, A.; Gupta, D.; Patra, A.K.; Saha, B.; Bhaumik, A. Synthesis of 5-Hydroxymethylfurural from Carbohydrates using Large-Pore Mesoporous Tin Phosphate. *ChemSusChem* **2014**, *7*, 925–933. [CrossRef]

37. Crisci, A.J.; Tucker, M.H.; Lee, M.Y.; Jang, S.G.; Dumesic, J.A.; Scott, S.L. Acid-Functionalized SBA-15-Type Silica Catalysts for Carbohydrate Dehydration. *ACS Catal.* **2011**, *1*, 719–728. [CrossRef]

38. Ghosh, A.; Chowdhury, B.; Bhaumik, A. Synthesis of Hollow Mesoporous Silica Nanospheroids with O/W Emulsion and Al(III) Incorporation and Its Catalytic Application. *Catalysts* **2023**, *13*, 354. [CrossRef]

 *catalysts*

Article

Pretreatment and Nanoparticles as Catalysts for Biogas Production Reactions in Pepper Waste and Pig Manure

Ana Isabel Parralejo Alcobendas *, Luis Royano Barroso, Juan Cabanillas Patilla and Jerónimo González Cortés

Instituto de Investigaciones Agrarias Finca la Orden-Valdesequera, Centro de Investigaciones Científicas y Tecnológicas de Extremadura, Scientific and Technological Research Centre of Extremadura (CICYTEX), Guadajira 06187, Badajoz, Spain; luis.royano@juntaex.es (L.R.B.); juan.cabanillas@juntaex.es (J.C.P.); jeronimo.gonzalez@juntaex.es (J.G.C.)
* Correspondence: ana.parralejoa@juntaex.es; Tel.: +34-924014013

Abstract: The circular economy is based on using waste generated from any process to obtain products with zero residues' criteria. This research was focused on pepper waste from the polyphenolic extraction method. Pepper waste was evaluated in batch and semi-continuous regime anaerobic digestion, adding, as catalysts, absorbent nanoparticles and/or using pretreatment strategies. The best methane yields were obtained from SB1 (assay without pretreatment in pepper waste): 464 ± 25 NL kg VS^{-1} for batch assays; and from period II (1.47 g VS L^{-1} d^{-1}) of S2 (assay of pig manure and pepper waste with thermal pretreatment): 160 NL/kg VS^{-1} for semi-continuous experiments. However, a kinetic study showed a methane production rate higher for SB2 (assay with nanoparticles as catalyst) than SB1 in batch assays.

Keywords: nanoparticles; pretreatment pepper waste; kinetic; anaerobic process

Citation: Parralejo Alcobendas, A.I.; Royano Barroso, L.; Cabanillas Patilla, J.; González Cortés, J. Pretreatment and Nanoparticles as Catalysts for Biogas Production Reactions in Pepper Waste and Pig Manure. *Catalysts* **2023**, *13*, 1029. https://doi.org/10.3390/catal13071029

Academic Editors: José María Encinar Martín, Sergio Nogales Delgado and Ioannis V. Yentekakis

Received: 17 May 2023
Revised: 9 June 2023
Accepted: 19 June 2023
Published: 21 June 2023

1. Introduction

Nowadays, the biorefinery concept is related optimizing the use of waste to obtain biofuels, energy, and high-added-value subproducts. Through this concept, it is possible fight against the climate change. Concretely, the European Commission has identified some priority areas where the European Directive from Renewables Energies [1] must act. A total elimination of waste generated by industries and the introduction of renewable energies in their processes can be regarded as appropriate measures undertaken to achieve some of the specific objectives against the climate change [2]. Particularly, in a region of Spain (Extremadura) where this research has been developed, there is a Regional Plan of Research, Technological Development, and Innovation [3]. This plan is focussed on economic priorities in the agrifood sector. For this reason, if waste from the agrifood sector is considered to be put to optimum use, we will be moving towards a more sustainable economy. According to the Spanish Statistics National Institute [4], there was 636,116 t of vegetable waste in Spain in 2020 from food and drink manufacturing industries and tobacco factories. A total of 26% in the production of fruits and vegetables in European countries in 2021 belong to Spain [5]. Furthermore, the amount of exported pepper from Spain exceeded 800,000 t in 2021, and in Spain, 1,500,000 t of pepper was produced in 2021 [6]. Normally, waste produced from pepper is around 50–60% of the total processed biomass [7]. The waste generated is usually employed as animal feed or discharged into landfills, leading to environmental degradation in the areas in which they are disposed of [8]. A strategic solution must be developed to manage the large amount of pepper waste produced in the country. Applying a biorefinery concept to pepper waste will be an excellent way to optimize the benefits. Different extraction methods can be carried out to achieve the valorization of this waste. High-added-value product can be obtained as polyphenolic and carotenoids compounds. A newly developed first-step extraction method offers an opportunity to obtain more degraded pepper waste to use as feed for microorganisms that produce biofuels—more concretely, biogas. The process to generate biogas is called anaerobic digestion (AD); it consists of degrading organic matter in the waste by specific microorganisms to produce biogas. Another product is generated in this process: digestate, which can be used as organic fertilizer. Digestate is a fertilizer containing odorless stabilized organic matter and NPK nutrients which have changed to mineral forms which are available for plants [9]. In this process, two or more residues can be employed, and this process is called anaerobic co-digestion (AC-D). There is a large number of studies concerning AC-D substrates employing vegetable waste with animal waste (i.e., slurry or cattle manure). An evaluation of pepper waste's addition in a co-digestion process with swine manure was developed by Riaño et al. [10]. In this study, the highest specific methane yield obtained under batch conditions was 309 N L CH_4 kg VS^{-1}, with a percentage of pepper waste in the mixture of 50% (on the VS basis). After AC-D under semicontinuous operation at different OLR values was studied, the method was shown to increase the specific methane yield by up to 86% compared to that obtained from a mono-digestion assay of swine manure (208 N L CH_4 kg VS^{-1} at 1.26 g VS $L^{-1}d^{-1}$). Li et al. [11] studied the AC-D of wood waste with pig manure and evaluated the methane production potential using a NaOH pretreatment in the wood waste. The obtained results showed that the methane yield was increased by 75.8% after NaOH pretreatment compared with the untreated wood waste. To improve the kinetics, different mechanisms can be employed, such as nanoparticles, bioelectrochemical applications, and nano-biochar. Madondo et al. [12] researched the application of bioelectrochemical systems and magnetite nanoparticles in sewage sludge for the improvement of organic content degradation. In this case, an enhanced methane percentage was obtained versus the control (88% versus 39%). A review [13] focused on the role of additive nano-biochar in the AC-D kinetic shows evidence of nano-biochar's value as a catalyst for enhancing biogas production. However, this review refers to the development of no-continuous operational modes. Semi-continuous operational modes using this kind of catalyst have not been so well studied. Magnetized nanoparticles (iron oxides and aluminum sulfate) were employed by Kweinor and Rathilal [14] to obtain a quicker reaction rate in the AC-D process of wastewater. The kinetic parameters calculated showed that the presence of these nanoparticles shortens the lag phase of the control system, with a kinetics rate of 0.285 d^{-1} for control and of 0.127 d^{-1} and 0.195 d^{-1} for iron oxides and aluminum sulfate nanoparticles, respectively.

Alkalinity is an important parameter in the AC-D process, mainly in a semi-continuous regime. High alkalinity (based on the equilibrium carbon dioxide–bicarbonate) provides an excellent buffer capacity of the digestion medium. VFA (Volatile Fatty Acids) accumulations or in pH values are avoided, according Smridhivej and Boyd [15]. When the feed of substrates added to the digester (organic load rate (OLR)) is increasing in the AC-D process, alkalinity and VFA must be controlled to avoid the inhibition of the process.

Due to the gap in the research related to the semi-continuous operational mode using diverse types of catalyst, the present study proposes to assess the performance and stability of assays employing pig manure and pepper waste in the AC-D, including absorbent nanoparticles and/or strategies of waste pretreatment, as catalysts. The obtained results are compared through different kinetic parameters calculated according to simulation models.

2. Results and Discussion

2.1. Chemical Characterization of Raw Materials

The assays developed in this research used different raw materials: pepper waste pretreatment (PWP) (with thermal pretreatment and/or nanoparticles), pig manure (PM), and pepper waste (PW). PWP with PM were studied in semi-continuous assays and PW with PM were employed in a pilot plant assay. PM, PW, and PWP were characterized before to start the studies, and results are shown in Table 1.

Table 1. Chemical parameters of raw materials determined.

Parameter	PM	PW	PWP
pH	7.70 ± 0.10	4.18 ± 0.04	4.10 ± 0.29
Redox potential, mV	-362 ± 23	-100 ± 1	206 ± 10
Alkalinity, mg $CaCO_3$ L^{-1}	9379 ± 75	-	-
N-NH_4, mg L^{-1}	1860 ± 85	870 ± 30	390 ± 80
C, %	2.23 ± 0.30	7.12 ± 0.19	7.63 ± 1.85
N, %	0.30 ± 0.03	0.36 ± 0.01	0.42 ± 0.09
C/N	7.32 ± 0.36	18.04 ± 1.37	18.71 ± 2.17
TS, %	5.71 ± 0.02	13.31 ± 0.29	17.17 ± 3.33
VS *, %	3.98 ± 0.12	12.32 ± 0.23	15.80 ± 3.14
Ca, ppm	2663 ± 48	1003 ± 21	1965 ± 18
Fe, ppm	209 ± 2	76 ± 1	120 ± 2
K, ppm	240 ± 30	2317 ± 25	1953 ± 4
Mg, ppm	1208 ± 11	439 ± 1	617 ± 4
Na, ppm	913 ± 2	64 ± 1	262 ± 3
P, ppm	1562 ± 7	473 ± 7	807 ± 46
Al, ppm	152 ± 7	81 ± 1	95 ± 5

* Total Volatile Solid over Total Solids.

As can be observed in Table 1, low values of pH from PW and PWP are presented. The alkalinity parameter value from PM is quite high to buffer the low pH values from pepper materials. If AC-D works with alkalinity values higher than 2000 mg $CaCO_3$ L^{-1}, as it does according to Flotats et al. [16], it indicates the stability of the process. The C/N proportion used in the feed must be close to 20–30 [17–20]; PW and PWP have values near to these values. The TS values of PW and PWP are very similar, and they are quite high. Moreover, the Total Volatile Solid is observed to be about 93 % of the TS. It entails a high potential of organic transformation of PW and PWP, as it is happened in research carried out by Arhoun et al. [21] (they developed AC-D of mixed sewage sludge and fruits and vegetable wholesale market waste).

2.2. Biochemical Methane Potential (BMP) of Different Strategies with PWP

Three assays were undertaken to find the most productive method: batch assay without pretreatment (SB1); batch assay with a determined absorbent nanoparticles dose (SB2); and batch assay with another determined absorbent nanoparticles dose and thermal pretreatment (SB3). Table 2 shows the BMP and the kinetic parameters. These results evidence that methane yield from SB1 is the highest of the studies carried out. Moreover, a thermal pretreatment of the PWP can be an adequate method to increase the methane average concentration in the biogas obtained. Gallego L. M. et al. [22] evaluated the empirical BMP through different models from some horticultural waste such as beet pulp and pear flesh; the obtained results (249 NL kg VS^{-1} and 318 NL kg VS^{-1} for beet pulp and pear flesh, respectively) were lower than the values obtained in this research. Kinetic parameters show higher R_{max} values for SB2 and SB3 than R_{max} for SB1. This probably means that the nanoparticles in the medium quickly support production methane rate.

Table 2. BMP and kinetic parameters for different studies developed with PWP.

Parameter	SB1		SB2		SB3	
Methane average yield, NL kg VS^{-1}	464 ± 25		331 ± 57		364 ± 49	
Methane average concentration, %	59 ± 2		56 ± 6		60 ± 1	
Replicates	R1	R2	R1	R2	R1	R2
R_{max}, Nm^3 kg VS^{-1} d^{-1}	0.64	0.75	0.82	1.51	0.85	1.00
l, d	2.82	1.56	0.77	1.29	-	-
R^2	0.9888	0.9757	0.9387	0.9727	0.9686	0.9766

Figures 1–3 illustrate the kinetic model fitting for the three studies. Two replicates were developed for each study. All of them are perfectly fitted to the modified Gompertz model because the regression coefficients are too elevated.

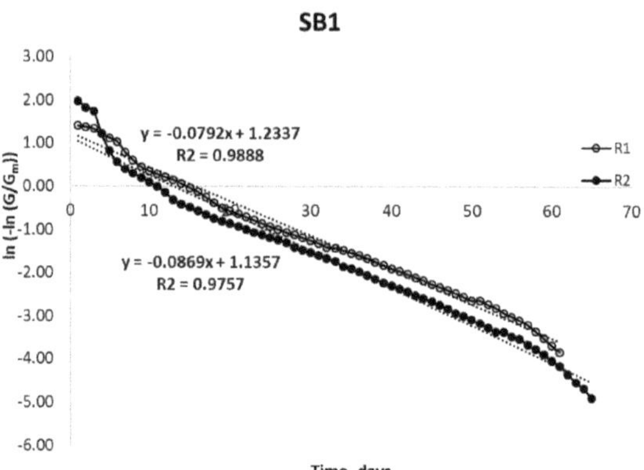

Figure 1. Experimental results fitted to the modified Gompertz model for SB1.

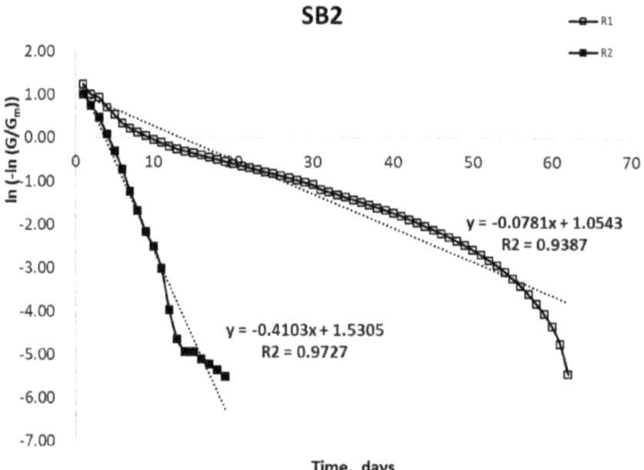

Figure 2. Experimental results fitted to the modified Gompertz model for SB2.

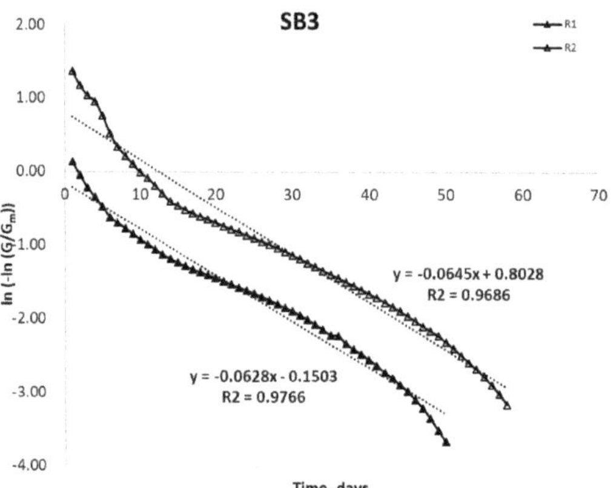

Figure 3. Experimental results fitted to the modified Gompertz model for SB3.

There is no lag phase (*l*) in SB3 because a previous pretreatment has been developed, and the experimental results fitted to the kinetic model were taken after this pretreatment. Regarding the lag phase from SB1 and SB2, the values obtained present more elevated values in SB1 than SB2. This fact seems to indicate that the nanoparticles' presence improves the methane production in the first stage. The lag phase average values of sorghum and corn stover (0.190 d and 2.648 d, respectively) obtained by González et al. [23] are the lowest in this research and very close the experimental to values of this work. Chiappero et al. [24] employed different biochars as catalysts during AD of mixed wastewater sludge, and the kinetic parameter R_{max} for the modified Gompertz model ranged between 0.014 and 0.034 $Nm^3 \, kg \, VS^{-1} \, d^{-1}$, i.e., lower values than results obtained in this work.

2.3. Different Pretreatment for Assays Semi-Continuous with PM and PWP

Assays were developed with three different conditions: S1: PM with PWP; S2: PM with PWP and thermal pretreatment; and S3: PM with PWP developing thermal pretreatment and nanoparticles use (the employed dose (0.015 g g VS^{-1}) was the most productive for batch assay). In Figure 4, we observe influence of the treatment carried out to increase the methane production for OLR (period I to III). The methane volume represented in Figure 4 seems very similar for periods II and III. A light difference can be seen in period I, obtaining the highest production for S2 and the lowest production for S1; this means that the nanoparticles are not increasing the methane production in the AC-D for period I (Table 3 does not show rises of methane production when nanoparticles are employed neither). Chen et al. [25] studied two types of magnetic nanoparticles (Ni ferrite nanoparticles and Ni Zn ferrite nanoparticles). They found a stimulation of anaerobic digestion in synthetic municipal wastewater with a certain type of nanoparticles but an inhibition of another type of nanoparticles was added to the anaerobic digestion medium. The values of methane yields and kinetic parameters are represented in Table 3.

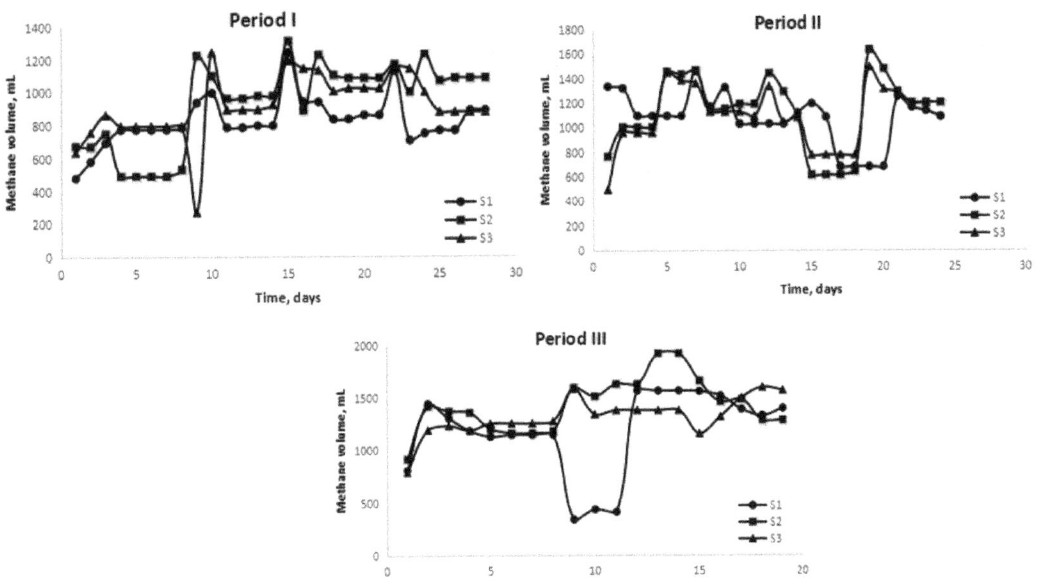

Figure 4. Methane volume evolution for different periods.

Table 3. Methane yield and kinetic parameters obtained in semi-continuous assays.

Period	Period I	Period II	Period III
		S1	
Methane yield, NL kg VS^{-1}	139	149	149
k_B, g COD L^{-1}		4.42	
U_{max}, d^{-1}		1.03	
k_2, d^{-1}		0.73	
		S2	
Methane yield, NL kg VS^{-1}	158	160	156
k_B, g COD L^{-1}		4.98	
U_{max}, d^{-1}		1.49	
k_2, d^{-1}		0.90	
		S3	
Methane yield, NL kg VS^{-1}	153	156	155
k_B, g COD L^{-1}		4.33	
U_{max}, d^{-1}		1.31	
k_2, d^{-1}		0.64	

According with the catalyst effect of the nanoparticles and the thermal pretreatment, the kinetic parameters' reaction constants—k_B and k_2—for each kinetic model fitted are shown in Table 2. The highest values of reaction constants belong to S2. This corresponds to the assay with the most elevate methane yield (S2). In any case, the reaction constant obtained in this work are higher than values of reactions constant obtained by other authors; 0.25 g COD g VS^{-1} d^{-1} has been reported by De la Lama D. et al. [26] for "alperujo" in semi-continuous anaerobic digestion of the thermally pretreated medium.

As can be seen in Figures 5 and 6, experimental results from the cumulative methane production of the S1 and S2 experiments are perfectly fitted to the Stover–Kincannon and second-order models because their regression coefficients are elevated (0.9999, 0.9822, and 0.9772).

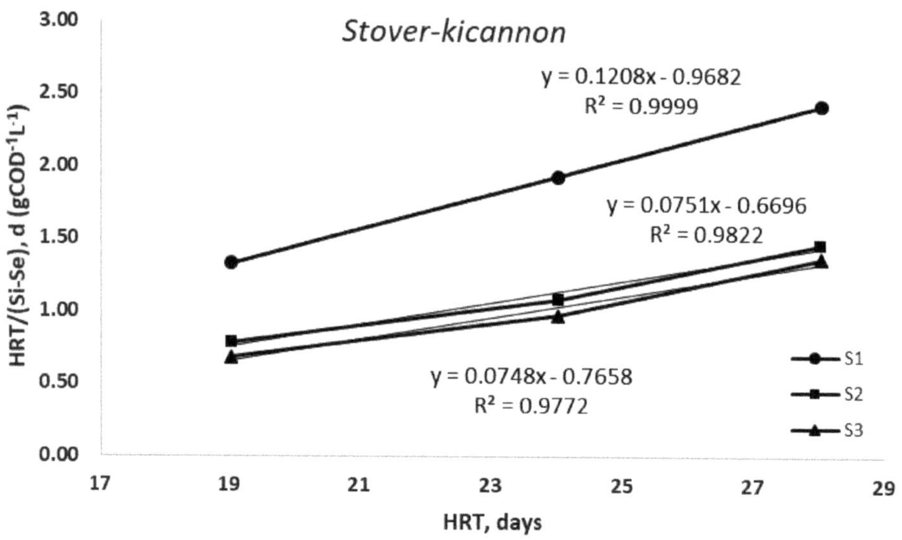

Figure 5. Experimental results fitted to the Stover–Kincannon model for semi-continuous experiments.

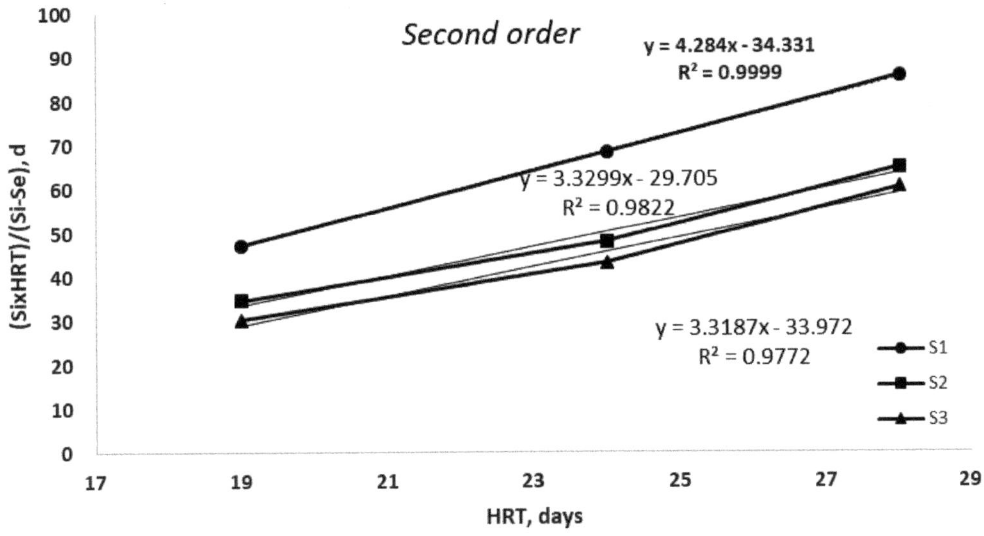

Figure 6. Experimental results fitted to the second-order model for semi-continuous experiments.

2.4. Effect of OLR on Different Parameters

The interactions study of certain parameters in the AC-D process and the studied values of OLR are represented in Figures 7 and 8. An expected, direct interaction is presented in alkalinity when the OLR is increased; this means that the process stability is increasing until 1.88 g VS L_D^{-1} d^{-1} for the three assays developed. This behavior was found in a work carried out by Parralejo et al. [27], where the AC-D process in semi-continuous assays was evaluated for OLR ranged 1.2–1.8 g VS L_D^{-1} d^{-1} for different mixtures of animal manure or nitrogen-rich biomass. However, the VS parameter evolution with OLR shows a light decrease in the second period (and small increases in methane yield (Table 3)). VFA and ammonia nitrogen parameters exhibit a direct interaction with the OLR evaluated in the most of assays developed. This is a normal behavior when the organic matter is enhanced. Nevertheless, the values of VFA and ammonia nitrogen are below the threshold values for the stability of the processes (4000 mg L^{-1} and 5000 mg L^{-1} for ammonia nitrogen and VFA, respectively) [28].

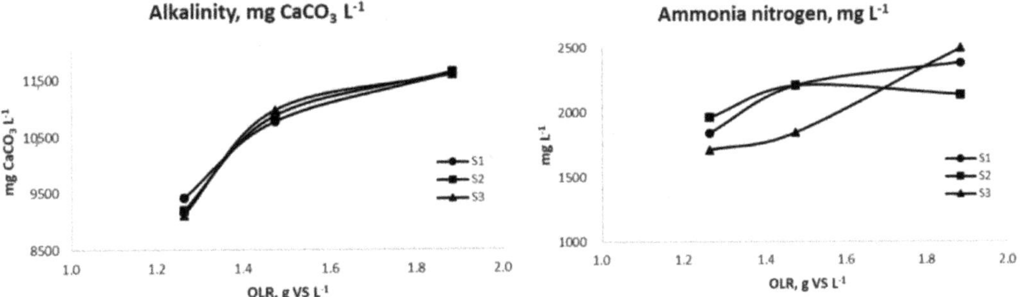

Figure 7. Alkalinity (**left**) and ammonia nitrogen (**right**) effects on OLR developed in semi-continuous assays.

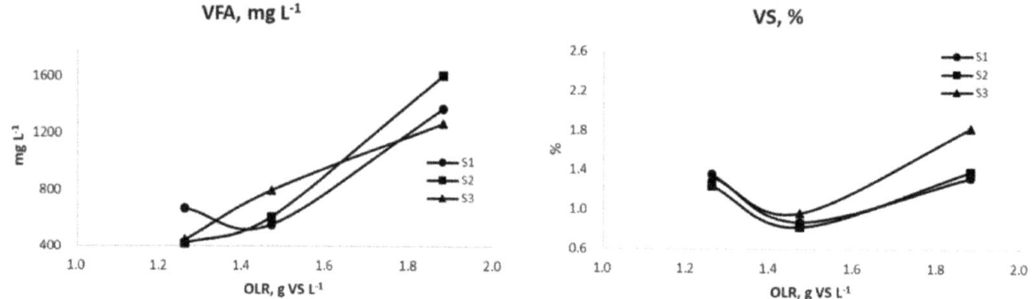

Figure 8. VFA (**left**) and VS (**right**) effects on OLR developed in semi-continuous assays.

2.5. Digestate Pilot Plant Experiment

An experiment in a pilot plant was carried out for two OLR values (period I and II). In this experiment, a semi-continuous AC-D for PW and PM was assessed. Methane yield and digestate composition were evaluated for each OLR studied, and the results are shown in Table 4.

Table 4. Methane yield and digestate composition for experiment carried out in pilot plant.

Parameter	Period I	Period II
pH	7.91 ± 0.10	7.95 ± 0.05
Redox potential, mV	-388 ± 12	-409 ± 9
Alkalinity, mg CaCO$_3$ L^{-1}	$10{,}892 \pm 2$	$10{,}340 \pm 37$
Methane yield, NL kg VS^{-1}	173 ± 45	264 ± 55
C, %	1.06 ± 0.03	2.20 ± 0.20
N, %	0.21 ± 0.01	0.37 ± 0.02
C/N	4.98 ± 0.15	6.02 ± 1.14
Ca, ppm	779 ± 4	2259 ± 9
Fe, ppm	75 ± 2	230 ± 2
K, ppm	1317 ± 28	2467 ± 29
Mg, ppm	526 ± 2	1121 ± 3
Na, ppm	669 ± 2	596 ± 6
P, ppm	359 ± 10	1052 ± 15
Al, ppm	40 ± 2	157 ± 2
Zn, ppm	34 ± 1	141 ± 1
Cu, ppm	10 ± 1	38 ± 1
Cr, ppm	<5	<5
Ni, ppm	<5	<5

Higher methane yield is obtained for period II (elevate OLR employed) than for period I. This difference in S1 was observed in assays carried out in laboratory conditions. It could be due to the pepper substrate being pretreated and the organic matter being degraded in the pretreatment process. All values of the elements showed in Table 4 are higher for period II than the values of elements for period I, and in any case, the values for period II are correct for the development of the AC-D process because the methane yield has an adequate value. If the digestates obtained for two periods are evaluated as fertilizer, N, P, and K nutrients are the most important. Normally, P and K are often expressed as P$_2$O$_5$ and K$_2$O, respectively. In Table 5, the nutrient compositions assimilable by plants are exposed along with a fertilizer classification for digestates according to the Spanish standard [29].

In the classification of the Spanish standard, "fertilizers A" are those that have the lowest amount of Zn, Cu, Cr, and Ni. The plant nutrients' availability amounts to 55%, 64%, and 92% for N, P$_2$O$_5$, and K$_2$O, respectively [30].

Table 5. Nutrient composition assimilable by plants of digestates from period I and II.

Parameter	Period I	Period II
Assimilable N content, %	0.12	0.20
Assimilable P_2O_5 content, %	0.80	2.33
Assimilable K_2O content, %	1.01	1.89
Fertilizer classification [28]	A	A

3. Materials and Methods

3.1. Evaluated Raw Materials

This research employed pig manure (PM) and pepper waste as raw materials, both without pretreatment (PW) and pretreated (PWP). PM was collected from a pig farm located in Guadajira (Badajoz, Spain) (+38°51′9.6768″, −6°40′15.5418″). PW was provided by a frozen vegetable factory. PW was composed of stem, peduncle, and seeds, so the heterogeneous waste was mixed and chopped via mechanical pretreatment to obtain a homogeneous paste. Moreover, PWP was subjected to the polyphenols extraction method to achieve optimization in the waste valorization. Polyphenols extraction employed water as solvent and ultrasound bath to be as close as possible to the most environmentally friendly techniques. PM was stored at room temperature; PW homogeneous paste was frozen, and the extraction method of PW to obtain PWP was carried out weekly. An inoculum was employed to help the development of the specific microorganisms. The inoculum used in assays consisted of a mixture of completely degraded organic material with a high content of methanogenic microorganisms. The inoculum was composed of a mixture of prickly pear and pig manure.

3.2. Digester Used and Experimental Design

Laboratory and pilot plant digesters were employed in this research, both made of stainless steel, with a central agitator electrically operated and adjustable by a potentiometer to obtain the mixture of substrates, and a thermostat to control the temperature inside the digesters. Laboratory digesters are coated with an outer jacket through which hot water circulates to maintain the constant temperature of the substrate, and pilot plant digester has an inner coil surrounding the walls for the hot water. The total volumes of laboratory and pilot plant digesters are 6 L and 2000 L, respectively, but the used volumes for these experiments were 4.5 L and 1500 L, respectively. In this study, a mesophilic temperature range (38 °C) was employed. At the beginning, three batch assays were developed (Table 6) to establish the influence of the nanoparticles' dose and the thermal pretreatment presence: SB1 batch assay of PWP; SB2 batch assay of PWP with a determined absorbent nanoparticles dose (0.064 g g VS^{-1}); and SB3 batch assay of PWP with another determined absorbent nanoparticles dose (0.015 g g VS^{-1}) and thermal pretreatment.

The ratio of inoculum to PWP was 1:2 on VS basis. The nanoparticles used belong to a small factory (Smallops), located in Badajoz (Extremadura, Spain), that manufactures the product from organic waste. Three semi-continuous assays were carried out (S1 to S3). The studied fed were S1: PM with PWP; S2: PM with PWP developing thermal pretreatment; S3: PM with PWP developing thermal pretreatment and nanoparticles use. The working procedure from semi-continuous regime assays consisted of a daily feeding of the substrate mixture. A hopper on the top of the digesters with a ball valve was employed to introduce the substrate mixture, and another valve located on the side of the digester was used to extract the digestate. Three different OLRs (1.26 g SV L_D^{-1} d^{-1}; 1.47 g SV L_D^{-1} d^{-1}; and 1.88 g SV L_D^{-1} d^{-1}) were studied for each assays set. Each OLR evaluated was considered a study period. Table 6 shows the experimental design. Finally, a pilot plant experiment was carried out, studying a mixture of 50% of PM and 50% of PW (on VS basis).

Table 6. Experimental design in assays sets evaluated.

Assay	OLR, g VS LD−1 d−1	Mixture Composition/Feed	Hydraulic Retention Time (HRT), d
SB1	-	Inoculum–PWP (ratio: 1:2)	-
SB2	-	Inoculum–PWP (ratio: 1:2) with nanoparticles (0.064 g g VS^{-1})	-
SB3	-	Inoculum–PWP (ratio: 1:2) with nanoparticles (0.015 g g VS^{-1}) and thermal pretreatment	-
S1	Period I: 1.26	50% PM and 50% PWP (on VS basis)	28
	Period II: 1.47		24
	Period III: 1.88		19
S2	Period I: 1.26	50% PM and 50% PWP (on VS basis) with thermal pretreatment	28
	Period II: 1.47		24
	Period III: 1.88		19
S3	Period I: 1.26	50% PM and 50% PWP (on VS basis) with thermal pretreatment and nanoparticles (0.015 g g VS^{-1})	28
	Period II: 1.47		24
	Period III: 1.88		19

3.3. Analytical Methods

APHA standard methods [31] were employed to characterize the substrates used. Drying the sample to a constant weight in an oven (JP Selecta Digitheat, Barcelona, Spain) at 105 °C for 48 h (2540 B method) and at 550 °C for 2 h in a muffle oven (Hobersal 12PR300CCH, Hobersal Furnaces & Ovens Technology, Barcelona, Spain) using an inert atmosphere (2540 E method) were the means employed to determine total solids (TS) and volatile solids (VS) in the samples analyzed. Specific electrodes were employed to measure the pH and redox potential values of the digestion medium connected to a pH meter (Crison Basic 20, Hach lange Spain S.L.U., Barcelona, Spain). To determine the alkalinity of the medium, method 2320 was employed; for the chemical oxygen demand (COD), method 410.4 was employed [32]; for ammonia nitrogen (N-NH$_4$) by volumetric titration, the E4500-NH$_3$ B method was employed; and for total volatile fatty acids (VFA), to Buchauer's volumetric method was employed [33]. The ratio between N and C nutrients was analyzed by a True-Spec CHN Leco 4084 elementary analyzer (ECO empowering results, Madrid, Spain, according to the UNE-EN 16948 standard for biomass analysis C, N, H [34]. A constant monitoring of the biogas volume and its composition was carried out with an Awite System of Analysis Process series 9 analyzer (Bioenergie GmbH, Awite Bioenergie GmbH, Langenbach, Germany) (composed of different sensors to detected methane, carbon dioxide, hydrogen, hydrogen sulfide, and oxygen concentration). The gas meter (Ritter model MGC-1 V3.2 PMMA, Awite Bioenergie GmbH, Langenbach, Germany) was employed to measure the biogas produced, which was stored in Tedlar bags. The biogas volume produced was corrected at standard conditions (0 °C, 101,325 kPa). The digestate was featured by spectroscopy using an ICP-OES Varian 715 ES (Agilent Technologies, Santa Clara, CA, USA).

3.4. Evaluation of Substrate Removal Kinetic Models

For batch assays, experimental results have been fitted to a kinetic model called a modified Gompertz [14]. For semi-continuous assays, based on the substrate removal rate, Grau second-order multicomponent and modified Stover–Kincannon models have been employed as kinetic models [26]. For the Grau second-order multicomponent model when multicomponent substrates are evaluated, the substrate removal rate can be expressed according to Equation (1):

$$\frac{-dS}{dt} = k_{n(s)} \cdot \mathrm{X} \cdot \left(\frac{S_e}{S_i}\right)^n \tag{1}$$

where $-dS/dt$ is the substrate removal rate; k_n (s) is the reaction constant; X is the concentration of the microorganisms, which ca be assumed as constant; S_e is the substrate concentration at any time; and S_i is the initial substrate concentration.

Integrating Equation (1) for $n = 2$ and linearizing it, the following linear expression is obtained (Equation (2)):

$$\frac{(S_i \cdot HRT)}{(S_i - S_e)} = HRT + \frac{S_i}{k_s \cdot X} \tag{2}$$

The value of the second-order reaction constant can be obtained by the plot of the $(S_i \cdot HRT)/(S_i - S_e)$ versus HRT. The term HRT is the hydraulic retention time value for each set assay.

In the modified Stover–Kincannon model, the substrate removal rate is expressed as a function of the OLR as follows in Equations (3) and (4):

$$\frac{dS}{dt} = \frac{(S_i - S_e)}{HRT} \tag{3}$$

$$\frac{dS}{dt} = \frac{U_{\max} \cdot \left(\frac{S_i}{HRT}\right)}{k_B + \left(\frac{S_i}{HRT}\right)} \tag{4}$$

where dS/dt is the substrate removal rate; k_B is the reaction constant; $U_{\max}$ is the maximum substrate removal rate; S_i and S_e are the substrate concentrations explained above; and HRT the hydraulic retention time, as has been specified before. When Equations (3) and (4) are equalized and integrated and the resulting expression is linearized, Equation (5) is obtained, as follows:

$$\frac{HRT}{(S_i - S_e)} = \frac{k_B}{U_{\max}} \cdot \frac{HRT}{S_i} + \frac{1}{U_{\max}}. \tag{5}$$

Experimental results fitted to Equation (5) give a linear expression where the reaction constant can be obtained from the slope.

4. Conclusions

A comparison between batch and semi-continuous assays have been developed concerning AC-D processes in pig manure and pepper waste, including absorbent nanoparticles and/or pretreatment strategies as catalysts. For batch assays, kinetic parameters specify that the presence of nanoparticles in the medium quickly supports the methane production rate (higher $R_{\max}$ values for SB2 and SB3 than $R_{\max}$ for SB1, and the lag phase is lowest for SB2). The studied influence of the pretreatment carried out in the methane production for the OLRs the evaluated OLRs (period I to III) via semi-continuous assays shows slightly more elevated values for the thermal pretreatment assay (160 NL kg VS^{-1}). Direct interaction among alkalinity, VS, ammonia nitrogen and VFA parameters, and the OLR has been found. Finally, digestates from experimental pilot plants evaluated for two OLR values have been assessed and classified, according the Spanish standard, as the fertilizer with the lowest heavy metal concentration and assimilable N, P_2O_5, and K_2O content of 0.12%, 0.80%, and 1.01% for period I, and 0.20%, 2.33%, and 1.89% for period II, respectively. However, this kind of fertilizer must be extensively studied before being applied to different crops.

Author Contributions: Conceptualization: J.G.C.; Methodology: A.I.P.A. and J.C.P.; Validation: J.G.C.; Investigation: A.I.P.A. and L.R.B.; Resources: J.C.P.; Writing—review and editing: A.I.P.A., L.R.B. and J.G.C.; Supervision: J.G.C. All authors have read and agreed to the published version of the manuscript.

Funding: The authors are grateful to the funding support by the National Institute of Research and Agro-Food Technology (INIA), co-financed with FEDER funds (PID2019-105039RR-C42).

Data Availability Statement: Any additional information will be provide upon reasonable request.

Conflicts of Interest: The authors declare no conflict of interest.

References

1. European Union. Directiva (UE) 2018/2001 del Parlamento Europeo y del Consejo de 11 de Diciembre de 2018 Relativa al Fomento del Uso de Energía Procedente de Fuentes Renovables. In *Diario Oficial de la Unión Europea*; European Union: Strasbourg, France, 2018.
2. *Manual Sobre las Biorrefinerías en España*; Ministerio de Economía, Industria y Competitividad: Madrid, Spain, 2017.
3. Consejería de Economía, Ciencia y Agenda Digital. *VII Plan Regional de Investigación, Desarrollo Tecnológico e Investigación de Extremadura 2022–2025*; Consejería de Economía, Ciencia y Agenda Digital: Mérida, Spain, 2022.
4. Instituto Nacional de Estadística. Available online: https://www.ine.es/jaxi/Datos.htm?path=/t26/p067/p02/residuos/serie/l0/&file=02001.px (accessed on 3 January 2023).
5. Ministerio de Agricultura, Pesca y Alimentación. *Cifras del Sector de Frutas y Hortalizas*; Ministerio de Agricultura, Pesca y Alimentación: Madrid, Spain, 2022.
6. Available online: https://es.statista.com/estadisticas/510920/produccion-de-pimiento-fresco-en-espa~na-por-comunidad-autonoma/ (accessed on 3 June 2023).
7. Ros, M.; Pascual, J.A.; Ayuso, M.; Morales, A.B.; Miralles, J.R.; Solera, C. Salidas valorizables de los residuos y subproductos orgánicos de la industria de los transformados de frutas y hortalizas: Proyecto LIFE + Agrowaste. *Residuos Rev. Técnica* **2012**, *22*, 28–35.
8. Vulic, J.; Seregelj, V.; Kalusevic, A.; Levic, S.; Nedovic, V.; Saponjac, V.T.; Canadanovic-Brunet, J.; Cetkovic, G. Bioavailability and bioactivity of encapsulated phenolics and carotenoids isolated from red pepper wastes. *Molecules* **2019**, *24*, 2837. [CrossRef] [PubMed]
9. Mórtola, N.; Romaniuk, R.; Consentino, V.; Eiza, M.; Carfagno, P.; Rizzo, P.; Bres, P.; Riera, N.; Roba, M.; Butti, M.; et al. Potential Use of a Poultry Manure Digestate as a Biofertiliser: Evaluation of Soil Properties and *Lactuca sativa* Growth. *Pedosphere* **2019**, *29*, 60–69. [CrossRef]
10. Riaño, B.; Molinuevo-Salcés, B.; Parralejo, A.; Royano, L.; González, J.; García-González, M.C. Techno-economic evaluation of anaerobic co-digestion fo pepper waste and swine manure. *Biomass Convers. Biorefin.* **2021**, *13*, 7763–7774. [CrossRef]
11. Li, R.; Tan, W.; Zhao, X.; Dang, Q.; Song, Q.; Xi, B.; Zhang, X. Evaluation on the Methane Production Potential of Wood Waste Pretreated with NaOH and Co-Digested with Pig Manure. *Catalysts* **2019**, *9*, 539. [CrossRef]
12. Madondo, N.I.; Tetteh, E.K.; Rathilal, S.; Femi Bakare, B. Application of Bioelectrochemical System and Magnetite Nanoparticles on the Anaerobic Digestion of Sewage Sludge: Effect of Electrode Configuration. *Catalysts* **2022**, *12*, 642. [CrossRef]
13. Goswami, L.; Kushwaha, A.; Singh, A.; Saha, P.; Choi, Y.; Maharana, M.; Patil, S.V.; Kim, B.S. Nano-Biochar as a Sustainable Catalyst for Anaerobic Digestion: A Synergetic Closed-Loop Approach. *Catalysts* **2022**, *12*, 186. [CrossRef]
14. Tetteh, E.K.; Rathilal, S. Kinetics and Nanoparticle Catalytic Enhancement of Biogas Production from Wastewater Using a Magnetized Biochemical Methane Potential (MBMP) System. *Catalysts* **2020**, *10*, 1200. [CrossRef]
15. Somridhivej, B.; Boyd, C.E. An assessment of factors affecting the reliability of total alkalinity measurements. *Aquaculture* **2016**, *459*, 99–109. [CrossRef]
16. Flotats, X.; Bonmatí, A.; Campos, E.; Antúnez, M. Ensayos en discontinuo de co-digestión anaerobia termofílica de purines de cerdo y lodos residuales. *Inf. Tecnol.* **1999**, *10*, 79–85.
17. Zhang, W.; Kong, T.; Xing, W.; Li, R.; Yang, T.; Yao, N.; Lv, D. Links between carbon/nitrogen ratio, synergy and microbial characteristics of long-term semi-continuous anaerobic co-digestion of food waste, cattle manure and corn straw. *Bioresour. Technol.* **2022**, *343*, 126094. [CrossRef]
18. Xiao, Y.; Zan, F.; Zhang, W.; Hao, T. Alleviating nutrient imbalance of low carbon-to-nitrogen ratio food waste in anaerobic digestion by controlling the inoculum-to-substrate ratio. *Bioresour. Technol.* **2022**, *346*, 126342. [CrossRef]
19. Slopiecka, K.; Liberti, F.; Massoli, S.; Bartocci, P.; Fantozzi, F. Chemical and physical characterization of food waste to improve its use in anaerobic digestion plants. *Energy Nexus* **2022**, *5*, 100049. [CrossRef]
20. Samoraj, M.; Mironiuk, M.; Izydorczyk, G.; Witek-Krowiak, A.; Szopa, D.; Moustakas, K.; Chojnacka, K. The challenges and perspectives for anaerobic digestion of animal waste and fertilizer application of the digestate. *Chemosphere* **2022**, *295*, 133799. [CrossRef]
21. Arhoun, B.; Villen-Guzman, M.; Gomez-Lahoz, C.; Rodriguez-Maroto, J.M.; Garcia-Herruzo, F.; Vereda-Alonso, C. Anaerobic co-digestion of mixed sewage sludge and fruits and vegetable wholesale market waste: Composition and seasonality effect. *J. Water Process Eng.* **2019**, *31*, 100848. [CrossRef]
22. Gallego, L.M.; Portillo, E.; Navarrete, B.; González, R. Estimation of methane production through the anaerobic digestion of greenhouse horticultural waste: A real case study for Almeria region. *Sci. Total Environ.* **2022**, *807*, 151012. [CrossRef]
23. González, J.F.; Parralejo, A.I.; González, J.; Álvarez, A.; Sabio, E. Optimization of the production and quality of biogas in the anaerobic digestion of different types of biomass in a batch laboratory biodigester and pilot plant: Numerical modeling, kinetic study and hydrogen potential. *Int. J. Hydrogen Energy* **2022**, *43*, 39386–39403. [CrossRef]
24. Chiappero, M.; Cillerai, F.; Berruti, F.; Masek, O.; Fiore, S. Addition of Different Biochars as Catalysts during the Mesophilic Anaerobic Digestion of Mixed Wastewater Sludge. *Catalysts* **2021**, *11*, 1094. [CrossRef]
25. Chen, J.L.; Steele, T.W.J.; Stuckey, D.C. The effect of Fe_2NiO_4 and Fe_2NiO_4Zn magnetic nanoparticles on anaerobic digestion activity. *Sci. Total Environ.* **2018**, *642*, 276–284. [CrossRef]

26. De la Lama, C.; Borja, R.; Rincón, B. Performance evaluation and substrate removal kinetics in the semi-continuous anaerobic digestion of thermally pretreated two phase olive pomace or "Alperujo". *Process Saf. Environ. Prot.* **2017**, *105*, 288–296. [CrossRef]
27. Parralejo, A.I.; Royano, L.; Cabanillas, J.; González, J. Biogas from Nitrogen-Rich Biomass as an Alternative to Animal Manure Co-Substrate in Anaerobic Co-Digestion Processes. *Enegies* **2022**, *15*, 5978. [CrossRef]
28. Drosg, B. *Process Monitoring in Biogas Plant*; Task 37—Energy from Biogas and Landfill Gas; International Energy Agency (IEA) Bioenergy: Paris, France, 2017.
29. *RD 535/2017*; Sobre Productos Fertilizantes de 26 de Mayo. Ministerio de la Presidencia y para las Administraciones Territoriales: Madrid, Spain, 2017.
30. Agriculture & Horticulture Development Board (AHDB). *Nutrient Management Guide (RB209)*; Section 2: Organic Materials; Agriculture & Horticulture Development Board (AHDB): Kenilworth, UK, 2017; p. 35.
31. APHA. *Standard Methods for Examination of Water and Wastewater*, 22nd ed.; American Public Health Association: Washington, DC, USA, 2012.
32. *EPA-600/4-79-020*; Methods for Chemical Analysis of Water and Wastes. Environmental Protection Agency: Washington, DC, USA; Environmental Monitoring and Support Laboratory: Des Plaines, IL, USA, 1983.
33. Buchauer, K. Titrations verfahren in der Abwasserund Schlam-manalytik zur Bestimmung von flüchtigen organischen Säuren. Das Gas- und Wasserfach (gfw). *Wasser Abwasser* **1997**, *138*, 313–320.
34. *Norma UNE-EN 16948*; Biocombustibles Sólidos, Determinación de Contenido Total de C, H y N, Método Instrumental. University of New England: Biddeford, UK, 2015.

 *catalysts*

Review

A Review on the Use of Catalysis for Biogas Steam Reforming

Sergio Nogales-Delgado [1,*], Carmen María Álvez-Medina [1], Vicente Montes [2,3] and Juan Félix González [1]

[1] Department of Applied Physics, University of Extremadura, Avda. De Elvas s/n, 06006 Badajoz, Spain; carmenmaam@unex.es (C.M.Á.-M.); jfelixgg@unex.es (J.F.G.)

[2] Department of Organic and Inorganic Chemistry, University of Extremadura, Avda. De Elvas s/n, 06006 Badajoz, Spain; vmontes@unex.es

[3] University Institute for Water Research, Climate Change and Sustainability (IACYS), Avda. De Elvas s/n, 06006 Badajoz, Spain

* Correspondence: senogalesd@unex.es

Abstract: Hydrogen production from natural gas or biogas, at different purity levels, has emerged as an important technology with continuous development and improvement in order to stand for sustainable and clean energy. Regarding biogas, which can be obtained from multiple sources, hydrogen production through the steam reforming of methane is one of the most important methods for its energy use. In that sense, the role of catalysts to make the process more efficient is crucial, normally contributing to a higher hydrogen yield under milder reaction conditions in the final product. The aim of this review is to cover the main points related to these catalysts, as every aspect counts and has an influence on the use of these catalysts during this specific process (from the feedstocks used for biogas production or the biodigestion process to the purification of the hydrogen produced). Thus, a thorough review of hydrogen production through biogas steam reforming was carried out, with a special emphasis on the influence of different variables on its catalytic performance. Also, the most common catalysts used in this process, as well as the main deactivation mechanisms and their possible solutions are included, supported by the most recent studies about these subjects.

Keywords: methane; Ni-based catalysts; sintering; coking; poisoning; promoters; hydrogen; syngas; catalyst support

Citation: Nogales-Delgado, S.; Álvez-Medina, C.M.; Montes, V.; González, J.F. A Review on the Use of Catalysis for Biogas Steam Reforming. *Catalysts* **2023**, *13*, 1482. https://doi.org/10.3390/catal13121482

Academic Editor: Binlin Dou

Received: 31 October 2023
Revised: 25 November 2023
Accepted: 26 November 2023
Published: 29 November 2023

1. Introduction

1.1. Biogas: Production, Characteristics, and Upgrading

In a global context where sustainability, green chemistry, and a circular economy are highly demanded by society, companies, and governmental agencies (who are encouraging the implementation of facilities to produce biomethane, for instance), the implementation of green technologies is becoming more and more important to contribute to a lower dependency on geopolitical changes or the energy market.

Thus, regarding the replacement of petrochemical products, whose treatment and processing are equally unsustainable from an environmental point of view, the use of green technologies could be an interesting alternative. In a sense, these practices could counteract the abovementioned negative effects, influencing considerably the current energy and geopolitical scenarios, which are always in constant evolution [1–3].

In this respect, the role of anaerobic digestion is gaining more and more importance, as it is a very versatile technology where different feedstocks with different characteristics can be used under different operating conditions that could be easily adapted to current green policies and standards.

Indeed, a considerable increase in biogas production worldwide has taken place (it has quadrupled in the last two decades), with Europe as the leading region with more than 50% of biogas global production and Germany with the most facilities for this purpose, although there are still opportunities for the European biogas industry [4,5].

In any case, due to its versatility by treating different wastes, this technology presents a great opportunity for the implementation of green technologies all over the world, especially in developing countries, where the management of some wastes is a challenge and, equally, an opportunity.

In that sense, biogas production seems to be a promising and strategic sector in the future energy scenario, and its consolidation in developed countries, as well as its possible implementation in developing areas, is a clear example of the potential of this technology in the near future.

As can be seen from Figure 1, biogas is produced by anaerobic bacteria that degrade organic material to biogas in four steps: hydrolysis, acidification, the production of acetic acid, and the production of methane.

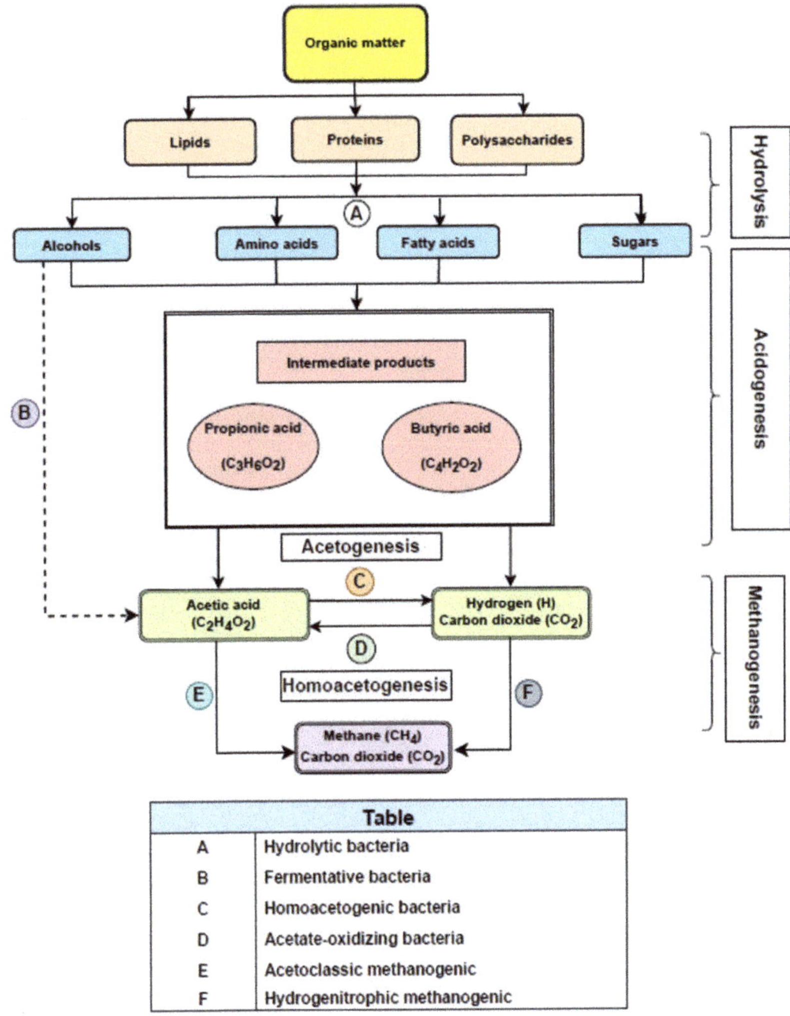

Table		
	A	Hydrolytic bacteria
	B	Fermentative bacteria
	C	Homoacetogenic bacteria
	D	Acetate-oxidizing bacteria
	E	Acetoclassic methanogenic
	F	Hydrogenitrophic methanogenic

Figure 1. Steps and products obtained during anaerobic digestion.

Initially, hydrolytic bacteria hydrolyze, that is, break down polymers (polysaccharides, proteins, and lipids) into monomers (fatty acids, amino acids, alcohols, and sugars); and solubilize the particulate material, and then fermenting bacteria ferment the resulting monomers into a wide range of end products. The end products of the acidogenic stage include acetic acid, hydrogen, and carbon dioxide. However, most of the products are volatile fatty acids (VFA) with higher carbon numbers, such as propionate, butyrate, and alcohols.

During acetogenesis, alcohols (ethanol) and volatile fatty acids (VFAs) like butyric acids (VFAs) are converted into acetate by acetate-producing bacteria, obtaining hydrogen and carbon dioxide as the main byproducts. This is an important process, as H_2 and CO_2 are reduced to acetate by homoacetogenic microorganisms, reducing excess hydrogen that may negatively affect the performance of acetogenic bacteria. Low hydrogen partial pressures (between 10.4 and 10.6 atm) are required for a suitable acetogenic reaction [6]. This is due to the fact that acetogenic bacteria can survive in a very-low-hydrogen-concentration environment.

Conversely, an increase in H_2 partial pressure may result in a lower acetate production by acetogens. To ensure that low pressure is maintained during this stage, a mutually symbiotic relationship between the acetogens and the hydrogenotrophic methanogens should take place, so that acetogens can produce acetate that can be used as substrate by methanogens [7].

Methanogenesis is a critical stage in AD, as in the case of hydrolysis. It has a major impact on the AD process because approximately 70% of the methane used in AD is generated. In this stage, carbon-dioxide-reducing and hydrogen-oxidizing methanogens produce CH_4 from H_2 and CO_2, whereas acetoclastic methanogens produce CH_4 from acetate [7].

Methanogens (Archaea) mainly use acetate, H_2, and CO_2 (also methanol, methylamines, and formate to a lesser extent), to generate CH_4 and CO_2. These are the main substrates for methanogenic bacteria to produce biogas, which consists of 50–75% CH_4, 50–25% CO_2, and lower amounts of N_2, H_2, and H_2S, which has a negative effect on the steam reforming of biogas, as explained in further sections.

In conclusion, methanogenesis indicates the extent of biological activity in an anaerobic system and the state of digestion. The more methane produced, the more stable and efficient the system is. Subsequently, in the context of this review, these steps are crucial to obtain a high-quality biogas, which should have a high methane percentage in order to carry out further upgrading processes in an optimized manner. Also, the presence of impurities (as previously mentioned) could worsen the performance of biogas steam reforming, from a catalytic point of view (for instance, through poisoning due to H_2S) or the requirement of purification steps once the biogas is processed (in order to obtain high-purity hydrogen, for instance).

In a more complex context, biogas production is linked to multiple steps to valorize this product through upgrading or energy use. As observed in Figure 2, biogas is obtained through anaerobic digestion from different wastes, such as agricultural wastes, sewage sludge, solid municipal waste, and manure, etc. [8–10]. In some cases, depending on the properties of the feedstock, the co-digestion of several wastes can be recommended so that some properties in biodigesters (such as acidity, organic matter, and the presence of contaminants, etc.) are balanced to obtain a high biogas production [4,11,12]. In this stage, biogas and digestate are obtained and usually reused as fertilizers or in other processes like active carbon production through pyrolysis or hydrothermal carbonization. Depending on the quality of the biogas (that is, moisture levels, methane content, and the absence of hydrogen sulfide, etc.), this product can be used directly for energy purposes or undergo upgrading (for further treatments such as steam reforming). In any case, even for energy purposes (for instance, its direct use in stoves, gas engines, or its introduction in the natural gas grid), this biogas upgrading could be recommended, which includes steps such as drying (to remove moisture), depuration (H_2S cleaning among other contaminants, such as siloxane, CO, or NH_3), and CH_4 separation from CO_2 [13,14].

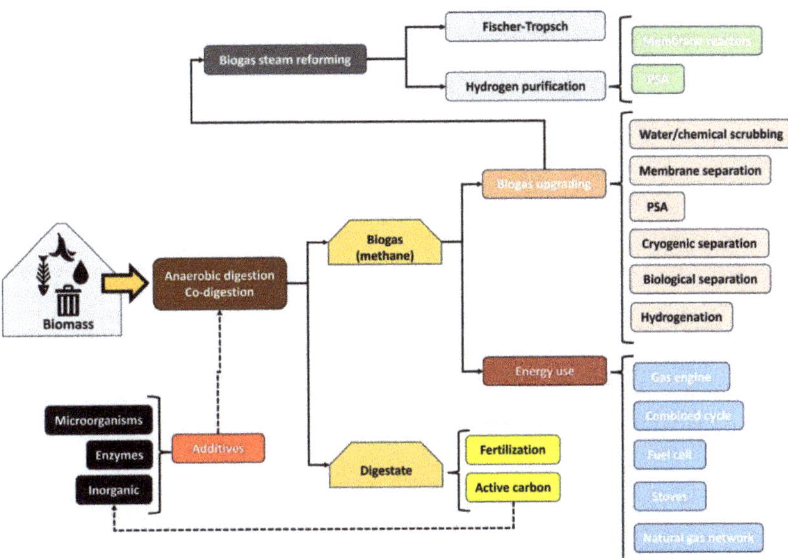

Figure 2. Scheme of processing chain from biomass to commercial products through biodigestion.

Once biogas is upgraded, with good enough quality for its use in further steps, its treatment in processes such as methane steam reforming can be considered to produce a gas phase enriched with H_2, typically between 50 and 80%. In order to generate a gas with a higher range of purity in hydrogen, a further purification step is needed, such as membrane reactors or pressure swing adsorption. If these are used, high-purity hydrogen (normally exceeding 98%) is obtained, with subsequent use in energy production or chemical synthesis (for instance, production of ammonia, methanol, hydrogen peroxide, acetic acid, aldehydes, dyes, hydrochloric acid, polyols, nylon, and polyurethane, etc.) [15]. Otherwise, a mixture of hydrogen with CO (which constitutes synthesis gas), among other components, can be obtained, which could be suitable for further processes such as Fischer–Tropsch synthesis to produce liquid fuels, among others [16,17]. It should be noted that biogas production and treatment offer a wide variety of opportunities (see Figure 2) for research and companies, including the aim of our review work, that is, biogas steam reforming and the use of catalysts.

It should be noted that the technologies applied to biogas are not limited to those observed in this figure, as further treatments of the derived products might be carried out. Regarding the subject of this review, two factors should be considered according to biogas processing:

- The role of catalysts is present in many aspects of biogas processing (specifically heterogeneous catalyst) when it comes to energy production (for instance in Fischer–Tropsch or steam reforming processes, the main topic of this review).
- In addition, many stages related to biogas can present an influence on catalytic performance in biogas steam reforming. For example, digestate can be transformed into active carbons, which can be used as additives for a better biogas production or biogas upgrading (with mercaptans or H_2S removal, for instance, as they are toxic and corrosive components). In turn, it can present a positive effect on catalytic steam reforming (as H_2S provokes poisoning and the subsequent deactivation of catalysts) [18,19]. Another example would be the use of membrane reactors to improve hydrogen yield during steam reforming, with the subsequent improvement in methane conversion. Many of these aspects (with a considerable influence on the catalytic steam reforming of biogas) will be covered in this review in following sections.

Considering the fact that, for every industrial process, the role of economics is essential, the valorization of some by-products or wastes during biogas production and treatment is important to make every process involved as efficient (and feasible) as possible, as it will be discussed in following sections.

1.2. Hydrogen Production from Methane (or Biogas) Steam Reforming

Hydrogen production is a perfect example of green chemistry, contributing to the transition to more renewable energies and the sustainable growth of population areas [20,21]. Pure hydrogen, as well as syngas (a mixture of hydrogen with CO), have been gaining importance recently, as they can be used as an energy carrier or in interesting industrial processes, like methanol synthesis (or more complex compounds) through Fischer–Tropsch reactions [22,23].

There are different chemical routes to produce hydrogen, such as thermochemical water decomposition, electrolysis, coal gasification, or fossil fuel reforming [21]. One of the main chemical routes to produce hydrogen or syngas from natural gas or biogas is methane steam reforming, which has been widely studied across the board, as explained in following sections. However, in many cases, it presents some advantages and disadvantages, as observed in Table 1 in a comparison with dry reforming. In the case of steam reforming, depending on feeding, operating conditions, or further/consecutive steps like the use of membrane reactors or pressure swing adsorption, different purity levels of hydrogen can be obtained [24].

Table 1. Advantages and disadvantages of SRM and DRM [25–27].

Methodology	Advantages	Disadvantages
SRM	Very developed industry, high H_2/CO ratio, O_2 is not required.	High temperatures required, high energy input, high CO_2 production, requirement of catalyst regeneration.
DRM	Syngas produced can be useful for downstream processes like Fischer–Tropsch.	Promotion of coke deposition on catalyst, reducing its activity.

From a chemical point of view, the steam reforming of methane (see Equation (1)) is an endothermic reaction that usually takes place at high temperatures, between 750 and 950 °C, and a wide range of pressure (5–20 bar) [25].

$$CH_4 + H_2O \leftrightarrow CO + 3H_2 \quad \Delta H^0_{298} = +206 \text{ kJ/mol} \tag{1}$$

With an enthalpy of 206 kJ/mol, reforming is a highly endothermic process. Consequently, a significant amount of external energy is required to carry it out. For this reason, the most common process takes place in a tubular reactor inside a furnace that provides the energy required for the reaction, with the subsequent economic costs.

Equally, a water–gas shift reaction (WGS) (see Equation (2)) can simultaneously take place. This way, both chemical reactions contribute to a higher yield in hydrogen production and, subsequently, a higher hydrogen concentration in the resulting gas.

$$H_2O + CO \leftrightarrow CO_2 + H_2 \quad \Delta H^0_{298} = -41 \text{ kJ/mol} \tag{2}$$

This exothermic reaction occurs at low rates in the reforming reactor, which explains the presence of CO_2 at the outlet of the reforming process. High carbon dioxide levels are undesirable from an environmental point of view, and different aspects such as catalyst design and consecutive separation processes. There are other possible side reactions, like a direct reaction between CH_4 and CO_2, CH_4 and CO, and methane decomposition, etc. These side reactions are more abundant if heterogeneous catalysts are used in the process [28,29]. The production of H_2 depends on the equilibrium of the reforming and adjustment reactions (Equations (1) and (2)), and it can be maximized with low pressures, high temperatures in reforming, and a high excess of steam [30,31].

Regarding pressure, it usually presents two contrary effects. Thus, high pressure promotes the interaction of molecules, whereas high pressure would shift the chemical balance, especially in methane steam reforming (Equation (1)) towards the reagent generation. That is the reason why the optimization of this parameter is necessary, considering the rest of the chemical conditions.

On the other hand, the amount of steam to be added to the feed is quantified by the steam to carbon (S/C) ratio, which represents the moles of steam introduced per mole of carbon in the hydrocarbon stream. This ratio takes values between 2.5 and 6 depending on the feed and process optimization conditions. High values of the S/C ratio promote H_2 formation while preventing catalyst deactivation due to carbon deposition, which is favored at low S/C ratios. However, there is a limitation on the maximum amount of steam that can be introduced, both from an energy penalty perspective, as it represents steam that will not generate power, and from an economic standpoint, as the investment cost increases. When the feed is natural gas, it is common to work with S/C ratios around 2.5–5.

Obviously, the physicochemical characteristics of biogas are essential for understanding the global performance of steam reforming. As observed in Table 2, depending on the kind of feedstock, the biogas composition might be different, although some general similarities are observed.

Table 2. Biogas composition (from different feedstocks).

Feedstock	CH_4, %	CO_2, %	H_2S, %	N_2, %	H_2, %	O_2, %	CO, %	NH_3, ppm	Reference
Agricultural waste	45–80	20–50	0.3–1	0–1	0–2	0–1	0–1	100	[32–35]
Industrial waste	50–70	30–50	0.8	0–1	0–2	0–1	0–1	0	[32,36]
Landfill	30–80	20–50	0.05–0.1	0–3	0–5	0–3	0–1	5	[32,34,36,37]
Sewage sludge	58–75	19–33	0.1–0.9	0–1	-	-	-	-	[34,35,38]
Slurry	40–62.3	58.7–37.7	-	0–13.9	0–1.4	0–1	-	-	[33]

In general, biogas contains between 45 and 80% CH_4 and around 20–60% CO_2, including other secondary compounds such as H_2O or N_2 and residual percentages of H_2S and H_2, among others. Biogas composition might vary depending on many different factors (mainly due to feedstock heterogeneity on account of the sampling date, but also due to changing anaerobic digestion conditions), but the majority component is methane, which could be the starting point to produce hydrogen through steam reforming, among other processes.

Consequently, the methane content in the final biogas will determine its further uses or treatments. In that sense, as explained in the literature, the use of methane as a hydrogen carrier could be an interesting alternative for hydrogen storage, which is difficult or expensive (implying the use of costly technologies such as liquefaction and compression) [39]. In that sense, it is important to note that the catalyst for reforming can be deactivated by the presence of certain contaminants in the gases entering the reformer (sulfur, copper, vanadium, and lead, etc.).

In particular, hydrogen sulfide content deserves a special mention, as it will imply a negative effect on the catalytic performance of biogas steam reforming, drastically affecting the catalyst's activity, even at very low concentrations (ppm). Therefore, it is common to use prior desulphurization systems based on activated carbons as adsorbents at room temperature.

1.3. Catalytic Biogas Steam Reforming in the Literature

Apart from the obvious industrial development of methane steam reforming (SRM), including those gases (like biogas) with a considerable amount of this compound, there has been an increase in the scientific interest in this subject in the last two decades [40]. Figure 3 shows the main trends observed in the literature for the search criterion "biogas steam reforming catalyst".

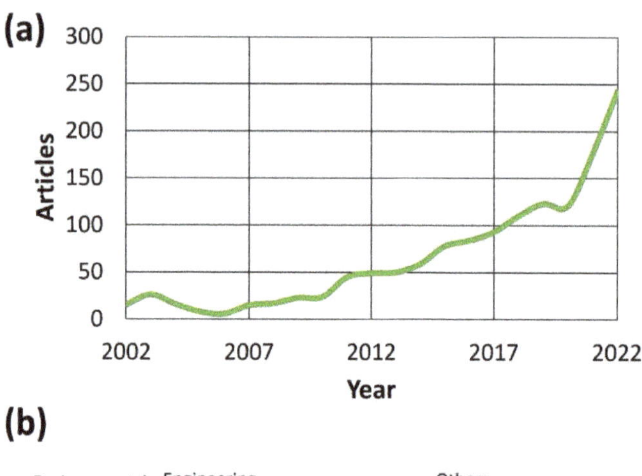

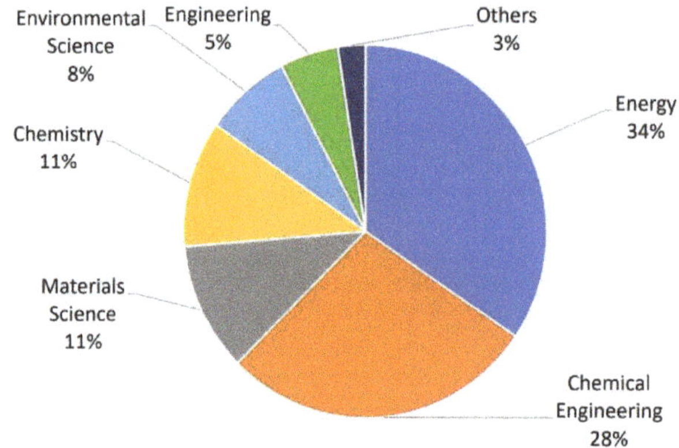

Figure 3. Articles published over the past two decades (**a**) and the main scientific fields of the journals where they were published (**b**) [41]. The search keywords were the following: "Catalyst" and "steam" and "reforming" and "biogas".

As observed in Figure 3a, there was an increasing trend in published articles related to the subject of this review, that is, the catalytic steam reforming of biogas. In that sense, especially from 2010, this increase was steady, whereas there was an exponential growth in the 2020–2022 stage, reaching up to 243 published articles in 2022. These figures confirm the increasing interest in this subject by the scientific community, where this subject has attracted the attention of diverse scientific fields (reflected in Figure 3b, including the main fields of the journals where articles devoted to catalytic biogas steam reforming were published). Thus, and as expected, energy was the field with most journals, pointing out the relevance of the products obtained during biogas steam reforming. Nevertheless, other fields like chemical engineering were equally important, as the implementation of new catalytic approaches at an industrial scale is vital in technologically advanced and mature industries. Also, materials science and chemistry are important in this sense, as the characterization of catalysts and the understanding of their main action mechanisms are important for understanding their effectiveness and durability during steam reforming. Finally, and not least, fields like environmental sciences and engineering (the latter related to the objectives explained for chemical engineering) are of interest in this review subject, as methane conversion to hydrogen and syngas could be a suitable way of reducing environmental impact (especially related to the greenhouse gas effect), pointing out the sustainability of this process.

There are numerous reviews on this subject among these articles published since 2000 [42–44], which have focused on several aspects such as methodologies, technologies, general processes, and catalysts, etc. Regarding research papers, according to Table 3, where the most cited articles about this subject are included, it should be noted that they are relatively new articles, proving the interest in this subject by the scientific community. In any case, only articles considering the main keywords in the title were included, as there are plenty of works (including reviews) that deal with this subject, requiring a stricter search criterion to obtain these results.

Table 3. Top 10 cited articles with the following search criteria for the title: catalyst and biogas and steam and reforming (source: [41]).

Ref.	Title	Authors	Year	Citations
[45]	Catalyst development for steam reforming of methane and model biogas at low temperature	Angeli et al.	2016	122
[46]	Experimental study of model biogas catalytic steam reforming: 2. Impact of sulfur on the deactivation and regeneration of Ni-based catalysts	Ashrafi et al.	2008	82
[47]	Deactivation and regeneration of Ni catalyst during steam reforming of model biogas: An experimental investigation	Appari et al.	2014	79
[48]	A detailed kinetic model for biogas steam reforming on Ni and catalyst deactivation due to sulfur poisoning	Appari et al.	2014	75
[49]	Influence of Ce-precursor and fuel on structure and catalytic activity of combustion synthesized Ni/CeO_2 catalysts for biogas oxidative steam reforming	Vita et al.	2015	53
[50]	H2 production from sorption enhanced steam reforming of biogas using multifunctional catalysts of Ni over Zr-, Ce- and La-modified CaO sorbents	Phromprasit et al.	2017	45
[51]	Activity and stability performance of multifunctional catalyst (Ni/CaO and $Ni/Ca_{12}Al_{14}O_{33}$-CaO) for bio-hydrogen production from sorption enhanced biogas steam reforming	Phromprasit et al.	2016	39
[52]	Bio-hydrogen production by oxidative steam reforming of biogas over nanocrystalline Ni/CeO_2 catalysts	Italiano et al.	2015	39
[53]	Syngas production by steam and oxy-steam reforming of biogas on monolith-supported CeO_2-based catalysts	Vita et al.	2018	38
[54]	Steam-biogas reforming over a metal-foam-coated $(Pd-Rh)/(CeZrO_2-Al_2O_3)$ catalyst compared with pellet type alumina-supported Ru and Ni catalysts	Roy et al.	2015	34

If the most cited authors with research work about catalytic biogas steam reforming are considered (see Table 4), some interesting remarks can be made, like the following:

- They have contributed with at least 30 articles each about this subject, which proves the endless possibilities of catalytic biogas steam reforming.
- Most authors are reputed scientists, with a h index from 30 to 76 and many citations. The fact that such prestigious scientists have dealt with this field to a certain extent proves its relevance in the scientific community.
- As expected, these authors are equally focused on other subjects and fields (with a considerable percentage of published articles about the subject of this review), pointing out the multidisciplinarity of their research teams, which could enrich the research in this field.

Table 4. Top 10 cited authors with the following search criteria: catalyst and biogas and steam and reforming (source: [41]).

Author Name	Citations	Published Articles about This Subject	Total Articles	h Index
Kawi, S.	21,760	62	421	76
Goula, M. A.	4222	44	106	36
Charisiou, N. D.	2990	41	88	30
Polychronopoulou, K.	7204	40	197	50
Chen, W. H.	24,413	39	598	83
Rezaei, M.	8326	38	250	51
Park, Y. K.	21,486	38	835	67
Fakeeha, A. H.	3145	38	162	31
Assabumrungrat, S.	9287	38	431	47
Haghighi, M.	9621	37	276	57

If the published articles about biogas steam reforming are arranged by country, an interesting outlook on scientific works can be found. Thus, as observed in Figure 4, there is a relatively homogeneous distribution of published articles about biogas steam reforming worldwide, with countries like Italy (83 articles) or China (53 articles) leading. Even though Europe has been traditionally focused on biogas upgrading, there are other regions like America and Asia where the role of this research is also representative. Equally, some works from African countries have been devoted to this subject, which could be an encouraging starting point for the implementation of these technologies on this continent.

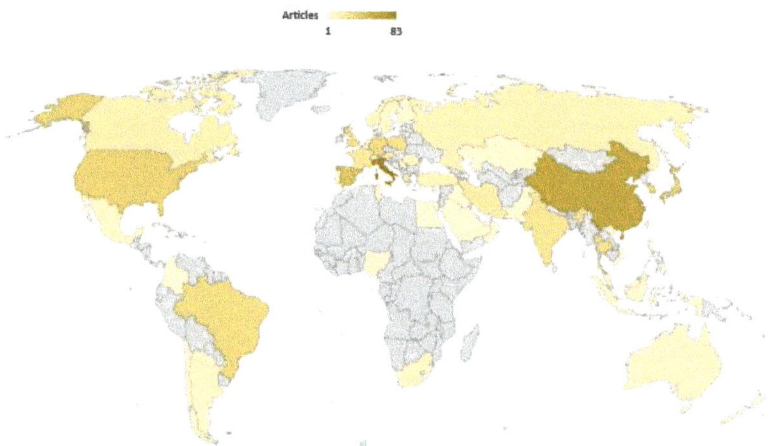

Figure 4. Worldwide distribution of published articles about this subject. Search criteria: biogas and steam and reforming [41].

Finally, taking into account the relationship between the keywords of the works dealing with the use of catalysts in the steam reforming of biogas, interesting findings can be found, as observed in Figure 5:

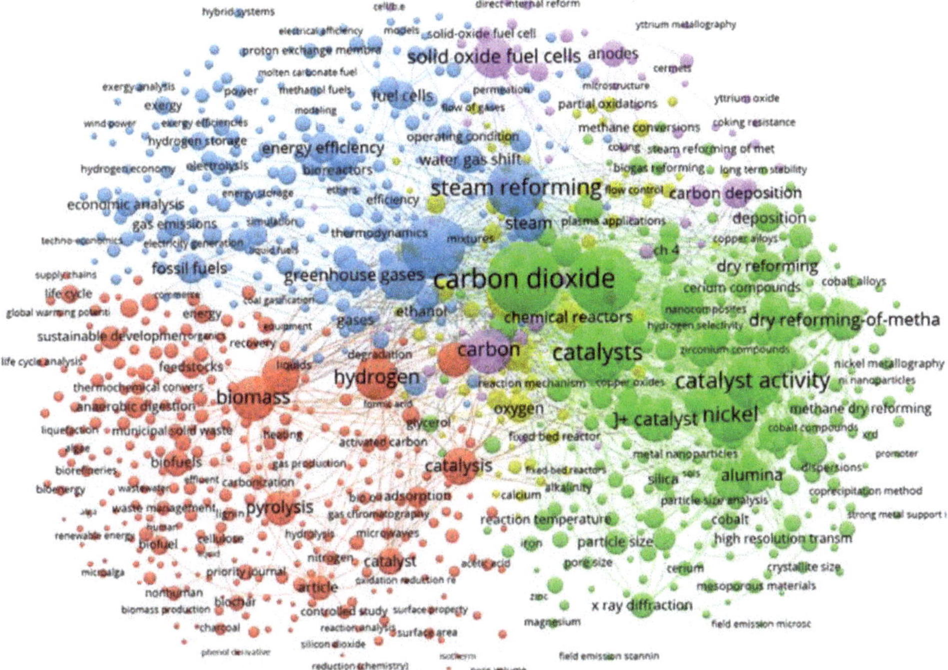

Figure 5. Keyword co-occurrence map based on Scopus, obtained by VOS Viewer. Search criteria used: catalyst and biogas and steam and reforming (7688 documents in total) [41].

Thus, five interesting clusters can be found (represented in different colors), which cover every aspect intended to be included in this review work, like the following:

- Red cluster: it is mainly focused on feedstocks for biogas generation (biomass, municipal solid waste, and sewage sludge, etc.), as well as biogas production through anaerobic digestion.
- Yellow cluster: focused on reaction mechanisms, methane conversion, and flow control, etc. In that sense, this is the most interesting cluster for engineering purposes.
- Blue cluster: the main subjects included in this cluster are steam reforming, water gas shift, fuels, energy efficiency and storage, operating conditions, thermodynamics, gas emissions, and economic analysis, among others. It is similar to the previous cluster but focused on energy conversion.
- Green cluster: mainly devoted to catalysts, where the role of nickel (and cerium, among other promoters) as an active phase and alumina and silica as supports seems to be important. Also, some factors such as particle and pore size or reaction temperature are relevant, which are vital to increasing the useful life of catalysts (avoiding carbon deposition or sintering, terms included in this cluster too). In other words, this is the cluster where everything about catalysts is covered.
- Purple cluster: as in the previous case, it is mainly focused on catalysts and some inhibitory effects such as carbon deposition and the corresponding properties of the catalysts to avoid it to a certain extent (coking resistance). In that sense, the role of yttrium seems to be important.

In that sense, as explained in this figure, most of the main aspects covered by these clusters will be included in this review, which points out the interrelation of many subjects (from feedstock to catalyst characterization or from energy efficiency to catalyst activity, for instance) and proving the interest in this field.

1.4. Aim of This Review

According to the above, the aim of this review work was to carry out a review of biogas steam reforming, especially focused on catalytic conversion to obtain hydrogen. Thus, the following points will be covered:

- Biogas production in context, including the main feedstocks and innovative technologies for improving methane yield and quality, which is essential for better performances of catalysts during steam reforming.
- Biogas steam reforming and the main factors affecting its performance, which can be improved by the use of catalysts and, at the same time, can affect the performances of catalysts in some respects.
- The role of catalysts in biogas steam reforming, including the main catalysts used, the foundations, mechanisms, and main deactivation processes (and how to avoid them or, at least delay them).
- The main techno-economic analyses and patents carried out on this subject, paying attention to the role of catalysts.

Thus, a thorough review was carried out, which aims to clarify the role of catalysts in biogas steam reforming and all the details affecting their performance, paying attention to the research carried out in the last five years.

1.5. Scope and Bibliometric Analysis

In order to carry out this review work, Scopus was investigated for all entries in the literature on the topics of biogas (including keywords such as catalysts and steam reforming) for the last 20 years, with special attention to the last 5-year period (2018–2023), where there has been a considerable increase in published papers and innovative research in this field. The search, which was made from May to October 2023, returned 6968 results, from which up to 281 articles were considered for their inclusion in this work, including the information of about 148 published works (mainly research works, reviews, and, to a lesser extent, proceeding papers and patents) in the final paper.

2. Use of Catalysts in Biogas Steam Reforming

As briefly explained in the previous section, the role of catalysts in steam reforming in general, and in methane or biogas (whose majority compound is methane) in particular, is essential for the implementation of a competitive technology at an industrial level.

2.1. Main Considerations

In general, the catalyst is located inside a tubular reactor, which can be arranged as a fixed or fluidized bed. The most studied/used configuration is the fixed bed, due to its simplicity and reproducibility, However, it is possible that the deactivation in this configuration is higher than that of the fluidized bed [55], because carbon depositions (between catalyst particles/pellets) are less prevented as the catalyst is immobile. To highlight the recent appearance of new types of reactors in this type of reactions, the photo-thermal ones [56], which can reach H_2 production velocities of 17.4 μmol s^{-1} with an STH efficiency of 22.5% and CO selectivity of 1% in the optimal design under concentrated light irradiance of 16 kW·m^{-2} in the lab, these reactors are positioned as an alternative to be developed to solve the great disadvantage of energy input.

When studying SMR, it is important to keep in mind that the diffusion of feed biogas and water is homogeneous. For this, the water feed can be performed in liquid form and vaporize inside the reactor, a fact that can lead to a gradient in the concentration of water along the catalytic bed. Another way to introduce water is in the vapor phase, a fact that implies having a vaporizer and a steam flow controller prior to entering the reforming reactor, in addition to avoiding condensation in the steam conduction.

What is perhaps one of the most influential parameters in the effectiveness of the reforming process is the activity of the catalyst. The order of catalytic activities on active metals for SRM has been reported: Rh > Ru > Ni > Ir = Pd = Pt > Co > Fe. In addition, other studies have calculated the TOF in DRM for various metals and showed that the order differs for Al_2O_3 support and SiO_2 support. The order of TOF of the methane reaction rate on each support is presented below [57]: Ni > Ru > Rh, Ir (SiO_2 support); Rh > Ni > Ir > Pt, Ru > Co (Al_2O_3 support). Recent studies have assessed the activity of several composite structured catalysts, showing the following decreasing activity: Rh > Ru > Pt > Ni [58]. In any case, for SRM and DRM, it is common to use Ni catalysts, which are active and inexpensive, supported on metal oxides such as Al_2O_3, which have a high heat resistance.

Generally, the catalyst consists of an active phase dispersed on a support. The active phase in catalysts for biogas reforming is commonly composed of nickel, whereas the support is usually an aluminosilicate. These materials are normally used due to their catalytic activity, resistance to operating conditions, commercial availability, and versatility. Within aluminosilicates, many types can be found commercially, and in different forms, like powder, spheres of various sizes, and pellets, etc. This variety allows for a great adaptability to the type of reactor, since it is possible to adjust some parameters such as the contact time, charge loss within the reactor, and deactivation, etc. In the following sections, these aspects will be covered.

2.2. Kinds of Catalysts and Their Preparation

Different catalysts can be used in biogas steam reforming, like the following [59]:

- Monometallic catalysts: they are mainly Ni-based catalysts, which are very popular in the literature due to their great catalytic activity and relatively low cost compared to other equivalent catalysts. However, they have some negative effects (which will be explained in detail in following sections) during steam reforming, such as deactivation due to coke deposition or poisoning.
- Catalysts with promoters: the abovementioned catalysts can be considerably improved by adding promoters (such as B, Ir, La, or Mg) that can help to improve the global performance during SRM thanks to the improvement of metal–support interaction or the ability to promote a higher dispersion compared to traditional catalysts. Recent works point out the relevance of adding some promoters (La and Mg) to typical Ni/Al_2O_3 catalysts, in order to improve their catalytic performance. Thus, these additives improved the stability and dispersion of the active phase, with a better deactivation resistance [60].
- Bimetallic or polymetallic catalysts: to avoid deactivation derived from sintering or coke deposition, the use of combined metallic catalysts could present a positive effect. Bimetallic catalysts are mainly based on Ni or Co combined with noble metals, non-noble metals, or metalloids, whereas polymetallic catalysts are combinations of different metals, like: Ni, Cu, and Zn; Ni, Co, and Ce; and Ni, Ru, and Mg. These combinations can present not only additive, but synergistic effects [30,61].

The main characteristics of these catalysts will be explained in further detail in the following sections, paying attention to different factors such as the catalyst support, active phase, and the interaction between them, which will determine the catalytic performance during methane or biogas steam reforming. For instance, the activity of the resulting catalyst and its resistance to sintering will vary depending on the use of different promoters. In that sense, the preparation of a certain catalyst is vital to understanding some of the final properties of this product. There are different ways to prepare catalysts for this purpose, such as impregnation, co-precipitation, and the sol–gel method.

In the impregnation method (Figures 6 and 7a), a precursor solution is combined with an active solid support phase, and then the solvent is removed by drying. In the application of this method, the solid and the solution are contacted in two ways: wet impregnation (WI) and incipient wet impregnation (IWI) (see Figure 6). In addition, there are dual mechanisms depending on the impregnation method used. WI involves a diffusion process, whereas IWI uses a capillary action method that allows the solution to penetrate pores in the support.

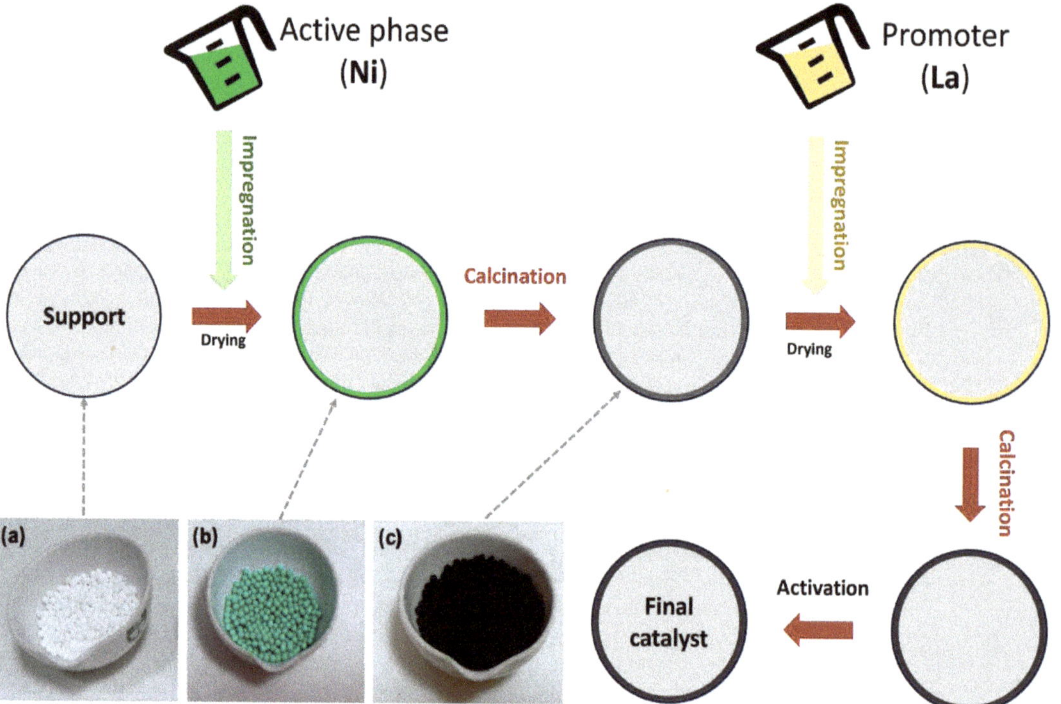

Figure 6. Different stages for catalyst preparation, including images of the most representative ones: (**a**) catalyst support; (**b**) wet impregnation with a Ni solution; and (**c**) calcination.

In wet impregnation, an excess solution is used which is separated from the solid by drying for a certain period of time. During the diffusion process, the composition of the solution changes, forming a residue of impurities and releasing heat due to adsorption in a short time interval. The wet impregnation method is effective for the preparation of the metal catalysts used in methane reforming processes, achieving a high yield and catalyst microstructure. This is the reason why this is the majority catalyst preparation according to the literature. However, a drawback of this method is sintering, especially on catalysts with high oxide loading. In wet impregnation, the metals are dispersed on the surface because the precursor is distributed on the support. This results in high use rates but a low dosage of active metal on the surface, which can result in a non-uniform dispersion of the catalysts [62].

Impregnation by incipient wetting, also known as dry or capillary impregnation, is a method in which the pore volume of the active phase/support is approximately equal to or slightly larger than the volume of the solution. To explain the mechanism of the capillary action of the incipient wetting impregnation method, several reactions occur at different rates. The selective adsorption of charged or uncharged species occurs via H-bonds, van der Waals, or Coulomb forces. Ions are then exchanged between the electrolyte and the charged surfaces, resulting in the polymerization/depolymerization of the ions deposited on the surface. This is followed by the partial dilution of the solid on the surface. After the impregnation of the catalyst into the solid support/active phase, drying and calcination (at different temperatures according to the nature of the active phase) are performed to obtain the desired catalyst material. The products of the impregnation processes are highly dependent on the precursors used, and the parameters that can influence the final mixture include the pH of the solution, the type of solvent, the concentration, and the nature of the dissolved solids [63].

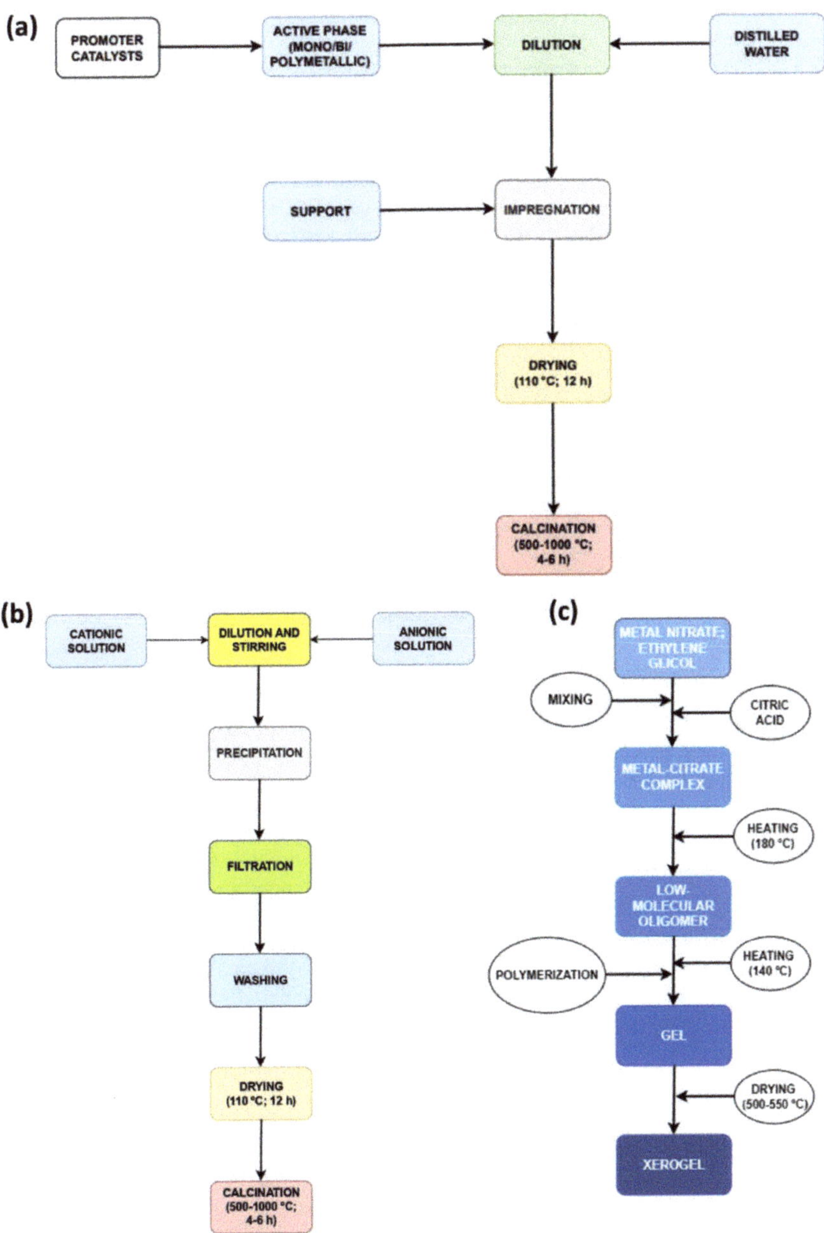

Figure 7. Main steps for different catalyst developments: (**a**) impregnation; (**b**) precipitation; and (**c**) sol–gel.

Co-precipitation (Figure 7b) is a versatile method that can be applied to the synthesis of simple, mixed, or supported catalysts [62,64]. Coprecipitation, often referred to as the "one-pot method", is a conventional approach for synthesizing catalysts in the context of methane steam reforming. The formation of metallic precipitates occurs from oversaturated solutions of their salts. Consequently, all precipitation methods share common components, specifically a soluble source of divalent or trivalent cations and a strong base, such as sodium hydroxide (NaOH), which promotes the precipitation of ions. This process involves mixing metal salts in an aqueous phase with alkaline solutions, resulting in the formation of insoluble metal hydroxides and/or carbonates. During the combination stage, reaction parameters such as temperature, evaporation, salt concentration, and pH stimulate the precipitation process.

These parameters can modify the growth and size of crystallites. The precipitation method consists of five steps: dissolution, precipitation, filtration, drying, and calcination. During the dissolution stage, the active-phase precursors (in salt form) dissolve or hydrolyze in a medium (normally water) to obtain hydroxides in a homogeneous solution. Subsequently, filtration and drying steps are needed, allowing the solids to filter and dry at the boiling temperature of the medium. The dried sample is then crushed, and a binder is added. The appropriate binder is selected to promote easy conversion into vapor and CO_2 during calcination and the subsequent activation. The calcination stage uses air (at an optimal temperature) to convert the material from its hydroxide or salt form into oxides. Several publications have resorted to the precipitation method to synthesize the catalyst support, and the catalyst for methane reforming processes [65–68].

The sol–gel method (Figure 7c) presents a different approach to prepare new materials. Conventional sol preparation involves the hydrolysis and condensation of metal precursors, resulting in a colloid suspension comprising various systems. Colloids are generated when one phase disperses into another, and the dispersed molecules have a dimension between 1 nm and 1 µm [62]. Depending on the kind of solvent, there are two pathways for using the sol–gel method: the aqueous sol–gel method, which refers to the use of water in the reaction, and the non-aqueous sol–gel method, which refers to the use of an organic solvent. In the aqueous method, O_2 from water decomposition is necessary for metallic oxide formation.

This method is advantageous due to the high affinity of most precursors for water. However, the main reactions (hydrolysis, condensation, and drying) occur simultaneously, making it difficult to control the particle morphology and process reproducibility. However, this disadvantage is insignificant when preparing metal oxides in bulk.

Thus, the aqueous method can be utilized for preparing bulk metal oxides as opposed to small-scale preparation [69]. In the non-aqueous method, also known as the non-hydrolytic method, the required O_2 is provided by solvents (like ketones and alcohols) or metal precursors. The organic solvent also contributes to modifying the process to refine the final properties of the material, such as the morphology, particle size, temperature, and humidity.

Most sol–gel processes use tetraethoxysilane (TEOS) in an aqueous solution, which forms SiO_2. This hydrolytic medium is required for hydrolysis and condensation reactions to occur. Hydrolysis is a chemical reaction where silanol (Si-OH) is generated from the reaction between water and an alkoxide (Si-OR), such as TEOS. Si-OH and Si-OR are responsible for the subsequent condensation reactions in the process, resulting in the formation of siloxane in a complex system of competition between hydrolysis and condensation during the intermediate steps of the sol–gel process [70].

Furthermore, the influence of acidic and basic conditions should be considered, as they compete and have their respective peculiarities. The acidic route allows for the syntheses of more branched compounds, whereas the basic route allows for the production of more spherical and compact materials. These parameters are defined around the point of zero charge (PZC), which is determined by the material's structure and porosity. The pH range of silica is between 1.5 and 4.5, and the condensation of silica species has a limited influence.

Pechini [71] patented a preparation method that adopted the principles used in the sol–gel method with modifications, which employs small molecules and chelating ligands. Initially, a homogeneous solution of metal/citrate complexes is formed in the method and the mixture is then converted into a covalently bonded polymeric matrix, thereby trapping the metal ions. The principle of the Pechini method is to slow down the thermal decomposition of the organic structure to control the resulting material. The primary reaction in this method is the transesterification that occurs between ethylene glycol and citric acid [72]. The Pechini method offers some benefits, including its simplicity, independence from process conditions due to the resulting material's ion positivity, and the use of a low temperature for precursor treatment, resulting in complete sintering elimination. However, its drawbacks involve the use of toxic ethylene glycol and a significant amount of organic reagents per mass unit [62].

There are other preparation methods that provide interesting catalysts, such as ion exchange, plasma synthesis, or the combination of solution combustion synthesis (SCS) with the wetness impregnation (WI) technique, offering high activities in Rh-based catalysts under typical SR operating conditions [53].

It should be noted that catalytic performance is highly influenced by the preparation method, as the catalyst dispersion and interaction with the support depend on the corresponding procedure, as observed in dry reforming [73]. Consequently, these methods aim to obtain a catalyst with specific characteristics, as explained in the following subsection.

Firstly, the selection of a suitable catalyst support is essential due to its surface characteristics, but also on account of its thermal or mechanical resistance. Also, strong interactions with the active phase are desired to delay deactivation processes.

Second, the active phase will play an essential role in biogas steam reforming, promoting this reaction, as explained in detail in following sections. In that sense, the distribution of this phase on the catalyst support is important, which will be determined by the surface characteristics of the support (pore size distribution) and the preparation method. For instance, in impregnation processes, the concentration of the active-phase precursor in the dilution will determine the final distribution of the active phase to a greater extent, as high concentrations could promote the agglomeration of the active phase, obtaining bigger active sites that usually imply a decrease in the surface area of the final catalyst, with a subsequent lower CH_4 conversion. On the other hand, lower concentrations would imply fewer active sites on the surface, which would decrease the hydrogen production. In a sense, an intermediate solution is suitable, taking into account the pore volume of the support and the concentration of the precursor required to cover the surface, avoiding agglomeration.

Finally, the use of bimetallic, trimetallic, or polymetallic catalysts is also advisable to complement the characteristics of typical active sites such as Ni. Also, the use of promoters (who are not directly involved in catalytic activity, but contribute to a suitable performance) is necessary, in order to promote a strong interaction between the active phase and the support. In this sense, the introduction of these components could change the development of the catalyst, as observed in Figure 6 in the case of impregnation, where successive steps should be carried out to introduce the promoter.

2.3. Characteristics of Catalysts

One of the key factors concerning the catalytic steam reforming of biogas is the main characteristics of the catalyst used. As in any field where catalysts are used, their properties should be perfectly adapted to the requirement of the corresponding conversion process. In this case, concepts such as the support (including shape or geometry), active phase (including the interaction with the support, which will determine the sintering or coke deposition resistance), or surface area should be taken into account, as observed in further subsections. It should be noted that the interaction between the active phase and the support is essential for understanding the catalytic performance during biogas steam reforming, as it will determine the resistance of the final catalyst to some factors such as poisoning, carbon deposition, or sintering, among others. Also, the combination of multiple metals in the active phase could improve some properties in the resulting heterogeneous catalyst, especially concerning some factors such as a longer useful life or selectivity towards hydrogen production.

2.3.1. Catalyst Support

The support plays an important role in a suitable catalytic design, as it holds the active phase where the catalytic conversion will take place. In that sense, concerning biogas steam reforming, the nature of this support (normally alumina or silica), its porosity, mechanical resistance, and geometry will allow for a maximum interaction of the gas phase with the solid catalyst, depending on operating conditions such as flow rate and pressure, etc. Figure 8 shows the different shapes of the catalyst supports used for these purposes, with a great interest in spheres and hole catalysts, according to the literature. In any case, other shapes are equally used, proving the versatility of catalytic steam reforming.

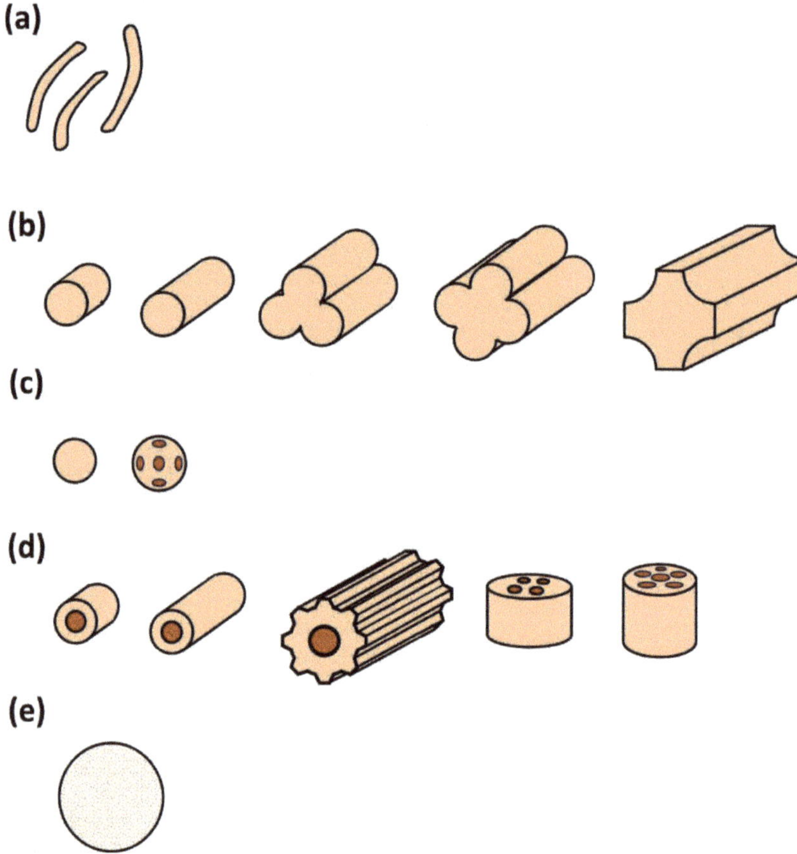

Figure 8. Different shapes of catalyst supports: (**a**) amorphous pellets; (**b**) extrudates (solid, hollow, trifolium, and quadrifolium, etc.); (**c**) sphere carriers, including holes; (**d**) hollow supports, with single hole cylinders, ribbed cylinders or with multiple holes; and (**e**) ceramic foam.

As commented in previous subsections, and according to Table 5, impregnation and co-precipitation seem to be the most popular ways of preparing catalysts for methane steam reforming, offering a wide range of surface areas, from around 100 to 500 m^2/g.

Table 5. Catalysts for SRM, with typical supports and BET surface area.

Catalyst	Preparation	BET Surface, m^2/g	Use	Reference
Ni/Al_2O_3 (10%)	Co-impregnation with cobalt on mesoporous alumina	103.84	SRM	[74]
Ni/Al_2O_3 (20%)	Impregnation on alumina spheres	188.80	SRM	[75]
Ni/Al_2O_3 (13%)	Co-impregnation with cobalt	171.20	SRM	[76]
Ni/C	Impregnation on commercial activated carbon	516	SRM	[77]
Ni/Mg-Al and Ni/La-Al	Co-precipitation	233 and 215	SRM and DRM	[60]

Also, there are other alternative supports, like CeO_2, which contribute to a better catalytic activity. Recent studies have carried out the combined steam and dry reforming of biogas using a Ni/CeO_2-Al_2O_3 catalyst with a bimodal porous structure. When the CeO_2 concentration was 5%, a great catalytic activity was found, thanks to the more intimate contact with alumina and the higher metal–support interaction, preserving it from carbon deposition by 70% [78].

These differences in shapes offer a wide range of SRM conditions. In any case, the maximum interaction between biogas and catalyst is highly desired, trying to avoid the free passage of gas as much as possible. Also, another aspect to be taken into account is the surface of the supports, as it plays a vital role both in catalyst preparation and their corresponding final performance.

Thus, the pore size distribution of different supports (see Figure 9 for different examples in SEM images) can influence the impregnation of active phases during catalyst preparation, whereas the performance of the final catalysts and (including some processes such as coke deposition or sintering) is highly determined by the pore size, whose profile should be selected to favorize a long useful life of the catalyst.

Equally, there are other characteristics of the support that should be considered, like its thermal and mechanical stabilities, which are essential in biogas steam reforming for different reasons. Firstly, due to the high temperatures taking place in this process, thermal stability is an ideal prerequisite to avoid surface or even structural changes in catalysts due to thermal shocks. Also, mechanical stability is important to avoid catalyst breakup due to different factors such as friction or high pressures, reducing the amount of detritus within the reactor and subsequent blockages. In that sense, the use of resistant materials like Al_2O_3 or SiO_2 is common, whereas other catalysts based on carbonaceous materials could present some challenges in that regard.

In this regard, innovative works have been carried out where the role of the support is essential. For example, metal-foam-coated Pd–Rh catalysts with variable $CeZrO_2$–Al_2O_3 support compositions were used in biogas steam reforming, resulting in higher CH_4 conversion with the extent of $CeZrO_2$ in the catalyst, a decreasing H_2/CO ratio, suppressed coke deposition due to oxygen storage, and an improvement in oxygen reducibility, with an improvement in resistance to the deterioration of surface area, pore structure, and active-phase dispersion [79,80]. Mesoporous catalysts prepared via a reverse precipitation method, $Ni_{2x}Ce_{1-x}O_2$ (x = 0.05, 0.13, 0.2), were compared with a commercial catalyst (R67), obtaining a higher H_2/CO ratio and excellent activity [81].

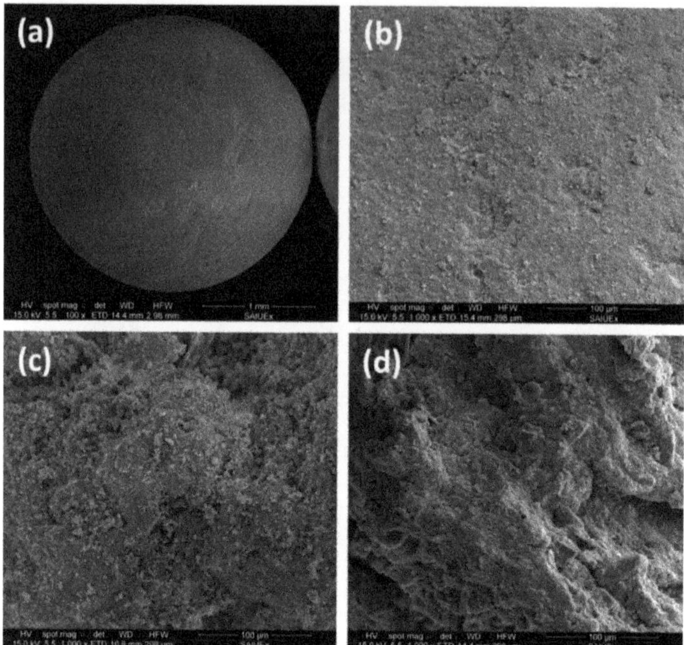

Figure 9. SEM images of a sphere catalytic support (**a**) and its surface (**b**), and comparison with other different surfaces: dried sewage sludge (**c**) and pyrolyzed sewage sludge (**d**).

2.3.2. Active Phase

Generally, catalysts consist of an active phase, usually a noble metal or acid/base site, deposited and dispersed on a porous support such as alumina, silica, or other material. The solid catalyst's active phase has a high affinity for molecules of specific reactants. Initially, the molecules chemically attach themselves to the active surface before reacting each other. This way, the activity of catalysts is normally proportional to the number of active sites on the surface. In the case of metal-supported catalysts, the active sites are represented by the exposed metal surface.

Nickel (a transition metal) is commonly used as an active phase in SRM processes due to its availability, low cost, and high activity. However, due to sintering and coke deposition, Ni-based catalysts are subject to rapid deactivation [82].

In addition to nickel, noble metals such as rhodium (Rh), ruthenium (Ru), palladium (Pd), platinum (Pt), and iridium (Ir) show promising potential as candidates for SMR due to their exceptional catalytic abilities and resistance to carbon deposition. Several experimental and numerical studies have reported that the catalytic activity of noble metals could be ordered as Rh~Ru > Ir > Pt~Pd [30].

It is important to consider different factors to make the addition of an active phase efficient. This way, catalysts with high activity, low concentrations of active phase, and subsequently high dispersion are desirable for carrying out high conversions in biogas upgrading into syngas. For this purpose, recent studies have proven different Ru-based catalysts (with different supports), with Ru/MgO showing an excellent catalytic performance in the bi-reforming of model biogas due to Ru dispersion with an ultra-small particle size [83]. Consequently, the use of nanoparticles seems to offer a promising outlook in this field. Also, the role of multi-metallic active phases is important for obtaining specific and interesting properties, as in the case of poisoning resistance. Thus, a catalyst (NiCeSnRh/Al$_2$O$_3$) was used in the bi-reforming of biogas, offering a high resistance to sulfur compounds [84].

2.3.3. Advantages and Disadvantages

There is such a wide range of catalysts that can be developed that it is difficult to cover the advantages and disadvantages related to their use. Nevertheless, there seems to be some common patterns, mainly to do with the active phase, which, in many cases, is the main limiting factor when it comes to producing an economically feasible catalyst. In that sense, the role of catalyst in techno-economic assessments in biogas steam reforming is important, as explained in the corresponding section. Essentially, it is a matter of cost–benefit analysis, paying attention to the benefits offered by a specific catalyst and its relative abundance. Regarding the active phase, Ni-based catalysts are popular for this reason, as they offer acceptable catalytic activity at a relatively low cost compared to other metals. However, there are other factors such as a higher propensity for deactivation processes, which could imply operational problems in the medium term, which could be solved with other more expensive catalysts based on Ru or the use of promoters such as La. In other words, life cycle assessments of catalysts for biogas steam reforming are, as in many other cases, essential for obtaining a cost–benefit balance.

Despite their advantages, noble-metal-based catalysts are limited due to their high prices. One way to keep the excellent performance of noble metals while maintaining a reasonable price is to combine two or more types of metals, using cheap transition metals (usually nickel or cobalt) as a base and noble metals as promoters, chemicals that are added to the catalyst in order to improve its catalytic properties. Bi/polymetallic catalysts have gained increasing attention in recent years, and the synergistic effect between commonly used metallic elements has been investigated experimentally and numerically. Numerical studies focus on the reaction pathway and the activation energies of certain reaction steps (specifically, C-H bond breaking during CH_4 decomposition), as well as the adsorption energies of atomic or molecular species on the catalyst surface, which are indicators of the catalytic activity and stability of the material [30].

Cobalt is also considered to be a promising promoter in SRM due to its good activity for the WGS reaction, which helps shift the equilibrium towards H_2 production. However, a problem related to the use of Co is its tendency to oxidize when the temperature and vapor partial pressure are in the range used for SMR. Alloying it with Ni is a possible solution to this problem while preserving the advantages of both elements [82].

Compared to the relatively simple mono and bimetallic systems, the application of catalysts containing three or more types of active metals in SMR has not been investigated in detail. The existing literature mainly examines Ni-based materials with the addition of two or three commonly used elements, such as Co, Cu, Pt, and Ru, etc. [30].

2.3.4. Catalytic Performance in Biogas Steam Reforming

Thus, after considering the relationship between the support and the active phase, the typical catalytic steam reforming of methane is explained in Figure 10, where the main steps that take place during the process are included.

Thus, the kinetic model of methane reforming, shown in Figure 8, is based on the following steps:

- H_2O is adsorbed on the catalyst and dissociates, giving rise to adsorbed oxygen atoms and H_2 in the gas phase.
- CH_4 is adsorbed on the catalyst and dissociates, generating CH_2 radicals and adsorbed H atoms.
- The adsorbed CH_2 radicals and oxygen react, with bonds being formed and breaking at the same time, generating a transition state (CHO) and H_2.
- The adsorbed CHO dissociates into adsorbed CO and H or reacts with adsorbed oxygen to produce CO_2 and H in parallel (controlling stage).
- The adsorbed CO reacts with adsorbed oxygen to form CO_2, or is desorbed to give gas-phase CO.

This way, and according to recent works found in the literature (see Table 6), different operating conditions with the subsequent methane conversion are included.

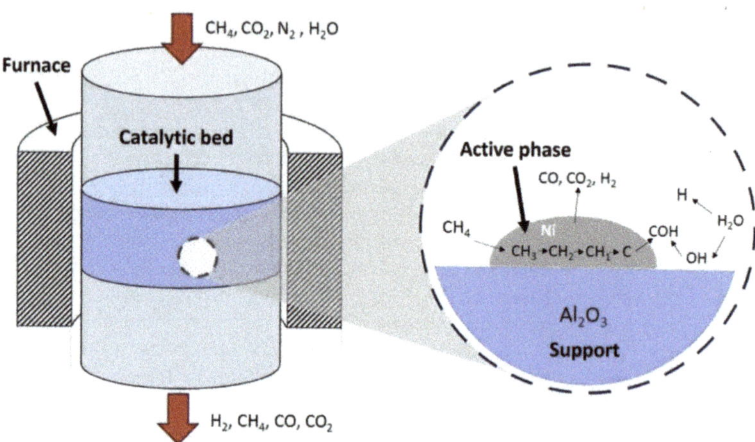

Figure 10. Stages taking place during catalytic steam reforming in a reactor. Example for a Ni-Al$_2$O$_3$ catalyst.

Table 6. Catalysts for SRM, including operating conditions and methane conversion.

Catalyst	T, °C	Pressure, bar	S/C	CH$_4$ Conversion, %	Reference
Ni/Al$_2$O$_3$	850	7	1.2	99	[85]
Ni-La/Si	800	1	0.8	90	[86]
Mo$_2$C-Ni/ZrO$_2$	700	1	0.8	74	[87]
Ni-hydrotalcite	900	5	2.0	98	[88]
Ni-Al$_2$O$_3$	800	1	1.77	<80	[89]
Ni-CaO/Al$_2$O$_3$	750	1	2.2	<90	[90]
Pt/Al$_2$O$_3$	800	1	3.0	<90	[91]
Ni/MgO	750	1	1	91.5	[92]
Ni/Al$_2$O$_3$	800	1.7	3	100	[93]
Ni based	700	1	3.5	98.3	[94]
Ni/γ-Al$_2$O$_3$	700	1	2	98.1	[47]
Ni/CeO$_2$	900	1	3	99.7	[95]
Pd-Rh/CeZrO$_2$-Al$_2$O$_3$	850	1	1.5	99.1	[96]
Ni/SiO$_2$	700	1	3.5	96	[97]
Ni/TiO$_2$	700	1	1.2	92	[98]
Pd/Ni-CaO-mayenite	750	1	4	97.8	[99]
Ni-MgO-CeZrO$_2$	750	1	1.5	>90	[100]
Ru/Mg$_x$AlO	800	1	2.6	>90	[83]

As observed in Table 6, the operating conditions allowed for methane conversions exceeding 90% in most cases, which usually implies a considerable hydrogen production in the obtained syngas. Ni and Pt catalysts are the most popular choices in these cases, adding, in some cases, promoters to improve the catalytic performance. In any case, these catalysts usually present long stabilities and a high coke resistance, with the possible softening of operating conditions within the range explained in previous sections. Regarding temperature, some studies have achieved very low values (under 800 °C), whereas pressure could be considerably decreases (up to 1 bar), with relatively low steam to carbon ratios, which could imply a considerable reduction in the fixed energy costs related to steam generation. All these improvements could mean an increase in the efficiency of biogas steam reforming. However, there are some challenges related to catalytic performance, mainly related to catalyst deactivation, which will be explained in the following section. Another emerging research line in catalytic steam reforming is the use of nano-catalysts supported on different materials. As previously explained, the dispersion of catalysts is essential to maximizing catalytic activity by increasing the effective active phase area, also avoiding negative effects like coke deposition. In that sense, as observed in Table 5, there are new research trends focused on the synthesis and performance of this kind of catalysts.

3. Catalytic Deactivation

One of the main aspects that should be taken into account for a suitable catalytic performance is the life cycle of catalysts during biogas steam reforming, especially in cases such as those with Ni-based catalysts and other heterogeneous catalysts. Indeed, deactivation can be caused by several factors due to mechanical, thermal, or chemical processes [101], that will be explained in following subsections. To a lesser extent, the durability of a catalyst can be affected by other factors, like its kind and shape, or operating conditions like steam to carbon ratio, pressure, or temperature [102].

3.1. Sintering

This is a process due to the agglomeration and growth of metal crystallites of the active phase, occurring at high temperatures. Considering that SRM usually requires high temperatures (above 600 °C), this is an event that should be considered. In that sense, Hüttig and Tamman temperatures determine the atom or crystallite migration for a certain metal, one third and one half of the melting point of the corresponding metal, respectively. Considering the typical operating conditions for biogas SR, the reaction temperatures for this process are high enough to provoke surface and bulk atom migration. Consequently, as included in Figure 11, there are three different stages in sintering on support's surface.

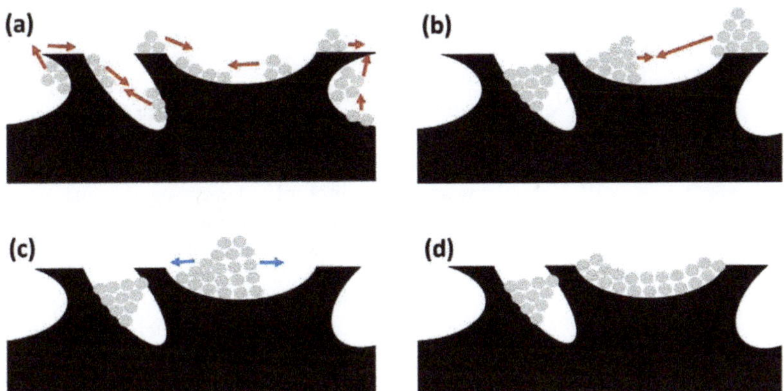

Figure 11. Different stages during sintering of active phase on catalyst support (catalyst support in black and metal particles in grey): (**a**) migration (red arrows); (**b**) collision (red arrows); (**c**) spreading (blue arrows); and (**d**) blockage (also in (**b**,**c**)).

First, atomic migration takes place, with a detachment from the crystallites and migration on the support's surface, generating bigger metal particles (Figure 11a). Afterwards, these bigger particles or crystallites can also migrate and collide, obtaining bigger particles (see Figure 11b). The process ends with particle spreading on the catalyst surface, as observed in Figure 11c, and blocking active sites.

This way, this phenomenon is related to a decrease in catalytic activity due to two main causes. First, sintering implies a decrease in the surface area of active sites, reducing the efficiency of the catalysts. Second, as observed in Figure 11d, crystallite growth can block pores on the catalyst support, containing further active sites that otherwise would be available for SRM [24,101].

Sintering can also be influenced by other aspects such as the catalyst structure and its porosity and metal–support interactions, promoting strong metal–support interactions that could decelerate the sintering [102].

3.2. Poisoning

Poisoning is due to a strong chemisorption of chemical species on catalytic sites, blocking them and avoiding SRM. There is a wide range of chemical products that can poison catalysts in biogas steam reforming. Among them, one of the most important ones is hydrogen sulfide (H_2S), whose content in biogas after biodigestion processes is not negligible, ranging from few ppm up to 1%. High H_2S content, apart from a poisoning effect, could be harmful or even deadly, promoting corrosion in industrial facilities, and decreasing the heating value of fuel gas. As a consequence, H_2S removal before biogas steam reforming is necessary to avoid a decrease in the global yield [103,104].

Depending on the kind of catalyst, this negative effect could be more noticeable. For instance, Ni is very sensitive to poisoning (see Equation (3)), whereas other metals like Co seem to offer a lower affinity for sulfur, with possible and interesting uses in bimetallic catalysts. Equally, other metals can react with H_2S, like Ag, Cu, Fe, Ru, or Pt [105].

$$M + H_sS \leftrightarrow MS + H_2 \tag{3}$$

where M can be any abovementioned metal. As a consequence of the interaction of hydrogen sulfide with the active site (through sulfidation), the catalyst cannot take part in steam reforming, partially or completely reducing its activity during the process.

In that sense, adsorption and absorption seem to be suitable techniques for removing hydrogen sulfide before biogas steam reforming, requiring low concentrations for this purpose (up to 5 ppm) before biogas processing in steam reforming facilities [102]. Specifically, the use of alkanolamines (such as methyl ethanolamine or methyl diethanolamine), alkaline salts, organic solvents, deep eutectic solvents, or ionic liquids for absorption, as well as zeolites, metal oxides, or carbon-based sorbents for adsorption, could be interesting treatments for removing H_2S under ambient pressure and low operating temperatures [104,106–108]. It must be borne in mind that, as previously explained, sewage sludge reuse as an active carbon (obtained through pyrolysis and gasification processes) to adsorb H_2S could be an interesting starting point for implementing a circular economy in wastewater treatment plants. Finally, recent studies have proposed a simplified heterogeneous fixed-bed reactor model to simulate the influence of H_2S poisoning on Ni-Al_2O_3 catalysts for methane steam reforming, with a good agreement between the simulated and experimental experiences if an order of deactivation (n = 1) is assumed [109]. These kinds of simulations are quite useful, as they can be easily adapted to different H_2S concentrations in biogas and GHSV. Also, the use of bimetallic catalysts to increase poisoning resistance is another interesting aspect to be taken into consideration, as explained in previous studies where the use of a Rh-Ni/Ce-Al_2O_3 catalyst showed a higher resistance to poisoning, being reversible by using regeneration processes, after which, the catalyst did not show selectivity to the reverse WGS reaction, allowing for high H_2 yields [110].

3.3. Carbon Deposition

Coking could represent another negative factor, related to the physical formation of carbon deposits due to gas-phase chemical reactions like methane cracking or CO disproportionation [102]. Carbon deposition can imply the deactivation or blockage of active sites, which decreases the effectiveness of the active phase over reaction time. As observed in Figure 12, successive phases take place during carbon deposition, with different effects depending on its degree of severity.

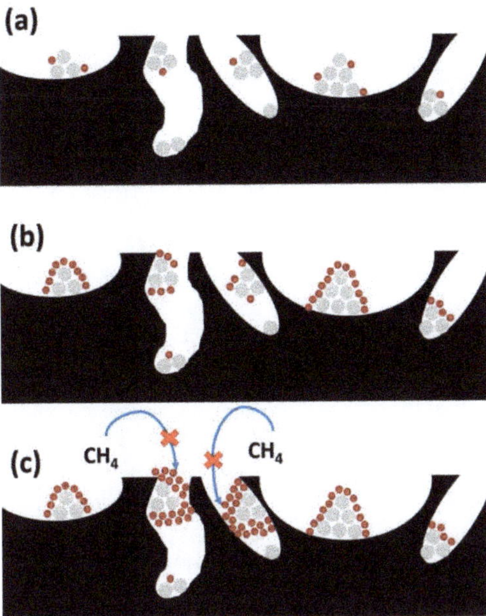

Figure 12. Different stages during coke deposition (coke in red, active phase in grey, and support in black): (**a**) coke production and chemisorption; (**b**) coke spreading with the subsequent encapsulation; and (**c**) active sites and pore blockage.

This way, coke chemisorption or adsorption takes place on active sites (Figure 12a), reducing their access to reactants. In further stages, coke diffusion or dispersion to generate active site encapsulation occurs (as observed in Figure 12b), completely blocking the active sites to reactants, and pore blockage takes place (Figure 12c), hampering the SRM reaction on the available active sites. This fact takes place especially when the CO and CH$_4$ decomposition is faster than the carbon removal.

To avoid this phenomenon, the use of promoters to strengthen the metal–support interactions could present a positive effect. Also, the particle size of the active phase could play an important role in controlling coke deposition [102].

4. Key Points to Improve the Performance of Biogas Steam Reforming

Considering the previous section, it is essential to take steps to solve deactivation processes, which hinders a suitable catalytic performance (with high methane conversions) and long service life. These steps could be taken, as observed in Figure 13, before, during, or after biogas steam reforming. As mentioned earlier (as it will be discussed throughout this text), every detail counts when it comes to contributing to a better catalytic performance in this process. Each stage is explained in the following subsections. It should be noted that these steps directly affect the performances of catalysts during steam reforming, but other aspects related to this process could be equally improved, such as the membrane reactor performance (by reducing the coke deposition or hydrogen sulfide content) or the maintenance of steam reforming facilities (delaying corrosion with a decrease in H$_2$S content). Nonetheless, other factors could be affected, implying efficiency loss and increases in costs.

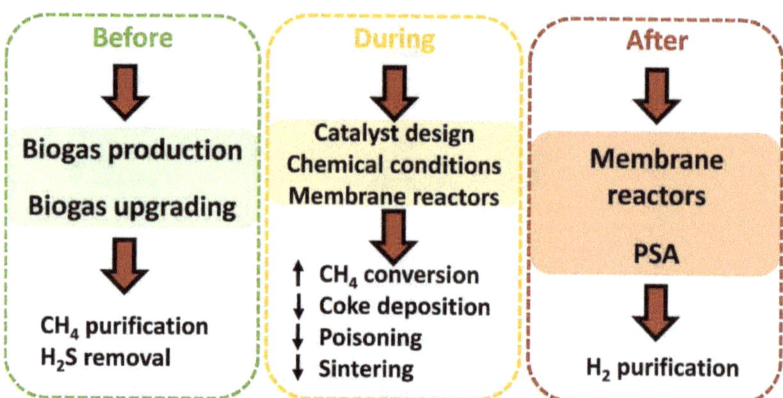

Figure 13. Steps carried out to improve the performance (through better conversion or longer useful life) of catalysts in steam reforming.

4.1. Steps before Steam Reforming (Biogas Production and Upgrading)

Regarding the previous steps, there are many processes that could improve methane production during the anaerobic digestion of biomass, as well as biogas upgrading to increase methane concentration, avoiding undesirable compounds such as hydrogen sulfide. As explained in the introduction section, there are plenty of measures for increasing the efficiency of biogas production and quality. Thus, the use of additives (such as microorganisms, enzymes, or inorganic compounds) could facilitate different steps during anaerobic digestion by removing inhibitors (such as ammonia, long-chain fatty acids, and acidification caused by VFAs, etc.) and creating suitable conditions for microorganism proliferation [4]. In this sense, the use of active carbons obtained from digestates could be interesting (such as sewage sludge, whose active carbon obtained through hydrothermal carbonization could be a promising starting point for valorizing this waste, as explained in previous works [38]), as it could be an example of an applied circular economy during biodigestion. On the other hand, biogas upgrading is essential for removing H_2S (which could deactivate catalysts through poisoning) and for increasing the CH_4 concentration in biogas (which is desirable for carrying out a more efficient SRM). For the former, adsorption systems are usually selected (for instance, activated carbons or nanoparticles), although there are many techniques for retaining H_2S, such as biological desulfurization, membrane separation (with polymeric membranes, normally capable of retaining CO_2 and H_2S [32]), or absorption with inorganic solutions (many of them based on iron) [103,104,111,112]. Also, some techniques such as PSA can be used to upgrade biogas by increasing methane content [113].

4.2. Steps during Steam Reforming (Catalyst Design, Chemical Conditions, Use of Membrane Reactors)

4.2.1. Catalyst Design (Promoters and Bi-Metallic Catalysts)

Regarding the catalyst design, if a certain catalyst is selected, with the aim of achieving the maximum conversion and stability, the catalyst should present as much dispersion of the active phase as possible, as there is not a minimum particle size from which a decrease in activity is found.

Another important factor is avoiding active-phase mobility and agglomeration, promoting that the active phase is dispersed enough to reduce the possibility of collision with other particles and the subsequent agglomeration. It is achieved by reducing the active phase concentration, whereas the number of active sites decrease. The typical Ni concentration in catalysts is 20% w/w, obtaining particle sizes from 10 to 100 nm.

In addition, feeding should be considered in catalyst design. In other words, catalyst deactivation by the pollutants included in biogas should be reduced. To reduce the effects related to coke deposition, the active phase surface on the support should be maximized, so that the diffusion of gas is enough to avoid a reducing atmosphere that allows for CO and CO_2 decomposition, producing carbon which is placed on the active phase surface and generating carbide and coke.

The use of promoters in the active phase has been studied and reviewed, with some common additives such as alkalis (K and Na), transition metals (La, Zr, and Zn), and non-metals (Al and B) [114,115].

An interesting issue with room for improvement is the resistance of catalysts to poisoning due to H_2S. This compound is usually removed through previous adsorption before steam reforming in a reactor [116], and few scientific articles have dealt with an increase in the resistance of catalysts to this pollutant. Some studies have focused on reviewing the influence of sulfur content in feeding for several processes [117]. As a conclusion, Ni seems to be a catalyst with a difficult direct protection from poisoning, with noble metals offering better results, except for Rh. On the other hand, there is a possibility of using metals from groups 4 to 6, through bifunctional catalysts, which could be a promising alternative.

4.2.2. Operating Conditions

There are several influences on catalytic performance depending on the operating conditions, mainly related to the promotion of deactivation processes [40]. Taking into account that methane steam reforming is an endothermic reaction (see Equation (1)), the effect of temperature is clear, with higher CH_4 conversions with temperature (as observed in specific cases such as the catalytic steam reforming of biogas with a Rd catalyst) [31]. However, intermediate solutions should be achieved, as extra energy costs associated with keeping the reactor temperature should be avoided, as explained in following sections.

Concerning steam addition, high S/C ratios (at least 1.5, achieving excess of feed vapor) are recommended to avoid coke deposition, among other factors like high pressure [100,118]. However, from an economic point of view, the production of large quantities of superheated vapor would imply a considerable increase in costs [42]. Additionally, some studies have pointed out the possible catalyst deactivation on time-on-stream, especially at high temperatures, observing a direct correlation between deactivation rates and high S/C, mainly due to the steam-induced metal–support interaction, resulting in an inactive spinel phase and not due to metal reoxidation [89].

Also, as previously explained, temperature presents opposite effects. On the one hand, high temperatures would imply a sintering effect, whereas low temperatures could promote coke deposition. In the case of steam and temperature, optimization and intermediate steps should be considered, because intermediate conditions to meet both low energy costs and high methane conversions should be obtained (apart from the obvious effects on catalytic performance in biogas steam reforming).

Equally, the CH_4/CO_2 ratio in biogas seems to present an influence on catalytic performance. In that sense, according to recent studies using two different catalysts (4% $Ni/NiAl_2O_4/Al_2O_3$ and 3.1% Ru/Al_2O_3), an increase in CO_2 implied a decrease in H_2/CO ratio and H_2 yield, finding an optimal CH_4/CO_2 ratio of 1.5/1 [119]. Therefore, CH_4/CO_2 ratios above 1 are advisable for a suitable catalytic performance, as explained in previous studies for Ni-based catalysts, possibly due to a CO_2-promoted Boudouard reaction, implying further coke deposition [120].

At this point, it is essential to consider the optimization and modelling of steam reforming for any specific case, as will be discussed in future sections.

4.2.3. Membrane Reactors

The use of membrane reactors is another interesting starting point for improving the performance of hydrogen production from biogas through steam reforming. Thus, the aim of this technology is to purify the hydrogen obtained during steam reforming using thin membranes (the selective layer is usually made of Pd and Ag on different kinds of supports, like stainless steel or alumina) where H_2 permeates, to obtain a final gas with a high purity (up to 95–99%) [121,122]. Figure 14 shows the different stages that take place during biogas (or methane, in a simplified form in this case), including the reactant inlet, H_2 permeation through the membrane, and, finally, the retentate and permeate outlets [123].

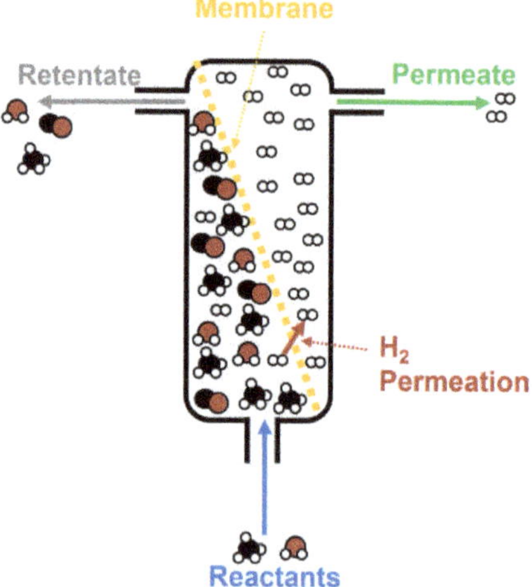

Figure 14. Scheme of a standard membrane reactor for hydrogen production during methane steam reforming.

It should be noted that the catalyst is usually put into contact with the membrane, in order to assure that the chemical reactions (and the subsequent products, like H_2) are as close as possible to the membrane.

This way, the chemical balance of the reactions observed in Equations (1) and (2) could be oriented towards product generation, as hydrogen is rapidly removed from the reaction medium to be delivered in a highly pure gas stream (permeate), whereas the rest of the products (mainly CO and CO_2, along with unreacted CH_4 and H_2O) are separated in another stream (retentate), which can be further treated to obtain higher yields in hydrogen [124]. This fact is very interesting, as some chemical conditions could be softened to obtain a higher efficiency during biogas steam reforming.

In that sense, temperature and pressure could be lowered, whereas the catalyst design can vary (for instance, the active phase in catalysts could be reduced, with the subsequent savings for this process). Regarding temperature decrease, it could imply a positive aspect for catalyst deactivation, as sintering effects could be delayed at lower temperatures, and hydrogen recovery in a membrane reactor is usually improved [125].

However, coke deposition could be promoted at low temperatures, which could present a reverse impact in methane steam reforming [126]. As in the case of many catalysts, H_2S present negative effects in membrane reactors, as Pd poisoning (the most popular element used in membranes) would provoke a progressive loss in separating performance [121,122].

4.3. Steps after Biogas Steam Reforming (Hydrogen Purification)

Even though there is not a clear and direct link between purification processes once biogas is converted into synthesis gas (a mixture of H_2 and CO, among other compounds) and their influence on catalyst performance, some indirect positive effects could be found.

As observed in Figure 15, there are different biogas steam reforming configurations to carry out hydrogen production with a high purity. In that sense, considering the first route (Figure 15a), the use of membrane reactors in situ could contribute, as explained in the previous subsection, to shifting the reaction balance towards product generation, with the subsequent possibility of decreasing chemical conditions such as temperature and, consequently, a delay in deactivation processes such as sintering.

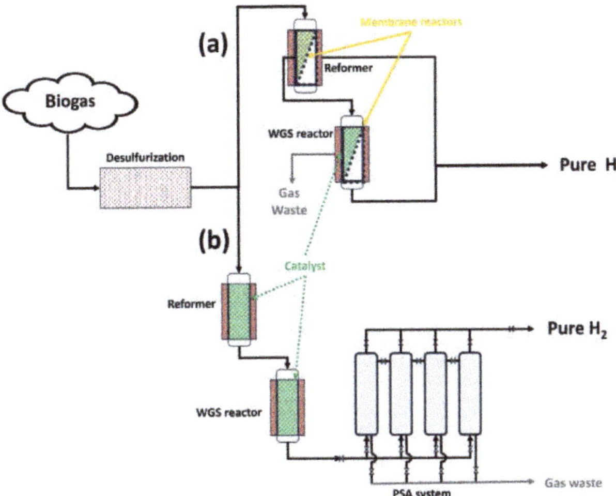

Figure 15. Different routes for biogas steam reforming to obtain high-purity hydrogen: (**a**) through membrane reactors; and (**b**) through PSA.

However, in Figure 15b, the purification process (in this case, PSA, where an adsorbent is used to bind molecules depending on the gas component, type of adsorbent material, partial pressure, and operating temperature [25,127]) takes place after the products are obtained, with no direct effect on the chemical balance.

Also, other configurations (due to economic adaptations of previous steam reforming facilities) could couple membranes after steam reforming to obtain pure hydrogen, presenting a similar situation compared to the use of PSA.

In these situations, an improvement in chemical conditions can be equally found, considering the high-quality product obtained (normally >98% hydrogen, which is highly valuable in the energy market). In this context, lower reaction temperatures or catalyst concentrations could be used, offsetting the lower conversion with a valuable final product and with the possibility of recirculating gas waste in the same process or other processes. Also, biogas can be converted to synthesis gas, which can be used in other processes such as Fischer–Tropsch, where different H_2/CO ratios (even at lower temperatures compared with normal steam reforming conditions to optimize hydrogen production) can be used to obtain, for instance, liquid fuels [128]. Under these circumstances, the catalyst could be less deactivated, especially if these steps are combined with the use of a suitable catalyst support or the use of promoters, which could equally improve the final conversion of methane in biogas.

5. Techno-Economic Analyses and Patents Derived from Catalytic Biogas Steam Reforming

As expected for such a mature technology, its implementation at an industrial level is highly extended, including research studies and patents for the application of steam reforming of methane (and biogas) about the feasibility of implementation (paying attention to techno-economic aspects) of this technology. In that sense, in general, there has been a considerable increase in hydrogen production patents focused on catalysts, especially at the beginning of this decade [26].

Firstly, the potential for the production of biogas, electricity, and heat from waste is important, as a wide range of electricity yields (from 52 to 850 kWh) could be obtained from

1 ton of biodegradable waste [35]. Previous studies recommend large-scale productions of biogas (at least 400 tons per day, and not recommending productions below 50 tons per day), especially for energy consumption due to steam generation [42]. It should be noted that, to make biogas steam reforming economically feasible, previous and further steps should be equally considered. As explained in the literature, biogas production and quality can be interrupted or reduced depending on some problems related to anaerobic digestion, such as over-acidification or foaming, which could lead to repeated and extended shutdowns of these units and the subsequent biogas production losses (up to 50%) [4]. Consequently, if biogas steam reforming is considered as a coupled technology related to biogas production (for instance, in WWTPs [38]), every aspect concerning the improvement of the stability and efficiency of biogas production will imply a better economic performance of biogas treatment on the whole. In that sense, the use of a circular economy (as seen in Figure 2, where the use of digestate to improve anaerobic digestion could be an interesting way of reusing waste with difficult management) could be resourceful, as well as the addition of macro, micronutrients enzymes, or carbon-based materials (for instance, hydrochar). Also, biogas upgrading should be considered, and an increase in the efficiency of contaminant removal, as well as an improvement in the service life of adsorbents (for instance), will have a positive repercussions for biogas steam reforming, allowing for the acquisition of more advanced equipment if global amortization takes place at an earlier economic stage.

Specifically, when it comes to the direct conversion of methane included in biogas, many factors are taken into account, especially the role of catalysts in this process, which will allow for softening some operating conditions (implying, for instance, a decrease in temperature reaction or steam addition) that could cheapen some permanent operating costs. Also, technological improvements in SRM are focused on the development of catalysts to contribute to an increase in hydrogen yield and a decrease in greenhouse gas emissions, with a reduction in energy consumption [26]. In other words, these works are devoted to minimize the main challenges related to SRM, like carbon dioxide production and energy costs [25].

Equally, regarding the treatment of gaseous gas obtained in biogas or methane steam reforming, there are plenty of choices to valorize biogas steam reforming. Obviously, in most cases, there must be a considerable initial investment, with the subsequent technological coupling or upgrading of biogas steam reforming facilities, as observed in Figure 2 in the case of Fischer–Tropsch synthesis, the use of membrane reactors, or PSA coupling (see also Figures 14 and 15). Nevertheless, the final added-value product (liquid fuels in the case of Fischer–Tropsch or pure H_2 in the case of membrane reactors and PSA) would offset this initial investment in the long run. Again, and even though these are technologies not directly related to biogas steam reforming by themselves, they can present a strong and positive influence, in a comprehensive manner, on the economic performance of this process.

Finally, a long, useful life of catalysts, as well as low concentrations (when possible), can positively contribute to an improvement in the economic performance of biogas steam reforming. As previously explained, the role of promoters is mainly focused on these purposes, as observed in previous studies where the use of nanosized Ni-Rh bimetallic clusters allowed for reducing the Ni content to 3% (as the size and distribution of the active phase was considerably improved compared to traditional Ni-based catalysts), with a considerable increase in catalytic stability and activity, which could contribute to a fixed cost reduction [129]. Thus, other works included in Table 7 follow a similar trend, where the use of promoters or even a change in the reactor disposition can make a reduction in catalyst addition (as well as other operating conditions like temperature) possible, with subsequent cost savings. As explained in previous sections, the role of some factors such as the use of membrane reactors can improve the economic performance of SR systems, including a reduction in catalyst amount or useful life. It should be noted that the works included in Table 7 are heterogeneous, dealing with facilities of different sizes (from the laboratory to semi-industrial level) and focusing on different aspects (from a direct improvement in catalysts through the use of promoters to the optimization of membrane reactors). In that sense, further studies about techno-economic assessments in this field will be really useful, enriching the current literature. Other works dealing with techno-economic research on biogas steam reforming are devoted to the use of PSA units [127,130,131], simulations and their comparisons with plant-scale experiences [132–134], or the whole process in general (in some cases. from biogas production to biogas steam reforming) [135,136], with an indirect role of the catalyst.

Table 7. Main techno-economic studies related to biogas steam reforming.

Study	Details	Comments	Reference
Performance and stability of doped Ceria-Zirconia catalyst for steam reforming	A quartz microreactor in a ceramic tube furnace was used, at atmospheric pressure, 800 °C, and 120,000 h^{-1} space velocity (GHSV). The gaseous mixture flows were adjusted for the different reactions: SR (steam to carbon mole fraction S/C = 2.5)	Doped catalysts doubled catalytic activity (from 55 to 100 h)	[137]
Performance of Ce-Ni catalyst	T= 600–850 °C; P = 1 atm; gas flow rate = 200 mlN/min; CH_4:CO_2:H_2O:He mole ratio = 1:0.8:0.4:2.8; 0.5 g of catalyst with a grain size of 0.3–0.8 mm was used	At 850 °C, 20 wt.% $Ce_{0.2}Ni_{0.8}O_{1.8}$/Al_2O_3-SG catalyst provided 100% hydrogen yield and full CH_4 conversion, with 85% CO_2 utilization, obtaining a more efficient catalyst	[138]
Catalyst rearrangement for improving hydrogen production in biogas SR	Study of hydrogen production through SR by improving the design of SR reactor and catalyst (nickel- and yttria-stabilized zirconia) disposition to increase efficiency	Higher effectiveness was achieved, allowing a decrease (above 40%) in the amount of catalyst used	[139]
Techno-economic analysis of swine manure biogas through SR	A complete study of a pilot-scale installation, covering the analysis of hydrogen production	After process flow modeling based on mass and energy balance, the plant produced 250 kg/h of H_2 from 1260 kg/h of biomethane from purified biogas (4208 m^3/h). Biomethane conversion offsets high investment costs	[140]
Optimization of a small-scale hydrogen production plant	The use of membrane reactors using a rhodium-based catalyst is optimized in order to assess the efficiency of SR at different levels	A reduced percentage of active phase (20%) was required, reducing 80% of the catalyst cost	[141]
Techno-economic analysis of a biogas SR plant	Membrane reactors as well as Ni-based catalysts are considered in the final configuration of the plant	Compared to biomethane, biogas is recommended from economic point of view at the current price	[142]
Design and operational considerations of a packed-bed membrane reactor for SRM	An experimental and computational study about the design of a membrane reactor for SRM was carried out, emphasizing geometrical scale of the reactor and catalytic activity	The optimum conditions were: GHSV = 1134 h^{-1}; S/C = 2; P = 30 atm; T = 773.15 K. The key limiting factor was catalytic activity, which points out the importance of improving its performance	[143]
Techno-economic analysis of an hydrogen production system including biogas reforming	Steam reforming of biogas was carried out to produce hydrogen from landfill gas by using PSA and CA	Simulation results were compared in terms of Levelized Cost of Hydrogen, which was below 2 €/kg_{H2} and feasible according to 2030 horizon	[144]

Table 7. *Cont.*

Study	Details	Comments	Reference
Energy and exergy analyses for hydrogen production from biogas	Biogas steam reforming and autothermal steam reforming are compared	The steam-reforming-based configuration achieved the best performance regarding H_2 production energy-based efficiency (59.8%). If co-production of heat and H_2 is considered, this value increased up to 73.5%	[145]
Technical, economical, and ecological aspects of biogas steam reforming to produce hydrogen	Investment, operation, and maintenance costs were analyzed	H_2 production cost achieved the value of 0.27 US\$/kWh with a payback period of 8 years. An ecological efficiency of 94.95% was obtained, proving the feasibility of the process	[146]
Optimization of biogas steam reforming	A steam reforming plant is optimized regarding electricity, freshwater, H_2, and cooling	The plant can generate 86.04 kW, 0.9662 kg/s, 958.2 kW, and 704.573 kg/h of electricity, fresh water, cooling, and hydrogen, respectively, with an improvement in exergy (19.9%) and energy efficiency (24.5%)	[147]
Multi-objective optimization of biogas steam reforming for multi-generation system	A design of a multi-generation integrated energy system powered by biogas energy is proposed, assessed, and optimized	The proposed plant generated 108.7 kW, 888.7 kW, and 703.3 kg/h, power, cooling load, and hydrogen, respectively. The energy and exergy efficiencies were 31.51% and 31.14%, respectively	[148]
Techno-economic modelling of a poly-generation system based on biogas	The system is devoted to power, H_2, freshwater, and ammonia production	The system provided 687.4 kW of power, 0.9662 kg/s of freshwater, 0.15 kg/s of hydrogen, and 1.149 kg/s of ammonia	[149]
Simulation of steam reforming of biogas in an industrial reformer	A simulation of steam reforming and its comparison with an industrial-scale performance was carried out	Total molar feed rate of 21 kmol/h, steam to methane ratio of 4.0, temperature of 973 K, and pressure of 25 bar, obtaining high CH_4 conversion (>93%) and H_2 yield	[150]

Another interesting proof of the feasibility of this process at the industry level is the proliferation of patents about this subject. As observed in Table 8, there are plenty of patents whose main objective is the improvement of biogas (or methane) steam reforming performance, in many cases focused on the role of catalysts, which is to be expected, as they usually reduce the activation energy of the process, allowing for optimizing methane conversion into hydrogen. In any case, different aspects related to catalytic performance should be considered, such as their durability and stability, which would determine the subsequent stable and efficient hydrogen production. Also, other patents are focused on the abovementioned technologies to improve SRM performance, like the use of biogas upgrading through PSA to increase methane concentration, the use of membrane reactors or PSA for biogas purification or syngas upgrading (which can contribute to a reduced use of catalysts or an increase in stability, for instance), or possible alternatives for the produced syngas, like Fischer–Tropsch synthesis to obtain hydrocarbon products. As observed, practically all the techniques explained in previous sections are directly or indirectly applied in these patents, which proves the correlation between scientific works and applied research. On the other hand, these patents are focused on an increase in SRM efficiency, which usually implies an improvement in economic and energy costs, as well as, for instance, an improvement in catalytic performance with a higher stability.

Table 8. Current patents focused on catalytic biogas steam reforming.

Description	Comments	Reference
Method for preparing hydrogen from biogas	Use of a reactor filled with catalyst, introducing biogas and water to obtain H_2 and CO. The catalyst is bifunctional, catalyzing methane steam conversion and CO_2 dry reforming reaction at the same time.	[151]
Method for producing hydrogen from biogas	A catalyst is used for steam reforming to produce syngas obtaining H_2 and CO. CO is oxidized in through carbon monoxide shift reaction, obtaining pure hydrogen in an adsorption device through pressure adjustment.	[152]
Biogas conversion to synthesis gas to produce hydrocarbons	This system offers different alternatives for biogas steam reforming, with a possible purification unit and an interesting alternative like the use of Fischer–Tropsch synthesis unit to produce hydrocarbons.	[153]
Method for producing hydrogen from biogas biomass	A method based on biogas steam reforming, with previous purification of biogas through desulfurization and PSA to obtain higher CH_4 concentration, and a final H_2 purification from syngas through another PSA system.	[154]
System for biomethane and syngas production from biogas stream	A plant for producing biomethane and syngas from biogas is proposed, with purification units to purify the biogas stream, steam reforming unit, and a membrane separation unit to produce biomethane stream.	[155]
Method for producing hydrogen from biogas	Production of hydrogen through biogas steam reforming, purifying biogas and final syngas produced by using PSA, offering a more efficient performance.	[156]
Method and plant for the production of synthesis gas from biogas	A more economic and efficient biogas steam reforming system is presented, with a previous H_2S and CO_2 removal.	[157]
Efficient use of biogas carbon dioxide in liquid fuel synthesis	Biogas is obtained through anaerobic digestion of different wastes, obtaining synthesis gas through steam reforming and converting CO_2 in biogas to synthesis gas by combining the CO_2 reforming reaction with steam reforming or partial oxidation.	[158]
Method and system for coupling biogas reforming with natural gas combined cycle based on solar energy driving	A method and system for coupling biogas reforming with natural gas combined cycle based on solar energy driving is presented, offering a higher efficiency thanks to the support of combined renewable energies.	[159]
Biogas and solar energy complementary two-stage preparation system and method for synthesis gas	The biogas steam reforming system consists of a desulfurization unit, and two SR reactor and heat regenerators, coupled to solar energy. Consumption of water was reduced, and the economic efficiency of biogas and solar energy use was improved.	[160]

6. Conclusions and Future Trends

Biogas steam reforming is a reality for obtaining high yields of hydrogen from wastes with difficult environmental management such as sewage sludge, with multiple emerging techniques that might be promising in the near future. Indeed, according to the scientific interest from many different and multidisciplinary research teams, it represents the future and, also, the present of biogas and methane processing for obtaining hydrogen.

In that sense, the role of catalysts is vital to make this process more effective, efficient and, therefore, suitable at an industrial scale. High yields and selectivity values for hydrogen production have been obtained according to the literature, but there are some challenges to be solved, especially concerning the durability of these catalysts, which can be diminished by factors such as poisoning, coke deposition, or sintering.

This way, cutting-edge research works, focused on a wide range of aspects related to biogas steam reforming, have recently been carried out.

With regard to the catalytic steam reforming of biogas, the use of innovative catalyst formulations offers a wide range of possibilities, including the use of bi-metallic, multi-metallic catalysts, or promoters. In that sense, a resistance to coke deposition, sintering, and poisoning is essential by enhancing the interactions between the active phase and support surface or improving the catalytic performance of the active phase. Equally, the role of nano-catalysts, as well as innovative supports such as ceramic foams, are interesting, as increases in surface interaction and selectivity are usually found.

Also, the optimization of the process is essential, as many factors can influence on the catalytic performance. In that sense, temperature and steam optimization, as well as biogas quality control (especially with a low H_2S content) should be taken into account.

Finally, concerning techno-economic assessments, recent studies seem to point out improvements to reduce costs related to catalyst addition (mainly reducing its amount and increasing its durability and useful life) and its performance (which could allow for softening chemical conditions, especially temperature). Equally, better hydrogen conversions could imply higher benefits (including softening of chemical conditions), as the final gas obtained (regardless of the subsequent purification processes) will be highly valued. Concurrently, patents are focused on similar topics, which points out the relevance of the catalytic steam reforming of biogas at industrial level.

Author Contributions: Conceptualization, S.N.-D.; methodology, S.N.-D.; investigation, S.N.-D., V.M. and C.M.Á.-M.; resources, J.F.G.; data curation, S.N.-D. and C.M.Á.-M.; writing—original draft preparation, S.N.-D., V.M. and C.M.Á.-M.; writing—review and editing, S.N.-D., V.M. and J.F.G.; visualization, S.N.-D., C.M.Á.-M. and J.F.G.; supervision, J.F.G.; project administration, J.F.G.; funding acquisition, V.M. and J.F.G. All authors have read and agreed to the published version of the manuscript.

Funding: Junta de Extremadura and FEDER funds through projects IB20042 and NextGeneration EU.

Acknowledgments: The authors thank the economic support to Junta de Extremadura and European FEDER funds through projects IB20042. S.N. and C.M. thank the NextGeneration EU funds, including investment lines C17.I1 "Planes complementarios con las Comunidades Autónomas", within line 17 "Reforma institucional y fortalecimiento de las capacidades del sistema nacional de Ciencia, Tecnología e Innovación" within the Recovery, Transformation and Resilience Plan.

Conflicts of Interest: The authors declare no conflict of interest.

Abbreviations

Abbreviation	Term
AD	Anaerobic digestion
BET	Brunauer–Emmett–Teller
CA	Chemical absorption
DRM	Dry reforming of methane
GHSV	Gas hourly space velocity
IWI	Incipient wet impregnation
PSA	Pressure swing adsorption
PZC	Point of zero charge
S/C	Steam to carbon ratio
SS	Sewage Sludge
SR	Steam reforming
SRM	Steam reforming of methane

VFAs	Volatile fatty acids
WGS	Water–gas shift
WI	Wet impregnation
WWTP	Wastewater treatment plant

References

1. Scholten, D.; Bosman, R. The Geopolitics of Renewables; Exploring the Political Implications of Renewable Energy Systems. *Technol. Forecast. Soc. Chang.* **2016**, *103*, 273–283. [CrossRef]
2. Sharma, N.K.; Govindan, K.; Lai, K.K.; Chen, W.K.; Kumar, V. The Transition from Linear Economy to Circular Economy for Sustainability among SMEs: A Study on Prospects, Impediments, and Prerequisites. *Bus. Strat. Environ.* **2021**, *30*, 1803–1822. [CrossRef]
3. Ramchuran, S.O.; O'Brien, F.; Dube, N.; Ramdas, V. An Overview of Green Processes and Technologies, Biobased Chemicals and Products for Industrial Applications. *Curr. Opin. Green Sustain. Chem.* **2023**, *41*, 100832. [CrossRef]
4. Leca, E.; Zennaro, B.; Hamelin, J.; Carrère, H.; Sambusiti, C. Use of Additives to Improve Collective Biogas Plant Performances: A Comprehensive Review. *Biotechnol. Adv.* **2023**, *65*, 108129. [CrossRef] [PubMed]
5. Bumharter, C.; Bolonio, D.; Amez, I.; García Martínez, M.J.; Ortega, M.F. New Opportunities for the European Biogas Industry: A Review on Current Installation Development, Production Potentials and Yield Improvements for Manure and Agricultural Waste Mixtures. *J. Clean. Prod.* **2023**, *388*, 135867. [CrossRef]
6. Mccarty, P.L.; Smith, D.P. Anaerobic Wastewater Treatment. *Environ. Sci. Technol.* **1986**, *20*, 1200–1206. [CrossRef]
7. Sawyerr, N.; Trois, C.; Workneh, T.; Okudoh, V. An Overview of Biogas Production: Fundamentals, Applications and Future Research. *Int. J. Energy Econ. Policy* **2019**, *9*, 105–116. [CrossRef]
8. Khawer, M.U.B.; Naqvi, S.R.; Ali, I.; Arshad, M.; Juchelková, D.; Anjum, M.W.; Naqvi, M. Anaerobic Digestion of Sewage Sludge for Biogas & Biohydrogen Production: State-of-the-Art Trends and Prospects. *Fuel* **2022**, *329*, 125416. [CrossRef]
9. Aromolaran, A.; Sartaj, M.; Abdallah, M. Supplemental Sewage Scum and Organic Municipal Solid Waste Addition to the Anaerobic Digestion of Thickened Waste Activated Sludge: Biomethane Potential and Microbiome Analysis. *Fermentation* **2023**, *9*, 237. [CrossRef]
10. Wang, Y.; Li, W.; Wang, Y.; Turap, Y.; Wang, Z.; Zhang, Z.; Xia, Z.; Wang, W. Anaerobic Co-Digestion of Food Waste and Sewage Sludge in Anaerobic Sequencing Batch Reactors with Application of Co-Hydrothermal Pretreatment of Sewage Sludge and Biogas Residue. *Bioresour. Technol.* **2022**, *364*, 128006. [CrossRef]
11. González, R.; Peña, D.C.; Gómez, X. Anaerobic Co-Digestion of Wastes: Reviewing Current Status and Approaches for Enhancing Biogas Production. *Appl. Sci.* **2022**, *12*, 8884. [CrossRef]
12. Nguyen, L.N.; Kumar, J.; Vu, M.T.; Mohammed, J.A.H.; Pathak, N.; Commault, A.S.; Sutherland, D.; Zdarta, J.; Tyagi, V.K.; Nghiem, L.D. Biomethane Production from Anaerobic Co-Digestion at Wastewater Treatment Plants: A Critical Review on Development and Innovations in Biogas Upgrading Techniques. *Sci. Total Environ.* **2021**, *765*, 142753. [CrossRef] [PubMed]
13. Mao, C.; Feng, Y.; Wang, X.; Ren, G. Review on Research Achievements of Biogas from Anaerobic Digestion. *Renew. Sustain. Energy Rev.* **2015**, *45*, 540–555. [CrossRef]
14. Karrabi, M.; Ranjbar, F.M.; Shahnavaz, B.; Seyedi, S. A Comprehensive Review on Biogas Production from Lignocellulosic Wastes through Anaerobic Digestion: An Insight into Performance Improvement Strategies. *Fuel* **2023**, *340*, 127239. [CrossRef]
15. Dragassi, M.-C.; Royon, L.; Redolfi, M.; Ammar, S. Hydrogen Storage as a Key Energy Vector for Car Transportation: A Tutorial Review. *Hydrogen* **2023**, *4*, 831–861. [CrossRef]
16. Mazurova, K.; Miyassarova, A.; Eliseev, O.; Stytsenko, V.; Glotov, A.; Stavitskaya, A. Fischer–Tropsch Synthesis Catalysts for Selective Production of Diesel Fraction. *Catalysts* **2023**, *13*, 1215. [CrossRef]
17. Al-Zuhairi, F.K.; Shakor, Z.M.; Hamawand, I. Maximizing Liquid Fuel Production from Reformed Biogas by Kinetic Studies and Optimization of Fischer–Tropsch Reactions. *Energies* **2023**, *16*, 7009. [CrossRef]
18. Kumar, M.; Dutta, S.; You, S.; Luo, G.; Zhang, S.; Show, P.L.; Sawarkar, A.D.; Singh, L.; Tsang, D.C.W. A Critical Review on Biochar for Enhancing Biogas Production from Anaerobic Digestion of Food Waste and Sludge. *J. Clean. Prod.* **2021**, *305*, 127143. [CrossRef]
19. Liu, M.; Wei, Y.; Leng, X. Improving Biogas Production Using Additives in Anaerobic Digestion: A Review. *J. Clean. Prod.* **2021**, *297*, 126666. [CrossRef]
20. Nikolaidis, P.; Poullikkas, A. A Comparative Overview of Hydrogen Production Processes. *Renew. Sustain. Energy Rev.* **2017**, *67*, 597–611. [CrossRef]
21. Dehghanimadvar, M.; Shirmohammadi, R.; Sadeghzadeh, M.; Aslani, A.; Ghasempour, R. Hydrogen Production Technologies: Attractiveness and Future Perspective. *Int. J. Energy Res.* **2020**, *44*, 8233–8254. [CrossRef]
22. Al-Qahtani, A.M. A Comprehensive Review in Microwave Pyrolysis of Biomass, Syngas Production and Utilisation. *Energies* **2023**, *16*, 6876. [CrossRef]
23. Parente, M.; Soria, M.A.; Madeira, L.M. Hydrogen and/or Syngas Production through Combined Dry and Steam Reforming of Biogas in a Membrane Reactor: A Thermodynamic Study. *Renew. Energy* **2020**, *157*, 1254–1264. [CrossRef]
24. Zhang, H.; Sun, Z.; Hu, Y.H. Steam Reforming of Methane: Current States of Catalyst Design and Process Upgrading. *Renew. Sustain. Energy Rev.* **2021**, *149*, 111330. [CrossRef]

25. Aziz, M.; Darmawan, A.; Juangsa, F.B. Hydrogen Production from Biomasses and Wastes: A Technological Review. *Int. J. Hydrogen. Energy* **2021**, *46*, 33756–33781. [CrossRef]
26. Arsad, S.R.; Ker, P.J.; Hannan, M.A.; Tang, S.G.H.; Norhasyima, R.S.; Chau, C.F.; Mahlia, T.M.I. Patent Landscape Review of Hydrogen Production Methods: Assessing Technological Updates and Innovations. *Int. J. Hydrogen Energy*, 2023; *in press*.
27. Gangadharan, P.; Kanchi, K.C.; Lou, H.H. Evaluation of the Economic and Environmental Impact of Combining Dry Reforming with Steam Reforming of Methane. *Chem. Eng. Res. Des.* **2012**, *90*, 1956–1968. [CrossRef]
28. Navarro, R.M.; Peña, M.A.; Fierro, J.L.G. Hydrogen Production Reactions from Carbon Feedstocks: Fossil Fuels and Biomass. *Chem. Rev.* **2007**, *107*, 3952–3991. [CrossRef]
29. Richter, J.; Rachow, F.; Israel, J.; Roth, N.; Charlafti, E.; Günther, V.; Flege, J.I.; Mauss, F. Reaction Mechanism Development for Methane Steam Reforming on a Ni/Al$_2$O$_3$ Catalyst. *Catalysts* **2023**, *13*, 884. [CrossRef]
30. Wang, S.; Nabavi, S.A.; Clough, P.T. A Review on Bi/Polymetallic Catalysts for Steam Methane Reforming. *Int. J. Hydrogen Energy* **2023**, *48*, 15879–15893. [CrossRef]
31. Ahmed, S.; Lee, S.H.D.; Ferrandon, M.S. Catalytic Steam Reforming of Biogas—Effects of Feed Composition and Operating Conditions. *Int. J. Hydrogen Energy* **2015**, *40*, 1005–1015. [CrossRef]
32. Roozitalab, A.; Hamidavi, F.; Kargari, A. A Review of Membrane Material for Biogas and Natural Gas Upgrading. *Gas Sci. Eng.* **2023**, *114*, 204969. [CrossRef]
33. Czekała, W.; Jasiński, T.; Dach, J. Profitability of the Agricultural Biogas Plants Operation in Poland, Depending on the Substrate Use Model. *Energy Rep.* **2023**, *9*, 196–203. [CrossRef]
34. Guerrero, F.; Espinoza, L.; Ripoll, N.; Lisbona, P.; Arauzo, I.; Toledo, M. Syngas Production From the Reforming of Typical Biogas Compositions in an Inert Porous Media Reactor. *Front. Chem.* **2020**, *8*, 145. [CrossRef] [PubMed]
35. Szyba, M.; Mikulik, J. Energy Production from Biodegradable Waste as an Example of the Circular Economy. *Energies* **2022**, *15*, 1269. [CrossRef]
36. Chen, X.Y.; Vinh-Thang, H.; Ramirez, A.A.; Rodrigue, D.; Kaliaguine, S. Membrane Gas Separation Technologies for Biogas Upgrading. *RSC Adv.* **2015**, *5*, 24399–24448. [CrossRef]
37. Abanades, S.; Abbaspour, H.; Ahmadi, A.; Das, B.; Ehyaei, M.A.; Esmaeilion, F.; El Haj Assad, M.; Hajilounezhad, T.; Hmida, A.; Rosen, M.A.; et al. A Conceptual Review of Sustainable Electrical Power Generation from Biogas. *Energy Sci. Eng.* **2022**, *10*, 630–655. [CrossRef]
38. González, J.F.; Álvez-Medina, C.M.; Nogales-Delgado, S. Biogas Steam Reforming in Wastewater Treatment Plants: Opportunities and Challenges. *Energies* **2023**, *16*, 6343. [CrossRef]
39. Xu, X.; Zhou, Q.; Yu, D. The Future of Hydrogen Energy: Bio-Hydrogen Production Technology. *Int. J. Hydrogen Energy* **2022**, *47*, 33677–33698. [CrossRef]
40. Kumar, R.; Kumar, A.; Pal, A. Overview of Hydrogen Production from Biogas Reforming: Technological Advancement. *Int. J. Hydrogen Energy* **2022**, *47*, 34831–34855. [CrossRef]
41. Scopus. Scopus. Available online: https://www.scopus.com/home.uri (accessed on 15 November 2023).
42. Boscherini, M.; Storione, A.; Minelli, M.; Miccio, F.; Doghieri, F. New Perspectives on Catalytic Hydrogen Production by the Reforming, Partial Oxidation and Decomposition of Methane and Biogas. *Energies* **2023**, *16*, 6375. [CrossRef]
43. Abou Rjeily, M.; Gennequin, C.; Pron, H.; Abi-Aad, E.; Randrianalisoa, J.H. Pyrolysis-Catalytic Upgrading of Bio-Oil and Pyrolysis-Catalytic Steam Reforming of Biogas: A Review. *Environ. Chem. Lett.* **2021**, *19*, 2825–2872. [CrossRef]
44. Nahar, G.; Mote, D.; Dupont, V. Hydrogen Production from Reforming of Biogas: Review of Technological Advances and an Indian Perspective. *Renew. Sustain. Energy Rev.* **2017**, *76*, 1032–1052. [CrossRef]
45. Angeli, S.D.; Turchetti, L.; Monteleone, G.; Lemonidou, A.A. Catalyst Development for Steam Reforming of Methane and Model Biogas at Low Temperature. *Appl. Catal. B Environ.* **2016**, *181*, 34–46. [CrossRef]
46. Ashrafi, M.; Pfeifer, C.; Pröll, T.; Hofbauer, H. Experimental Study of Model Biogas Catalytic Steam Reforming: 2. Impact of Sulfur on the Deactivation and Regeneration of Ni-Based Catalysts. *Energy Fuels* **2008**, *22*, 4190–4195. [CrossRef]
47. Appari, S.; Janardhanan, V.M.; Bauri, R.; Jayanti, S. Deactivation and Regeneration of Ni Catalyst during Steam Reforming of Model Biogas: An Experimental Investigation. *Int. J. Hydrogen Energy* **2014**, *39*, 297–304. [CrossRef]
48. Appari, S.; Janardhanan, V.M.; Bauri, R.; Jayanti, S.; Deutschmann, O. A Detailed Kinetic Model for Biogas Steam Reforming on Ni and Catalyst Deactivation Due to Sulfur Poisoning. *Appl. Catal. A Gen.* **2014**, *471*, 118–125. [CrossRef]
49. Vita, A.; Italiano, C.; Fabiano, C.; Laganà, M.; Pino, L. Influence of Ce-Precursor and Fuel on Structure and Catalytic Activity of Combustion Synthesized Ni/CeO$_2$ Catalysts for Biogas Oxidative Steam Reforming. *Mater. Chem. Phys.* **2015**, *163*, 337–347. [CrossRef]
50. Phromprasit, J.; Powell, J.; Wongsakulphasatch, S.; Kiatkittipong, W.; Bumroongsakulsawat, P.; Assabumrungrat, S. H$_2$ Production from Sorption Enhanced Steam Reforming of Biogas Using Multifunctional Catalysts of Ni over Zr-, Ce- and La-Modified CaO Sorbents. *Chem. Eng. J.* **2017**, *313*, 1415–1425. [CrossRef]
51. Phromprasit, J.; Powell, J.; Wongsakulphasatch, S.; Kiatkittipong, W.; Bumroongsakulsawat, P.; Assabumrungrat, S. Activity and Stability Performance of Multifunctional Catalyst (Ni/CaO and Ni/Ca$_{12}$Al$_{14}$O$_{33}$—CaO) for Bio-Hydrogen Production from Sorption Enhanced Biogas Steam Reforming. *Int J Hydrogen Energy* **2016**, *41*, 7318–7331. [CrossRef]
52. Italiano, C.; Vita, A.; Fabiano, C.; Laganà, M.; Pino, L. Bio-Hydrogen Production by Oxidative Steam Reforming of Biogas over Nanocrystalline Ni/CeO$_2$ Catalysts. *Int. J. Hydrogen Energy* **2015**, *40*, 11823–11830. [CrossRef]

53. Vita, A.; Italiano, C.; Ashraf, M.A.; Pino, L.; Specchia, S. Syngas Production by Steam and Oxy-Steam Reforming of Biogas on Monolith-Supported CeO_2-Based Catalysts. *Int. J. Hydrogen Energy* **2018**, *43*, 11731–11744. [CrossRef]
54. Roy, P.S.; Park, C.S.; Raju, A.S.K.; Kim, K. Steam-Biogas Reforming over a Metal-Foam-Coated (Pd–Rh)/($CeZrO_2$–Al_2O_3) Catalyst Compared with Pellet Type Alumina-Supported Ru and Ni Catalysts. *J. CO2 Util.* **2015**, *12*, 12–20. [CrossRef]
55. Abbas, H.F.; Wan Daud, W.M.A. Hydrogen Production by Methane Decomposition: A Review. *Int. J. Hydrogen Energy* **2010**, *35*, 1160–1190. [CrossRef]
56. Li, D.; Sun, J.; Zhang, Z.; Ma, R.; Wei, J. High-Efficient Sunlight-Driven Hydrogen Production from Methanol Steam Reforming on a Novel Photo-Thermo-Catalysis and Thermo-Catalysis Dual-Bed Reactor. *Fuel* **2024**, *357*, 129895. [CrossRef]
57. Ferreira-Aparicio, P.; Rodríguez-Ramos, I.; Anderson, J.A.; Guerrero-Ruiz, A. Mechanistic Aspects of the Dry Reforming of Methane over Ruthenium Catalysts. *Appl. Catal. A Gen.* **2000**, *202*, 183–196. [CrossRef]
58. Ruban, N.V.; Rogozhnikov, V.N.; Stonkus, O.A.; Emelyanov, V.A.; Pakharukova, V.P.; Svintsitskiy, D.A.; Zazhigalov, S.V.; Zagoruiko, A.N.; Snytnikov, P.V.; Sobyanin, V.A.; et al. A Comparative Investigation of Equimolar Ni-, Ru-, Rh- and Pt-Based Composite Structured Catalysts for Energy-Efficient Methane Reforming. *Fuel* **2023**, *352*, 128973. [CrossRef]
59. Schiaroli, N.; Battisti, M.; Benito, P.; Fornasari, G.; Di Gisi, A.G.; Lucarelli, C.; Vaccari, A. Catalytic Upgrading of Clean Biogas to Synthesis Gas. *Catalysts* **2022**, *12*, 109. [CrossRef]
60. Dan, M.; Mihet, M.; Borodi, G.; Lazar, M.D. Combined Steam and Dry Reforming of Methane for Syngas Production from Biogas Using Bimodal Pore Catalysts. *Catal. Today* **2021**, *366*, 87–96. [CrossRef]
61. Saeidi, S.; Sápi, A.; Khoja, A.H.; Najari, S.; Ayesha, M.; Kónya, Z.; Asare-Bediako, B.B.; Tatarczuk, A.; Hessel, V.; Keil, F.J.; et al. Evolution Paths from Gray to Turquoise Hydrogen via Catalytic Steam Methane Reforming: Current Challenges and Future Developments. *Renew. Sustain. Energy Rev.* **2023**, *183*, 113392. [CrossRef]
62. Osazuwa, O.U.; Abidin, S.Z.; Fan, X.; Amenaghawon, A.N.; Azizan, M.T. An Insight into the Effects of Synthesis Methods on Catalysts Properties for Methane Reforming. *J. Environ. Chem. Eng.* **2021**, *9*, 105052. [CrossRef]
63. Zhao, X.; Joseph, B.; Kuhn, J.; Ozcan, S. Biogas Reforming to Syngas: A Review. *iScience* **2020**, *23*, 101082. [CrossRef]
64. Perego, C.; Villa, P. Catalyst Preparation Methods. *Catal. Today* **1997**, *34*, 281–305. [CrossRef]
65. Rotaru, C.G.; Postole, G.; Florea, M.; Matei-Rutkovska, F.; Pârvulescu, V.I.; Gelin, P. Dry Reforming of Methane on Ceria Prepared by Modified Precipitation Route. *Appl. Catal. A Gen.* **2015**, *494*, 29–40. [CrossRef]
66. Świrk Da Costa, K.; Zhang, H.; Li, S.; Chen, Y.; Rønning, M.; Motak, M.; Grzybek, T.; Da Costa, P. Carbon-Resistant NiO-Y_2O_3-Nanostructured Catalysts Derived from Double-Layered Hydroxides for Dry Reforming of Methane. *Catal. Today* **2021**, *366*, 103–113. [CrossRef]
67. Mallikarjun, G.; Sagar, T.V.; Swapna, S.; Raju, N.; Chandrashekar, P.; Lingaiah, N. Hydrogen Rich Syngas Production by Bi-Reforming of Methane with CO_2 over Ni Supported on CeO_2-SrO Mixed Oxide Catalysts. *Catal. Today* **2020**, *356*, 597–603. [CrossRef]
68. Florea, M.; Matei-Rutkovska, F.; Postole, G.; Urda, A.; Neatu, F.; Pârvulescu, V.I.; Gelin, P. Doped Ceria Prepared by Precipitation Route for Steam Reforming of Methane. *Catal Today* **2018**, *306*, 166–171. [CrossRef]
69. Niederberger, M. Nonaqueous Sol–Gel Routes to Metal Oxide Nanoparticles. *Acc. Chem. Res.* **2007**, *40*, 793–800. [CrossRef]
70. Rex, A.; dos Santos, J.H.Z. The Use of Sol–Gel Processes in the Development of Supported Catalysts. *J. Sol-Gel Sci. Technol.* **2023**, *105*, 30–49. [CrossRef]
71. Pechini, M.P. Method of Preparing Lead and Alkaline Earth Titanates and Niobates and Coating Method Using the Same to Form a Capacitor. U.S. Patent 3,330,697, 11 July 1967.
72. Lin, J.; Yu, M.; Lin, C.; Liu, X. Multiform Oxide Optical Materials via the Versatile Pechini-Type Sol–Gel Process: Synthesis and Characteristics. *J. Phys. Chem. C* **2007**, *111*, 5835–5845. [CrossRef]
73. Moogi, S.; Hyun Ko, C.; Hoon Rhee, G.; Jeon, B.H.; Ali Khan, M.; Park, Y.K. Influence of Catalyst Synthesis Methods on Anti-Coking Strength of Perovskites Derived Catalysts in Biogas Dry Reforming for Syngas Production. *Chem. Eng. J.* **2022**, *437*, 135348. [CrossRef]
74. Zolghadri, S.; Honarvar, B.; Rahimpour, M.R. Synthesis, Application, and Characteristics of Mesoporous Alumina as a Support of Promoted Ni-Co Bimetallic Catalysts in Steam Reforming of Methane. *Fuel* **2023**, *335*, 127005. [CrossRef]
75. Meshksar, M.; Farsi, M.; Rahimpour, M.R. Hollow Sphere Ni-Based Catalysts Promoted with Cerium for Steam Reforming of Methane. *J. Energy Inst.* **2022**, *105*, 342–354. [CrossRef]
76. Zarei-Jelyani, F.; Salahi, F.; Farsi, M.; Reza Rahimpour, M. Synthesis and Application of Ni-Co Bimetallic Catalysts Supported on Hollow Sphere Al_2O_3 in Steam Methane Reforming. *Fuel* **2022**, *324*, 124785. [CrossRef]
77. Wang, Z.; Wei, B.; Lv, J.; Wang, Y.; Wu, Y.; Yang, H.; Hu, H. In-Situ Catalytic Upgrading of Tar from Integrated Process of Coal Pyrolysis with Steam Reforming of Methane over Carbon Based Ni Catalyst. *J. Fuel Chem. Technol.* **2022**, *50*, 129–142. [CrossRef]
78. Dan, M.; Mihet, M.; Barbu-Tudoran, L.; Lazar, M.D. Biogas Upgrading to Syngas by Combined Reforming Using Ni/CeO_2–Al_2O_3 with Bimodal Pore Structure. *Microporous Mesoporous Mater.* **2022**, *341*, 112082. [CrossRef]
79. Roy, P.S.; Kang, M.S.; Kim, K. Effects of Pd-Rh Composition and $CeZrO_2$-Modification of Al_2O_3 on Performance of Metal-Foam-Coated Pd-Rh/Al_2O_3 Catalyst for Steam Reforming of Model Biogas. *Catal. Lett.* **2014**, *144*, 2021–2032. [CrossRef]
80. Roy, P.S.; Song, J.; Kim, K.; Kim, J.M.; Park, C.S.; Raju, A.S.K. Effects of $CeZrO_2$–Al_2O_3 Support Composition of Metal-Foam-Coated Pd–Rh Catalysts for the Steam-Biogas Reforming Reaction. *J. Ind. Eng. Chem.* **2018**, *62*, 120–129. [CrossRef]

81. Lin, K.H.; Chang, H.F.; Chang, A.C.C. Biogas Reforming for Hydrogen Production over Mesoporous $Ni_{2x}Ce_{1-x}O_2$ Catalysts. *Int. J. Hydrogen Energy* **2012**, *37*, 15696–15703. [CrossRef]
82. Arbag, H.; Yasyerli, S.; Yasyerli, N.; Dogu, G.; Dogu, T. Enhancement of Catalytic Performance of Ni Based Mesoporous Alumina by Co Incorporation in Conversion of Biogas to Synthesis Gas. *Appl. Catal. B* **2016**, *198*, 254–265. [CrossRef]
83. Zhong, J.; Han, B.; Zhang, Z.; Bi, G.; Xie, J. Bi-Reforming of Model Biogas to Syngas over Ru Nano-Catalysts Supported on Mg-Al Oxide Derived from Layered Double Hydroxides. *Fuel* **2023**, *343*, 127941. [CrossRef]
84. Poggio-Fraccari, E.; Mariño, F.; Herrera, C.; Larrubia-Vargas, M.Á.; Alemany, L. Bi-Reforming of Biogas for Hydrogen Production with Sulfur-Resistant Multimetallic Catalyst. *Chem. Eng. Technol.* **2023**, *46*, 1176–1184. [CrossRef]
85. Park, D.; Lee, C.; Moon, D.J.; Kim, T. Design, Analysis, and Performance Evaluation of Steam-CO_2 Reforming Reactor for Syngas Production in GTL Process. *Int. J. Hydrogen Energy* **2015**, *40*, 11785–11790. [CrossRef]
86. Chen, C.; Wang, X.; Chen, X.; Liang, X.; Zou, X.; Lu, X. Combined Steam and CO_2 Reforming of Methane over One-Pot Prepared Ni/La-Si Catalysts. *Int. J. Hydrogen Energy* **2019**, *44*, 4780–4793. [CrossRef]
87. Ren, P.; Zhao, Z. Unexpected Coke-Resistant Stability in Steam-CO_2 Dual Reforming of Methane over the Robust Mo_2C-Ni/ZrO_2 Catalyst. *Catal. Commun.* **2019**, *119*, 71–75. [CrossRef]
88. Schiaroli, N.; Lucarelli, C.; Sanghez de Luna, G.; Fornasari, G.; Vaccari, A. Ni-Based Catalysts to Produce Synthesis Gas by Combined Reforming of Clean Biogas. *Appl. Catal. A Gen.* **2019**, *582*, 117087. [CrossRef]
89. Ponugoti, P.V.; Pathmanathan, P.; Rapolu, J.; Gomathi, A.; Janardhanan, V.M. On the Stability of Ni/γ-Al_2O_3 Catalyst and the Effect of H_2O and O_2 during Biogas Reforming. *Appl. Catal. A Gen.* **2023**, *651*, 119033. [CrossRef]
90. Kolbitsch, P.; Pfeifer, C.; Hofbauer, H. Catalytic Steam Reforming of Model Biogas. *Fuel* **2008**, *87*, 701–706. [CrossRef]
91. Azevedo, I.R.; da Silva, A.A.A.; Xing, Y.T.; Rabelo-Neto, R.C.; Luchters, N.T.J.; Fletcher, J.C.Q.; Noronha, F.B.; Mattos, L.V. Long-Term Stability of Pt/$Ce_{0.8}Me_{0.2}O_2$-γ/Al_2O_3 (Me = Gd, Nb, Pr, and Zr) Catalysts for Steam Reforming of Methane. *Int. J. Hydrogen Energy* **2022**, *47*, 15624–15640. [CrossRef]
92. Raju, A.S.K.; Park, C.S.; Norbeck, J.M. Synthesis Gas Production Using Steam Hydrogasification and Steam Reforming. *Fuel Process. Technol.* **2009**, *90*, 330–336. [CrossRef]
93. Dong, W.-S.; Roh, H.-S.; Liu, Z.-W.; Jun, K.-W.; Park, S.-E. Hydrogen Production from Methane Reforming Reactions over Ni/MgO Catalyst. *Bull. Korean Chem. Soc.* **2001**, *22*, 1323–1327.
94. Díez-Ramírez, J.; Dorado, F.; Martínez-Valiente, A.; García-Vargas, J.M.; Sánchez, P. Kinetic, Energetic and Exergetic Approach to the Methane Tri-Reforming Process. *Int. J. Hydrogen Energy* **2016**, *41*, 19339–19348. [CrossRef]
95. Chouhan, K.; Sinha, S.; Kumar, S.; Kumar, S. Utilization of Biogas from Different Substrates for SOFC Feed via Steam Reforming: Thermodynamic and Exergy Analyses. *J. Environ. Chem. Eng.* **2019**, *7*, 103018. [CrossRef]
96. Bkour, Q.; Marin-Flores, O.G.; Graham, T.R.; Ziaei, P.; Saunders, S.R.; Norton, M.G.; Ha, S. Nickel Nanoparticles Supported on Silica for the Partial Oxidation of Isooctane. *Appl Catal A Gen* **2017**, *546*, 126–135. [CrossRef]
97. Bej, B.; Pradhan, N.C.; Neogi, S. Production of Hydrogen by Steam Reforming of Methane over Alumina Supported Nano-NiO/SiO_2 Catalyst. *Catal. Today* **2013**, *207*, 28–35. [CrossRef]
98. Shinde, V.M.; Madras, G. Catalytic Performance of Highly Dispersed Ni/TiO_2 for Dry and Steam Reforming of Methane. *RSC Adv.* **2014**, *4*, 4817–4826. [CrossRef]
99. Dang, C.; Xia, H.; Yuan, S.; Wei, X.; Cai, W. Green Hydrogen Production from Sorption-Enhanced Steam Reforming of Biogas over a Pd/Ni–CaO-Mayenite Multifunctional Catalyst. *Renew. Energy* **2022**, *201*, 314–322. [CrossRef]
100. Park, M.-J.; Kim, H.-M.; Gu, Y.-J.; Jeong, D.-W. Optimization of Biogas-Reforming Conditions Considering Carbon Formation, Hydrogen Production, and Energy Efficiencies. *Energy* **2023**, *265*, 126273. [CrossRef]
101. Santamaria, L.; Lopez, G.; Fernandez, E.; Cortazar, M.; Arregi, A.; Olazar, M.; Bilbao, J. Progress on Catalyst Development for the Steam Reforming of Biomass and Waste Plastics Pyrolysis Volatiles: A Review. *Energy Fuels* **2021**, *35*, 17051–17084. [CrossRef]
102. Farooqi, A.S.; Yusuf, M.; Mohd Zabidi, N.A.; Saidur, R.; Sanaullah, K.; Farooqi, A.S.; Khan, A.; Abdullah, B. A Comprehensive Review on Improving the Production of Rich-Hydrogen via Combined Steam and CO_2 Reforming of Methane over Ni-Based Catalysts. *Int. J. Hydrogen Energy* **2021**, *46*, 31024–31040. [CrossRef]
103. Chan, Y.H.; Lock, S.S.M.; Wong, M.K.; Yiin, C.L.; Loy, A.C.M.; Cheah, K.W.; Chai, S.Y.W.; Li, C.; How, B.S.; Chin, B.L.F.; et al. A State-of-the-Art Review on Capture and Separation of Hazardous Hydrogen Sulfide (H_2S): Recent Advances, Challenges and Outlook. *Environ. Pollut.* **2022**, *314*, 120219. [CrossRef]
104. Pudi, A.; Rezaei, M.; Signorini, V.; Andersson, M.P.; Baschetti, M.G.; Mansouri, S.S. Hydrogen Sulfide Capture and Removal Technologies: A Comprehensive Review of Recent Developments and Emerging Trends. *Sep. Purif. Technol.* **2022**, *298*, 121448. [CrossRef]
105. Capa, A.; González-Vázquez, M.P.; Chen, D.; Rubiera, F.; Pevida, C.; Gil, M.V. Effect of H_2S on Biogas Sorption Enhanced Steam Reforming Using a Pd/Ni-Co Catalyst and Dolomite as a Sorbent. *Chem. Eng. J.* **2023**, *476*, 146803. [CrossRef]
106. Rahim, D.A.; Fang, W.; Wibowo, H.; Hantoko, D.; Susanto, H.; Yoshikawa, K.; Zhong, Y.; Yan, M. Review of High Temperature H_2S Removal from Syngas: Perspectives on Downstream Process Integration. *Chem. Eng. Process.-Process Intensif.* **2023**, *183*, 109258. [CrossRef]
107. Ozekmckci, M.; Salkic, G.; Fellah, M.F. Use of Zeolites for the Removal of H_2S: A Mini-Review. *Fuel Process. Technol.* **2015**, *139*, 49–60. [CrossRef]
108. Secco, C.; Fuziki, M.E.K.; Tusset, A.M.; Lenzi, G.G. Reactive Processes for H_2S Removal. *Energies* **2023**, *16*, 1759. [CrossRef]

109. Fabrik, M.; Salama, A.; Ibrahim, H. Modeling of Catalyst Poisoning during Hydrogen Production via Methane Steam and Dry Reforming. *Fuel* **2023**, *347*, 128429. [CrossRef]

110. Izquierdo, U.; García-García, I.; Gutierrez, Á.M.; Arraibi, J.R.; Barrio, V.L.; Cambra, J.F.; Arias, P.L. Catalyst Deactivation and Regeneration Processes in Biogas Tri-Reforming Process. The Effect of Hydrogen Sulfide Addition. *Catalysts* **2018**, *8*, 12. [CrossRef]

111. Cattaneo, C.R.; Muñoz, R.; Korshin, G.V.; Naddeo, V.; Belgiorno, V.; Zarra, T. Biological Desulfurization of Biogas: A Comprehensive Review on Sulfide Microbial Metabolism and Treatment Biotechnologies. *Sci. Total Environ.* **2023**, *893*, 164689. [CrossRef] [PubMed]

112. Costa, C.; Cornacchia, M.; Pagliero, M.; Fabiano, B.; Vocciante, M.; Reverberi, A. Pietro Hydrogen Sulfide Adsorption by Iron Oxides and Their Polymer Composites: A Case-Study Application to Biogas Purification. *Materials* **2020**, *13*, 4725. [CrossRef]

113. Abd, A.A.; Othman, M.R.; Majdi, H.S.; Helwani, Z. Green Route for Biomethane and Hydrogen Production via Integration of Biogas Upgrading Using Pressure Swing Adsorption and Steam-Methane Reforming Process. *Renew. Energy* **2023**, *210*, 64–78. [CrossRef]

114. Wang, X.; Su, X.; Zhang, Q.; Hu, H. Effect of Additives on Ni-Based Catalysts for Hydrogen-Enriched Production from Steam Reforming of Biomass. *Energy Technol.* **2020**, *8*, 2000136. [CrossRef]

115. Torimoto, M.; Sekine, Y. Effects of Alloying for Steam or Dry Reforming of Methane: A Review of Recent Studies. *Catal. Sci. Technol.* **2022**, *12*, 3387–3411. [CrossRef]

116. Becker, C.M.; Marder, M.; Junges, E.; Konrad, O. Technologies for Biogas Desulfurization—An Overview of Recent Studies. *Renew. Sustain. Energy Rev.* **2022**, *159*, 112205. [CrossRef]

117. Hulteberg, C. Sulphur-Tolerant Catalysts in Small-Scale Hydrogen Production, a Review. *Int. J. Hydrogen Energy* **2012**, *37*, 3978–3992. [CrossRef]

118. Pashchenko, D.; Makarov, I. Carbon Deposition in Steam Methane Reforming over a Ni-Based Catalyst: Experimental and Thermodynamic Analysis. *Energy* **2021**, *222*, 119993. [CrossRef]

119. Tuna, C.E.; Silveira, J.L.; da Silva, M.E.; Boloy, R.M.; Braga, L.B.; Pérez, N.P. Biogas Steam Reformer for Hydrogen Production: Evaluation of the Reformer Prototype and Catalysts. *Int. J. Hydrogen Energy* **2018**, *43*, 2108–2120. [CrossRef]

120. Papurello, D.; Chiodo, V.; Maisano, S.; Lanzini, A.; Santarelli, M. Catalytic Stability of a Ni-Catalyst towards Biogas Reforming in the Presence of Deactivating Trace Compounds. *Renew. Energy* **2018**, *127*, 481–494. [CrossRef]

121. Iulianelli, A.; Manisco, M.; Bion, N.; Le Valant, A.; Epron, F.; Colpan, C.O.; Esposito, E.; Jansen, J.C.; Gensini, M.; Caravella, A. Sustainable H2 Generation via Steam Reforming of Biogas in Membrane Reactors: H_2S Effects on Membrane Performance and Catalytic Activity. *Int. J. Hydrogen Energy* **2021**, *46*, 29183–29197. [CrossRef]

122. Amiri, T.Y.; Ghasemzageh, K.; Iulianelli, A. Membrane Reactors for Sustainable Hydrogen Production through Steam Reforming of Hydrocarbons: A Review. *Chem. Eng. Process. Process Intensif.* **2020**, *157*, 108148. [CrossRef]

123. Arratibel Plazaola, A.; Pacheco Tanaka, D.A.; Van Sint Annaland, M.; Gallucci, F. Recent Advances in Pd-Based Membranes for Membrane Reactors. *Molecules* **2017**, *22*, 51. [CrossRef]

124. Iulianelli, A.; Alavi, M.; Bagnato, G.; Liguori, S.; Wilcox, J.; Rahimpour, M.R.; Eslamlouyan, R.; Anzelmo, B.; Basile, A. Supported Pd-Au Membrane Reactor for Hydrogen Production: Membrane Preparation, Characterization and Testing. *Molecules* **2016**, *21*, 581. [CrossRef] [PubMed]

125. Ben-Mansour, R.; Haque, M.A.; Habib, M.A.; Paglieri, S.; Harale, A.; Mokheimer, E.M.A. Effect of Temperature and Heat Flux Boundary Conditions on Hydrogen Production in Membrane-Integrated Steam-Methane Reformer. *Appl. Energy* **2023**, *346*, 121407. [CrossRef]

126. Jokar, S.M.; Farokhnia, A.; Tavakolian, M.; Pejman, M.; Parvasi, P.; Javanmardi, J.; Zare, F.; Gonçalves, M.C.; Basile, A. The Recent Areas of Applicability of Palladium Based Membrane Technologies for Hydrogen Production from Methane and Natural Gas: A Review. *Int. J. Hydrogen Energy* **2023**, *48*, 6451–6476. [CrossRef]

127. Di Marcoberardino, G.; Vitali, D.; Spinelli, F.; Binotti, M.; Manzolini, G. Green Hydrogen Production from Raw Biogas: A Techno-Economic Investigation of Conventional Processes Using Pressure Swing Adsorption Unit. *Processes* **2018**, *6*, 19. [CrossRef]

128. Al-Zuhairi, F.K.; Azeez, R.A.; Mahdi, S.A. Synthesis of Long-Chain Hydrocarbons from Syngas over Promoted Co/SiO_2 Catalysts Using Fischer–Tropsch Reaction. *AIP Conf. Proc.* **2022**, *2443*, 030028. [CrossRef]

129. Schiaroli, N.; Lucarelli, C.; Iapalucci, M.C.; Fornasari, G.; Crimaldi, A.; Vaccari, A. Combined Reforming of Clean Biogas over Nanosized Ni–Rh Bimetallic Clusters. *Catalysts* **2020**, *10*, 1345. [CrossRef]

130. He, X.; Lei, L.; Dai, Z. Green Hydrogen Enrichment with Carbon Membrane Processes: Techno-Economic Feasibility and Sensitivity Analysis. *Sep. Purif. Technol.* **2021**, *276*, 119346. [CrossRef]

131. Cherif, A.; Nebbali, R.; Sen, F.; Sheffield, J.W.; Doner, N.; Nasseri, L. Modeling and Simulation of Steam Methane Reforming and Methane Combustion over Continuous and Segmented Catalyst Beds in Autothermal Reactor. *Int. J. Hydrogen Energy* **2022**, *47*, 9127–9138. [CrossRef]

132. Kenkel, P.; Wassermann, T.; Zondervan, E. Biogas Reforming as a Precursor for Integrated Algae Biorefineries: Simulation and Techno-Economic Analysis. *Processes* **2021**, *9*, 1348. [CrossRef]

133. Udemu, C.; Font-Palma, C. Modelling of Sorption-Enhanced Steam Reforming (SE-SR) Process in Fluidised Bed Reactors for Low-Carbon Hydrogen Production: A Review. *Fuel* **2023**, *340*, 127588. [CrossRef]

134. Avraam, D.G.; Halkides, T.I.; Liguras, D.K.; Bereketidou, O.A.; Goula, M.A. An Experimental and Theoretical Approach for the Biogas Steam Reforming Reaction. *Int. J. Hydrogen Energy* **2010**, *35*, 9818–9827. [CrossRef]

135. Cormos, C.C.; Cormos, A.M.; Petrescu, L.; Dragan, S. Techno-Economic Assessment of Decarbonized Biogas Catalytic Reforming for Flexible Hydrogen and Power Production. *Appl. Therm. Eng.* **2022**, *207*, 118218. [CrossRef]
136. Park, M.J.; Kim, J.H.; Lee, Y.H.; Kim, H.M.; Jeong, D.W. System Optimization for Effective Hydrogen Production via Anaerobic Digestion and Biogas Steam Reforming. *Int. J. Hydrogen Energy* **2020**, *45*, 30188–30200. [CrossRef]
137. Frontera, P.; Malara, A.; Macario, A.; Miceli, M.; Bonaccorsi, L.; Boaro, M.; Pappacena, A.; Trovarelli, A.; Antonucci, P.L. Performance and Stability of Doped Ceria–Zirconia Catalyst for a Multifuel Reforming. *Catalysts* **2023**, *13*, 165. [CrossRef]
138. Matus, E.; Kerzhentsev, M.; Ismagilov, I.; Nikitin, A.; Sozinov, S.; Ismagilov, Z. Hydrogen Production from Biogas: Development of an Efficient Nickel Catalyst by the Exsolution Approach. *Energies* **2023**, *16*, 2993. [CrossRef]
139. Pajak, M.; Brus, G.; Kimijima, S.; Szmyd, J.S. Enhancing Hydrogen Production from Biogas through Catalyst Rearrangements. *Energies* **2023**, *16*, 4058. [CrossRef]
140. Wodołażski, A.; Magdziarczyk, M.; Smoliński, A. Techno-Economic Analysis of Hydrogen Production from Swine Manure Biogas via Steam Reforming in Pilot-Scale Installation. *Energies* **2023**, *16*, 6389. [CrossRef]
141. Ongis, M.; Di Marcoberardino, G.; Baiguini, M.; Gallucci, F.; Binotti, M. Optimization of Small-Scale Hydrogen Production with Membrane Reactors. *Membranes* **2023**, *13*, 331. [CrossRef]
142. Ongis, M.; Di Marcoberardino, G.; Manzolini, G.; Gallucci, F.; Binotti, M. Membrane Reactors for Green Hydrogen Production from Biogas and Biomethane: A Techno-Economic Assessment. *Int. J. Hydrogen Energy* **2023**, *48*, 19580–19595. [CrossRef]
143. Yan, P.; Cheng, Y. Design and Operational Considerations of Packed-Bed Membrane Reactor for Distributed Hydrogen Production by Methane Steam Reforming. *Int. J. Hydrogen Energy* **2022**, *47*, 36493–36503. [CrossRef]
144. Lo Basso, G.; Pastore, L.M.; Mojtahed, A.; de Santoli, L. From Landfill to Hydrogen: Techno-Economic Analysis of Hybridized Hydrogen Production Systems Integrating Biogas Reforming and Power-to-Gas Technologies. *Int. J. Hydrogen Energy* **2023**, *48*, 37607–37624. [CrossRef]
145. Minutillo, M.; Perna, A.; Sorce, A. Green Hydrogen Production Plants via Biogas Steam and Autothermal Reforming Processes: Energy and Exergy Analyses. *Appl. Energy* **2020**, *277*, 115452. [CrossRef]
146. Braga, L.B.; Silveira, J.L.; Da Silva, M.E.; Tuna, C.E.; Machin, E.B.; Pedroso, D.T. Hydrogen Production by Biogas Steam Reforming: A Technical, Economic and Ecological Analysis. *Renew. Sustain. Energy Rev.* **2013**, *28*, 166–173. [CrossRef]
147. Bo, G.; Li, F.; Chauhan, B.S.; Ghaebi, H. Multi-Objective Optimization and 4E Analyses to Simultaneous Production of Electricity, Freshwater, H_2 and Cooling Based on Biogas-Steam Reforming. *Process Saf. Environ. Prot.* **2023**, *177*, 1307–1320. [CrossRef]
148. He, J.; Han, N.; Xia, M.; Sun, T.; Ghaebi, H. Multi-Objective Optimization and Exergoeconomic Analysis of a Multi-Generation System Based on Biogas-Steam Reforming. *Int. J. Hydrogen Energy* **2023**, *48*, 21161–21175. [CrossRef]
149. Farhang, B.; Ghaebi, H.; Javani, N. Techno-Economic Modeling of a Novel Poly-Generation System Based on Biogas for Power, Hydrogen, Freshwater, and Ammonia Production. *J. Clean. Prod.* **2023**, *417*, 137907. [CrossRef]
150. Chouhan, K.; Sinha, S.; Kumar, S.; Kumar, S. Simulation of Steam Reforming of Biogas in an Industrial Reformer for Hydrogen Production. *Int. J. Hydrogen Energy* **2021**, *46*, 26809–26824. [CrossRef]
151. Zhou, Y.; Zhang, S.; Zhou, H. Method for Preparing Hydrogen by Utilizing Biogas. China Patent CN113772628 (A), 10 December 2021.
152. Xu, S.; Pan, X.; Pehowa, P.; Lillabrig, A. Method for Producing Hydrogen from Biogas. China Patent CN113460963 (A), 1 October 2021.
153. Mortensen, P.M.; Aasberg-Petersen, K.; Nielsen, C.S. Biogas Conversion to Synthesis Gas for Producing Hydrocarbons. United States Patent A/S US 2023/0012800 A1, 19 January 2023.
154. Zhong, Y.; Liu, K.; Zhong, Y.; Cai, Y.; Chen, Y.; Mou, S. Method for Producing Hydrogen from Biogas Biomass. China Patent CN107758617 (A), 6 March 2018.
155. Xu, S.; Pan, X.; Bazan, M.; Pechova, P. Apparatus for Producing Biomethane and Syngas from Biogas Stream with Means for Regulating Quality Biogas. China Patent CN202210209672 20220304, 23 September 2022.
156. Allidieres, L. Method for Producing Hydrogen from Biogas. France Patent US2014186258 (A1), 3 July 2014.
157. Guenther, L. Method and Plant for the Production of Synthesis Gas from Biogas. Germany Patent US2011175032 (A1), 21 July 2011.
158. Offerman, J.D. Efficient Use of Biogas Carbon Dioxide in Liquid Fuel Synthesis. Germany Patent US2008220489 (A1), 11 September 2008.
159. Su, B.; Tang, S.; Huang, S. Method and System for Coupling Biogas Reforming with Natural Gas Combined Cycle Based on Solar Energy Driving. China Patent CN114836480 (A), 2 August 2022.
160. Han, W.; Su, B.; Jin, H. Biogas and Solar Energy Complementary Two-Stage Preparation System and Method for Synthesis Gas. China Patent CN110436413 (A), 12 November 2019.

MDPI
St. Alban-Anlage 66
4052 Basel
Switzerland
www.mdpi.com

Catalysts Editorial Office
E-mail: catalysts@mdpi.com
www.mdpi.com/journal/catalysts

Disclaimer/Publisher's Note: The statements, opinions and data contained in all publications are solely those of the individual author(s) and contributor(s) and not of MDPI and/or the editor(s). MDPI and/or the editor(s) disclaim responsibility for any injury to people or property resulting from any ideas, methods, instructions or products referred to in the content.